A. V. Boiko
G. R. Grek
A. V. Dovgal
V. V. Kozlov

The Origin of Turbulence in Near-Wall Flows

Springer-Verlag Berlin Heidelberg GmbH

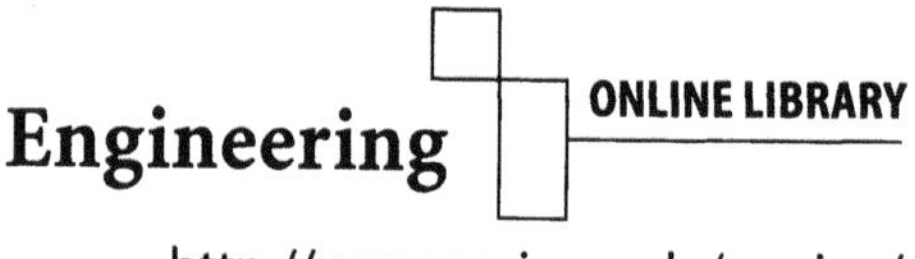

http://www.springer.de/engine/

A. V. Boiko
G. R. Grek
A. V. Dovgal
V. V. Kozlov

The Origin of Turbulence in Near-Wall Flows

Dr. Andrey V. Boiko
Professor Genrih R. Grek
Professor Alexander V. Dovgal
Professor Victor V. Kozlov
Institute of Theoretical and Applied Mechanics
Siberian Branch of the Russian Academy of Sciences
630090 Novosibirsk
Russia
e-mail: kozlov@itam.nsc.ru

Library of Congress Cataloging-in-Publication-Data is applied for

Die Deutsche Bibliothek – CIP-Einheitsaufnahme

The origin of turbulence in near wall flows / A. V. Boiko ...

DOI 10.1007/978-3-662-04765-1

http://www.springer.de

Originally published by Springer-Verlag Berlin Heidelberg **New York in** 2002
MyCopy version of the original edition 2002

Typesetting: Data delivered by author
Cover Design: de'blik, Berlin
Printed on acid free paper SPIN: 10839231 62/3020/M – 5 4 3 2 1 0
www.springer.com/mycopy

Preface

The Origin of Species
Charles Darwin

The origin of turbulence in fluids is a long-standing problem and has been the focus of research for decades due to its great importance in a variety of engineering applications. Furthermore, the study of the origin of turbulence is part of the fundamental physical problem of turbulence description and the philosophical problem of determinism and chaos.

At the end of the nineteenth century, Reynolds and Rayleigh conjectured that the reason of the transition of laminar flow to the 'sinuous' state is instability which results in amplification of wavy disturbances and breakdown of the laminar regime. Heisenberg (1924) was the founder of linear hydrodynamic stability theory. The first calculations of boundary layer stability were fulfilled in pioneer works of Tollmien (1929) and Schlichting (1932, 1933). Later Taylor (1936) hypothesized that the transition to turbulence is initiated by free-stream oscillations inducing local separations near wall. Up to the 1940s, skepticism of the stability theory predominated, in particular due to the experimental results of Dryden (1934, 1936). Only the experiments of Schubauer and Skramstad (1948) revealed the determining role of instability waves in the transition. Now it is well established that the transition to turbulence in shear flows at small and moderate levels of environmental disturbances occurs through development of instability waves in the initial laminar flow. In Chapter 1 we start with the fundamentals of stability theory, employing results of the early studies and recent advances.

As for a complete theory of the laminar–turbulent transition in shear layers, an adequate formal procedure should allow description of the complete instability evolution including the generation of flow perturbations, their development, the breakdown of the initial laminar state, and final transformation to a turbulent regime. The solution of the problem in such a complex formulation produces substantial mathematical difficulties, so the process is divided into a sequence of stages that can be examined by simplified models. The classical linear stability theory describes low-intensity shear-layer perturbations as the instability waves. The lack of a theory to account for the origin of the latter inevitably results in the problem of their excitation by external flow disturbances. Finally, the growth of perturbations beyond a certain amplitude threshold is accompanied by non-linear effects such as wave interactions, base-flow distortions and appearance of turbulent spots. Thus, the transition process initiated by low-level environmental disturbances has three milestones: amplification of small-amplitude instability waves, their generation, and non-linear destruction of the laminar flow. These aspects of the laminar–turbulent transition are considered step-by-step in Chapters 2 to 4.

In Chapter 5 we turn to boundary layer transition at high external flow perturbations, which is very different from that under 'quiet' conditions. In Chapter 6 the developed approaches are applied to the transition process of laminar flow separation. Finally, the prediction and control of the transition are focused on in Chapter 7.

Thus, in the present book we try to give a panoramic view of the origin of turbulence in near-wall shear layers by binding its different aspects for prototypical incompressible flows. Basically, the foregoing considerations reflect the authors' experience in experimental studies of the transition process. We expect that the result of our efforts will be of interest to researchers, high-school teachers, and students involved in hydro-aerodynamics, stability and transition problems.

To be in line with this general idea, there was a need to pick out the material for inclusion from plenty of available research data. In this way, our priority was a conceptual treatment of different – in particular modern – aspects of the problem, rather then delving too much into the details. An interested reader can find these details in recent publications on the transition problem which are referred to in the book.

The research work has been performed in collaboration with our colleagues from the Institute of Theoretical and Applied Mechanics, Siberian Branch of the Russian Academy of Sciences, as well as with Henrik Alfredsson, Barbro Klingmann, Masaharu Matsubara, Alfons Michalke, Oleg Ryzhov and William Saric, which provided exploration results included in the present book. Our thanks also go to Uwe Dallmann, Friedrich Grosche and Gerd Meier fot their cooperation.

We are grateful to Fabio Bertolotti, Hans Bippes, Alessandro Bottaro, Michael Gaster, Dan Henningson, Thorwald Herbert, Yasuaki Kohama, Paolo Luchini, Michio Nishioka, Ulrich Rist, Frank Smith, Jin Hyung Sung, Vassilios Theofilis and Israel Wygnanski, for useful contacts and discussions. Suggestions by Sergei Gaponov, Yuri Kachanov, Victor Levchenko, Anatoli Tumin and Nikolai Yavorski on the preparation of the manuscript were very helpful.

Special thanks are due to the Alexander von Humboldt Foundation for supporting the research of the first two authors.

We are grateful to our families for their patience and their recognition of our efforts.

Novosibirsk, April 2001

Andrey V. Boiko
Alexander V. Dovgal
Genrih R. Grek
Victor V. Kozlov

Contents

1 Fundamentals of stability theory

A general and indicative definition of stability was given by Betchov and Criminale (1967): 'the stability can be defined as quality of immunity to small disturbances.' An illustration of this general property to the stability of mechanical systems is served by the elementary examples shown in Fig. 1.1.

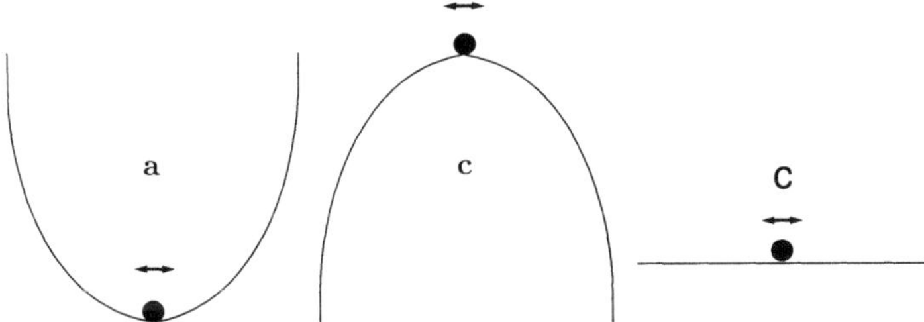

Fig. 1.1. Simple mechanical examples of equilibrium states: **a** stable state; **b** unstable state; **c** neutral (indefinite) state

Similarly, it is known that for certain parameters of a hydrodynamic system, the hydrodynamic equations with precise stationary (laminar) solutions at constant boundary conditions cannot be implemented in practice: the motion is unstable. Consequently, an important subject in the theory of hydrodynamic stability is the analysis of the development of disturbances in an initially laminar flow. The fundamentals of this theory are the subject of this chapter.

1.1 Concept of hydrodynamic stability

To be used in hydrodynamic applications, the definition of stability must be properly specified. Because of the complexity of the hydrodynamic equations of motion, it is obviously not possible to give a unique rational definition of the stability. In general, if at importation of certain disturbances in a flow, it returns to the initial state (speaking in the language of the theory of dynamic systems, is 'attracted') in time and/or space, the flow is stable to these disturbances. If the disturbances grow and the flow comes to another (not necessarily turbulent) state, it is unstable.

Sometimes the transition to turbulence, at least at an initial stage, occurs through a chain of (quasi-)stable states and consequent loss of stability (bifurcation) in each of them. The phenomenon is denoted as the change of stability and, usually, it is characterized by a consequent loss of symmetry in the flow motion. Such a laminar–turbulent transition is a characteristic of non-dissipative systems (i.e. closed flows in which there are no sources and sinks of energy), but it sometimes also occurs in open (dissipative) flows.

Frequently in the stability problems the asymptotic (after long period) response of a system affected by a disturbance is considered. However, situations are not excluded where the disturbance at the beginning, during its establishment, experiences transient growth and only then decays. It is a typical phenomenon that is found in many other branches of physics, such as at the breaking of an electric circuit that can lead to a bubble break. Similarly, if the disturbance becomes dangerously large during this transient growth, it can trigger the laminar–turbulent transition.

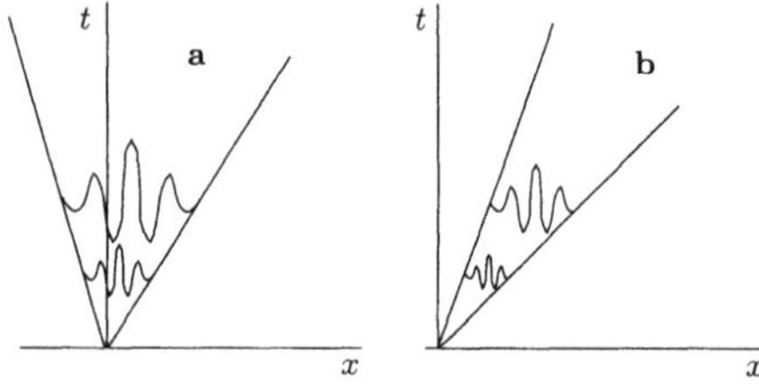

Fig. 1.2. Illustration of the propagation of disturbances in systems with different types of instability: **a** absolutely unstable system; **b** convectively unstable system

Two classes of physical problems regarding the propagation of disturbances in hydrodynamic systems can be selected (Fig. 1.2):

1. The problem of initial conditions, or stability in time. If the initial disturbance decays in time at each fixed point of space (or, at least, does not monotonically grow), the system is called stable to these disturbances. Otherwise, if the initial disturbance monotonically grows in time at a fixed point of space, the system is called absolutely unstable.
2. The problem of boundary conditions, or amplification in space. If an external signal at the entrance to the system decays whilst propagating in it, it is said that the spatial attenuation (non-transmission of a signal) takes place. Otherwise, there is a spatial amplification, and the system is called convectively unstable.

Obviously, the classification depends on the frame of reference chosen. The convective disturbances always can be transformed to 'developing in time' and vice versa by Galilei transformations. Since the velocity distributions and boundary conditions are not invariant with respect to the coordinate transformation, these problems are not identical. In practice, there is usually a 'natural' laboratory reference system, related to the walls of a wind tunnel, aerofoil, disturbance source, etc., in which the classification is conducted. When such a reference system is not obvious, it is always possible to formulate a problem of finding the reference system, in which the instability is, e.g., absolute with the maximum increment. In practice it is possible that the disturbance grows in the chosen reference system in both space and time simultaneously, as in certain regions of wakes behind bluff bodies or heated jets (Monkewitz 1988; Huerre and Monkewitz 1990).

1.2 Stability of fluid motion in time

We consider the instability in time first. Formally, the mathematical analysis of the spatial amplification is more difficult. The reasons are in the properties of the underlying equations and uncertainty in the selection of a disturbance measure (Henningson and Schmid 1994). Features related to the amplification in space are described in detail in Sect. 1.3.

The concept of stability in time can be defined using various positive-definite norms of parameters (measures) of the disturbances (Galdi and Padula 1990). However, the natural physical measure of the disturbance is usually its kinetic energy. Therefore, we give various formal definitions of the stability based on the kinetic energy of disturbance velocity $\boldsymbol{u}$, integrated over the whole volume V covered by the hydrodynamic system $E_{\mathrm{V}} = \int_{\mathrm{V}} (\boldsymbol{u}^2/2)\mathrm{dV}$ (Joseph 1976). This implies either a localization of the disturbance in the volume V which is large enough for open flows (the disturbance developing only inside the volume during the observation) or the spatial periodicity of the motion for closed flows, V covering the whole range of the disturbance motion.

Some definitions are helpful:

Asymptotic stability: A flow is (asymptotically) stable to disturbances, if

$$\lim_{t\to\infty} \frac{E_{\mathrm{V}}(t)}{E_{\mathrm{V}}(0)} \to 0,$$

where t is time.

Conditional stability: If there is a value $\delta > 0$, such that a solution of equations of motion is stable at $E(0) < \delta$, the solution is called *conditionally stable.* The value δ (the attraction radius) determines a set of initial conditions attracting to the undisturbed solution. If the disturbance energy $E(0) \geqslant \delta$, the disturbance grows or forms a new stable state (exchange of stability).

Global stability: If the value $\delta \to \infty$, the solution is *globally or unconditionally stable.*

Monotonic stability: If the flow is stable and $\mathrm{d}E/\mathrm{d}t \leqslant 0$ at all $t > 0$, the solution is *monotonically stable.*

As seen, each next definition imposes new restrictions on the stability.

1.2.1 Critical parameters for onset of instability

The instability is defined as a state when the corresponding stability conditions are broken. In accordance with the principle of similarity first discovered experimentally by Reynolds (1883) for flow in a channel, the conditions for the appearance of the instability depend on certain dimensionless ratios of problem parameters as viscosity, density, velocity, temperature and frequency. For many simple flows, such a fundamental quantity is the ratio between inertial and friction forces, the Reynolds number $\mathrm{Re}_l = Ul/\nu$, where U is a characteristic flow velocity, l is a characteristic length, and ν is the kinematic viscosity, which in turn is the ratio of dynamic viscosity of the fluid and density. Based on the definitions of stability given above, the *critical Reynolds numbers* separating the regions of stable and unstable motion (Fig. 1.5) are defined below. For more complex flows, such as those with curvature, other similarity parameters, e.g., the Görtler number Gö, which is a dimensionless measure of a curvature of a wall, appear. Then the same classification is applicable to them as well; see Sect. 2.3.2.

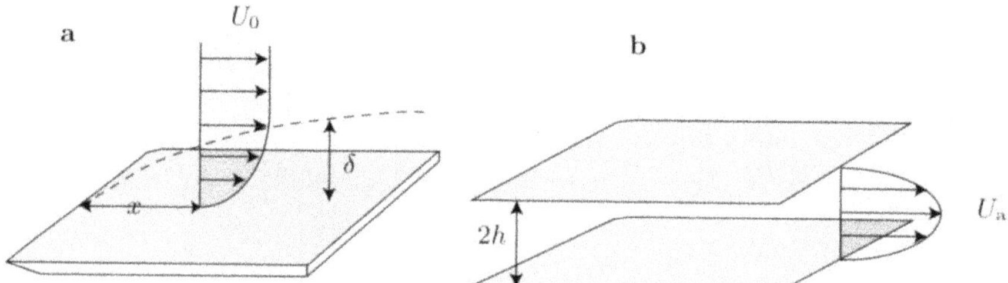

Fig. 1.3. Examples of two-dimensional shear flows: **a** Blasius boundary layer on a flat plate, $\mathrm{Re}_x = U_0x/\nu$; **b** Poiseuille flow in a plane channel, $\mathrm{Re}_h = U_a h/\nu$

At $\mathrm{Re} \geqslant \mathrm{Re}_\mathrm{E}$ the flow loses monotonic stability; i.e. the disturbances are possible, whose energy can experience a transient growth. For example, in plane channel flow (also called plane Poiseuille flow, see Fig. 1.3)[1] $\mathrm{Re}_\mathrm{E} = U_a h/\nu = 49.6$.

At $\mathrm{Re} \geqslant \mathrm{Re}_\mathrm{G}$ the flow loses global stability – it becomes conditionally stable. The sense of the conditional stability is easily seen from the simple

[1] In what follows we use Re with small letter subscripts to denote the characteristic length and with capital letter subscripts to denote some characteristic points in the Reynolds number range.

mechanical model in Fig. 1.4: the system is stable to infinitesimal disturbances, but unstable to disturbances exceeding a certain threshold value in amplitude (the 'condition'). This is in contrast to the 'global' stability case in Fig. 1.1a.

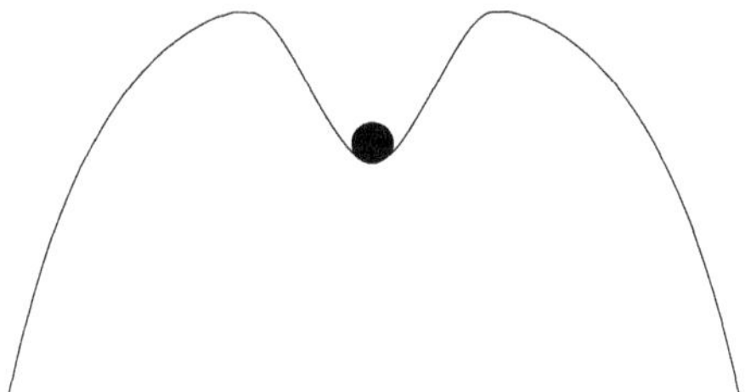

Fig. 1.4. An example of a mechanical system that is steady state stable to small unstable to large velocity disturbances

In other words, at $\mathrm{Re} \geqslant \mathrm{Re_G}$ there may be initial disturbances capable, as a minimum, not to decay in time and, as a maximum, cause transition to turbulence. For such flows, where the principle of exchange of stability holds, the transition Reynolds number $\mathrm{Re_T}$ obviously differs from $\mathrm{Re_G}$, i.e. stable-not-turbulent equilibrium solutions have to be taken into account. We will come back to the conditional stability in Sect. 4.1.2.

At $\mathrm{Re} \geqslant \mathrm{Re_L}$ the flow is linearly unstable, i.e. there is an infinitesimal disturbance which does not decrease in time. Plane Poiseuille flow and the Blasius boundary layer becomes linearly unstable at certain finite Reynolds numbers. However, there are flows such as the flow in a round pipe, which are linearly stable at any Re.

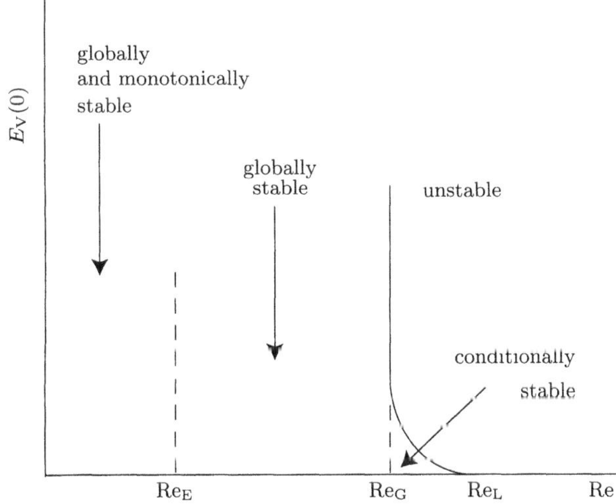

Fig. 1.5. Critical Reynolds numbers and flow stability

As we will see below, global and conditional stability relate closely to the linear stability of small disturbances for the flows of our interest.

1.2.2 Growth of disturbance energy

Let $\boldsymbol{U}(\boldsymbol{r})$ and $P(\boldsymbol{r})$ be distributions of velocity and pressure of a known stationary solution of the equations of motion (the Navier–Stokes equations) of an incompressible fluid (Schlichting and Gersten 2000):

$$(\boldsymbol{U}\nabla)\boldsymbol{U} = -\nabla P + \frac{1}{\mathrm{Re}}\nabla^2\boldsymbol{U}, \tag{1.1}$$

$$(\nabla\boldsymbol{U}) = 0, \tag{1.2}$$

with natural boundary conditions $\boldsymbol{U}|_{\mathrm{s}} = 0$ at boundaries S (walls and/or infinity). Let us impose a disturbance $\boldsymbol{u}(\boldsymbol{r},t)$ and $p(\boldsymbol{r},t)$, where $\boldsymbol{r}$ is a coordinate vector and t is time, so that the resultant motion $\boldsymbol{\mathcal{U}}(\boldsymbol{r}) = \boldsymbol{U}(\boldsymbol{r}) + \boldsymbol{u}(\boldsymbol{r},t)$, and $\mathcal{P}(\boldsymbol{r}) = P(\boldsymbol{r}) + p(\boldsymbol{r},t)$, also satisfy the Navier–Stokes equations and the boundary conditions:

$$\frac{\partial\boldsymbol{\mathcal{U}}}{\partial t} + (\boldsymbol{\mathcal{U}}\nabla)\boldsymbol{\mathcal{U}} = -\nabla\mathcal{P} + \frac{1}{\mathrm{Re}}\nabla^2\boldsymbol{\mathcal{U}}, \tag{1.3}$$

$$(\nabla\boldsymbol{\mathcal{U}}) = 0. \tag{1.4}$$

Substitution of (1.1)–(1.2) into (1.3)–(1.4) yields

$$\frac{\partial\boldsymbol{u}}{\partial t} + (\boldsymbol{U}\nabla)\boldsymbol{u} + (\boldsymbol{u}\nabla)\boldsymbol{U} + (\boldsymbol{u}\nabla)\boldsymbol{u} = -\nabla p + \frac{1}{\mathrm{Re}}\nabla^2\boldsymbol{u}, \tag{1.5}$$

$$(\nabla\boldsymbol{u}) = 0. \tag{1.6}$$

Equations (1.5)–(1.6) are fundamental hydrodynamic equations for disturbances. Non-conservation of energy in the system is a known feature of these equations. An obvious way exclude it consists in decomposing the initial flowfield to the *deformable mean* and perturbed flows, as is assumed frequently for turbulent flows (Waleffe 1995b; Henningson 1996). However, this approach leads to the celebrated closure problem and is not used here; that is we assume that the mean velocity deformation is negligible during the characteristic time of disturbance development in the system.

The equation of energy balance, *the Reynolds–Orr equation*, which serves as a basis for the energy theory, is one of the consequences of the above equations. It is obtained by multiplication of (1.5) by $\boldsymbol{u}$ and integration by parts over volume V using continuity equation (1.6). The procedure yields:

$$\frac{\mathrm{d}E_{\mathrm{V}}}{\mathrm{d}t} = -\int_{\mathrm{V}}(\boldsymbol{u}\boldsymbol{D})\boldsymbol{u}\mathrm{dV} - \frac{1}{\mathrm{Re}}\int_{\mathrm{V}}(\nabla\boldsymbol{u})^2\mathrm{dV}, \tag{1.7}$$

where $D = [\partial U_i/\partial x_j + \partial U_j/\partial x_i]/2 \neq 0$ is the symmetric deformation tensor of the mean laminar flow and $(\nabla\boldsymbol{u})^2 = (\partial u_i/\partial x_j)^2$ in Cartesian coordinates $\{x_1, x_2, x_3\}$.

The first right-hand side term of (1.7) describes the exchange of energy with the mean flow, the second term describes the energy dissipation due to viscosity. If the exchange term is positive, the energy is extracted from the mean flow. The dissipation term is always negative. The relative value of these two terms defines whether the energy of the disturbance decays or grows. The rate of these two terms at $\mathrm{d}E_\mathrm{V}/\mathrm{d}t = 0$ forms the global stability problem for Re_E, below which any disturbance monotonically decays:

$$\frac{1}{\mathrm{Re}_\mathrm{E}} = \max_{\boldsymbol{u}(\boldsymbol{r})\neq 0} -\frac{\int_V (\boldsymbol{u}\boldsymbol{D})\boldsymbol{u}\mathrm{d}V}{\int_V (\nabla \boldsymbol{u})^2 \mathrm{d}V}. \tag{1.8}$$

This is a variational problem to search for *kinematically allowable* velocities which maximize $1/\mathrm{Re}_\mathrm{E}$ – the rate of two quadratic forms (the Rayleigh quotient). It is known that such maximum is always exists and is positive. However, as is, this does not guarantee that the first unstable disturbance is physical, and so to make the prediction realistic, one needs to impose additional constrains on the dynamics of the velocity field. For particular procedures, an interested reader is referred to Joseph (1976).

The above analysis is limited to fluids in which the mean velocity field is independent of the field of disturbances and to a particular definition of the kinetic energy. As we see, the non-linear terms in this case vanish in the Reynolds–Orr equation, and the instantaneous disturbance growth rate $1/E_\mathrm{V}\mathrm{d}E_\mathrm{V}/\mathrm{d}t$ does not depend on the initial disturbance amplitude. This indicates that the possible energy growth of any disturbances *in incompressible Newtonian fluids* described by (1.1)–(1.2) is related to linear growth mechanisms of infinitesimal disturbances (Henningson and Reddy 1994; Waleffe 1995b). The same is true for any other fluids possessing the same conservative properties of the non-linear terms in their equations of motion. In particular, it means that there are implicit linear mechanisms for the transient subcritical growth of energy of disturbances at 'subcritical' $\mathrm{Re}_\mathrm{E} < \mathrm{Re} < \mathrm{Re}_\mathrm{L}$.

It appears that the structure of the most 'dangerous' subcritical disturbances in the energy analysis for a large majority of classical shear flows is a chain of streamwise vortex-like structures (Joseph 1976). More advanced studies based on linear stability theory give similar results (see Sect. 1.5). However, this subcritical growth of energy of disturbances is frequently not capable of producing the laminar–turbulent transition directly, but rather serves as a trigger for other mechanisms, e.g., the so-called *secondary instability* (see Sect. 4.5).

1.2.3 Formulation of linear hydrodynamic stability problems

Many streams stable to 'small' disturbances appear unstable to disturbances of a 'large' amplitude. Nevertheless, in a number of cases, the transition to turbulence begins as a result of instability to small perturbations. In such cases, it is possible to essentially simplify the problem of stability investigation by reducing the consideration of non-linear equations of motion to the analysis of linearized equations for disturbances. A physical substantiation of the linearization procedure was given by Yudovich (1965) and Sattinger (1970). It was shown that the stability to infinitesimal disturbances is the necessary and sufficient condition for stability to physically realizable small – but finite – disturbances.

Dropping the quadratic terms for perturbations in (1.3) yields

$$\frac{\partial \boldsymbol{u}}{\partial t} + (\boldsymbol{U}\nabla)\boldsymbol{u} + (\boldsymbol{u}\nabla)\boldsymbol{U} = -\nabla p + \frac{1}{\mathrm{Re}}\nabla^2 \boldsymbol{u}. \tag{1.9}$$

These equations, with relevant initial and boundary conditions as well as the continuity equation for disturbances (1.6) form the *linear stability problem.* A general theoretical study even of these linearized equations is impossible at present time. Therefore, simplified models that reflect real-life phenomena well enough are usually considered. For shear flows the flow *quasi-stationarity* is sometimes assumed – i.e. a weakness of changes of disturbance amplitude in time. A selection of appropriate coordinate system, which reflects possible symmetry features of the flow is also significant for its mathematical analysis.

In such a way, velocity vectors $\boldsymbol{U}$ and $\boldsymbol{u}$ for a flat flow in component-wise form in Cartesian coordinates are

$$\boldsymbol{U} = \{U(y), V = 0, W(y)\}, \tag{1.10}$$

$$\boldsymbol{u} = \{u(x,y,z,t), v(x,y,z,t), w(x,y,z,t)\}, \tag{1.11}$$

where x is a streamwise coordinate, y is a wall normal, and z is a spanwise coordinate normal to x and y.

Another useful approximation is *local parallelity* of flow streamlines, that is a dependence of velocity of a basic laminar flow $\boldsymbol{U}$ on the only coordinate y normal to the streamlines. If the flow is strictly parallel, e.g., when it is a plane channel flow – a developed internal flow in which fluid moves between two fixed planes at $y = y_1$ and $y = y_2$ (Fig. 1.3a) – the Navier–Stokes equations show that the stationary flow can only be a quadratic function of y. Removing the requirement that $\boldsymbol{U}$ is an exact solution of the stationary equations of motion, it is possible to consider $\boldsymbol{U}$ as a model – not necessarily quadratic – function of y, thereby making it possible to apply further analysis to various types of *quasi-parallel* flows. Such a reasoning can be applied also for flows described in other coordinate systems, e.g., for axisymmetric configurations.

The correctness of the approach must, however, be tested separately in each particular case. For example, the Blasius boundary layer is a classical case of growing external flow (Fig. 1.3a). Its thickness δ increases proportional to $\sqrt{\mathrm{Re}_x}$, where $\mathrm{Re}_x = U_\infty x/\nu$, U_∞ is the free-stream velocity and x is the distance from the leading edge of the plate. The normal velocity component V is small at large Re_x, and the approach of local parallelity works quite satisfactorily for two-dimensional disturbances observed in experiment (Klingmann et al. 1993).

Taking into account (1.10) and (1.11), the equations of motion (1.9) and the continuity equation can be written as

$$\frac{\partial u}{\partial t} + U\frac{\partial u}{\partial x} + W\frac{\partial u}{\partial z} + v\frac{\partial U}{\partial y} = -\frac{\partial p}{\partial x} + \frac{\nabla^2 u}{\mathrm{Re}}, \tag{1.12}$$

$$\frac{\partial v}{\partial t} + U\frac{\partial v}{\partial x} + W\frac{\partial v}{\partial z} = -\frac{\partial p}{\partial y} + \frac{\nabla^2 v}{\mathrm{Re}}, \tag{1.13}$$

$$\frac{\partial w}{\partial t} + U\frac{\partial w}{\partial x} + W\frac{\partial w}{\partial z} + v\frac{\partial W}{\partial y} = -\frac{\partial p}{\partial z} + \frac{\nabla^2 w}{\mathrm{Re}}, \tag{1.14}$$

$$\frac{\partial u}{\partial x} + \frac{\partial v}{\partial y} + \frac{\partial w}{\partial z} = 0. \tag{1.15}$$

These equations together with the boundary and initial conditions describe completely the evolution of an arbitrary infinitesimal disturbance in space and time. The boundary conditions are the non-slippage and non-permeability of the fluid at solid surfaces and the boundedness of $\boldsymbol{u}$ at infinity for external flows. In the latter case, the requirement of disappearance of disturbances at infinity guarantees that the flow relaxes to its undisturbed state in the free stream; i.e. the consideration is limited to disturbances of a shear layer of interest.

The classical approach to the solution of such stability problems is the method of normal modes, consisting of a reduction of the linear initial-boundary-value problem to an eigenvalue problem. Let us suppose that the full solution can be expressed as a sum of elementary solutions (modes), which have the form

$$\{u, v, w, p\} = \{\hat{u}(y), \hat{v}(y), \hat{w}(y), \hat{p}(y)\}\mathrm{e}^{\mathrm{i}[(\boldsymbol{\kappa r}) - \omega t]}, \tag{1.16}$$

where $\hat{u}$, $\hat{v}$, $\hat{w}$, $\hat{p}$ are complex amplitude functions of the disturbances; $\boldsymbol{\kappa}$ is a wave vector, so that $\kappa^2 = \alpha^2 + \beta^2$, where α and β are (generally complex) wave numbers in streamwise and spanwise directions x and z, respectively; and ω is the circular (complex) frequency of the disturbance. In a number of cases the problem can be further simplified to consider only symmetrical or antisymmetrical modes, if the flow has a symmetry in wall-normal direction as, for example, in plane Poiseuille flow.

The performed decomposition is equivalent to transformation from the physical space of Cartesian coordinates and time to the Fourier–Laplace spectral space of wave numbers and frequencies. The elementary solutions are the eigenmodes, i.e. (free, not forced) hydrodynamic waves of the corresponding spectral problem. Let us note that in reducing the initial-boundary-value problem, it is necessary to check the completeness of the obtained set of the waves, since it is possible part of the elementary solutions to be of other types; see Sect. 1.4.

To be specific, let us assume that an elementary disturbance develops in time, i.e. the values α and β are real, and $\omega = \omega_r + \mathrm{i}\omega_i$ is complex. Some further kinematic definitions are necessary. The value ω_i is the growth rate of the wave. The angle between the direction of propagation of the wave (direction of a wave vector $\boldsymbol{\kappa}$) and the streamwise axis x is $\gamma = \arctan(\beta/\alpha)$. The complex vector of the phase velocity of wave propagation (velocity of movement of a crest of the wave along the wave vector) is determined by the expression

$$\boldsymbol{c} = \boldsymbol{c_r} + \mathrm{i}\boldsymbol{c_i} = (\boldsymbol{\kappa}/\omega)^{-1} = \boldsymbol{\kappa} \cdot (\omega/\kappa^2). \tag{1.17}$$

Substituting (1.16) in (1.12)–(1.15), yields:

$$\begin{aligned} \mathrm{i}\,(\alpha U + \beta W - \omega)\,\hat{u} + U'\hat{v} &= -\mathrm{i}\alpha\hat{p} + \frac{1}{\mathrm{Re}}\left[\hat{u}'' - (\alpha^2+\beta^2)\hat{u}\right], \\ \mathrm{i}\,(\alpha U + \beta W - \omega)\,\hat{v} &= -\hat{p}' + \frac{1}{\mathrm{Re}}\left[\hat{v}'' - (\alpha 2+\beta^2)\hat{v}\right], \\ \mathrm{i}\,(\alpha U + \beta W - \omega)\,\hat{w} + W'\hat{v} &= -\mathrm{i}\beta\hat{p} + \frac{1}{\mathrm{Re}}\left[\hat{w}'' - (\alpha^2+\beta^2)\hat{w}\right], \\ \mathrm{i}\,(\alpha\hat{u} + \beta\hat{w}) + \hat{v}' &= 0, \end{aligned}$$

where a prime denotes a derivative in respect to y. In the matrix form:

$$\begin{pmatrix} \mathcal{L}_{SQ}, & \mathrm{d}U/\mathrm{d}y, & 0, & \mathrm{i}\alpha \\ 0, & \mathcal{L}_{SQ}, & 0, & \mathrm{d}/\mathrm{d}y \\ 0, & \mathrm{d}W/\mathrm{d}y, & \mathcal{L}_{SQ}, & \mathrm{i}\beta \\ \mathrm{i}\alpha, & \mathrm{d}/\mathrm{d}y, & \mathrm{i}\beta, & 0 \end{pmatrix} \cdot \begin{pmatrix} \hat{u} \\ \hat{v} \\ \hat{w} \\ \hat{p} \end{pmatrix} = 0. \tag{1.18}$$

Here

$$\mathcal{L}_{SQ} = \mathrm{i}(\alpha\overline{U} - \omega) - (\mathrm{d}^2/\mathrm{d}y^2 - \kappa^2)/\mathrm{Re}, \tag{1.19}$$

where $\overline{U} = U + W\beta/\alpha$ is the effective mean velocity perpendicular to the wavefront. Introducing an effective mean velocity parallel to the wavefront $\overline{W} = (\beta U - \alpha W)/\kappa$, the equations are reduced by successive elimination of variables to the following system:

$$\mathcal{L}_{\mathrm{OS}}\hat{v} = 0, \tag{1.20}$$

$$\mathcal{L}_{\mathrm{SQ}}\hat{u} = \beta\hat{v}\overline{W}'/\kappa + \mathrm{i}\alpha\mathcal{L}_{\mathrm{SQ}}\hat{v}_y/\kappa^2, \tag{1.21}$$

$$\mathcal{L}_{\mathrm{SQ}}\hat{w} = \alpha\hat{v}\overline{W}'/\kappa + \mathrm{i}\beta\mathcal{L}_{\mathrm{SQ}}\hat{v}_y/\kappa^2, \tag{1.22}$$

$$\hat{p} = (\mathrm{i}\alpha\hat{v}\overline{U}' - \mathcal{L}_{\mathrm{SQ}}\hat{v}_y)/\kappa^2. \tag{1.23}$$

The first of these equations is called the (three-dimensional) Orr–Sommerfeld equation with the linear operator:

$$\mathcal{L}_{\mathrm{OS}} = \left(\overline{U} - \frac{\omega}{\alpha}\right)\left(\frac{\mathrm{d}^2}{\mathrm{d}y^2} - \kappa^2\right) - \overline{U}'' + \frac{\mathrm{i}}{\alpha\mathrm{Re}}\left(\frac{\mathrm{d}^2}{\mathrm{d}y^2} - \kappa^2\right)^2. \tag{1.24}$$

Equations (1.21) and (1.22) can be simplified ever more by transforming to 'natural' in this case variables $\overline{u} = (\alpha\hat{u} + \beta\hat{w})/\kappa$ and $\overline{w} = (\beta\hat{u} - \alpha\hat{w})/\kappa$; i.e. to the effective horizontal velocities of the disturbance perpendicular and parallel to the wavefront, respectively. This yields

$$\overline{u} = \mathrm{i}\hat{v}_y/\kappa, \tag{1.25}$$

$$\mathcal{L}_{\mathrm{SQ}}\overline{w} = -\hat{v}\overline{W}'. \tag{1.26}$$

The linear operators $\mathcal{L}_{\text{OS}}$, $\mathcal{L}_{\text{SQ}}$ are called the Orr–Sommerfeld and Squire operators, respectively. Note that it is easy to reformulate (1.26) in terms of normal vorticity of the perturbed motion $\eta = \partial u/\partial z - \partial w/\partial x$, since its amplitude component $\hat{\eta} = \mathrm{i}(\beta\hat{u} - \alpha\hat{w}) = \mathrm{i}\overline{w}\kappa$:

$$\mathcal{L}_{\text{SQ}}\hat{\eta} = -\mathrm{i}\hat{v}\overline{W}'\kappa. \tag{1.27}$$

Therefore, the Squire operator $\mathcal{L}_{\text{SQ}}$ is known also as the normal vorticity operator.

The Orr–Sommerfeld equation (1.20) at homogeneous boundary conditions constitutes the eigenvalue problem for the normal velocity of the disturbance $\hat{v}$. The other components are found then from (1.25) and (1.26). Equation (1.25) shows that the horizontal velocity perpendicular to the wavefront is determined by the continuity condition in this direction. In the second equation, the right-hand term serves as a 'driving force' for the disturbances of the horizontal velocity parallel to the wavefront. Note that the horizontal velocities have opposite properties of symmetry with respect to $\hat{v}$; i.e. if $\hat{v}$ is symmetric, $\hat{u}$ and $\hat{w}$ are antisymmetrical and vice versa.

Physically, the normal velocity modes are eigenoscillations inside the 'hydrodynamic resonator' (the shear flow). The absolute value of the amplitude function (waveform) of the wave can be considered as its 'shape' or amplitude profile along coordinate y, whereas the phase, which depends on y too, denotes time shifts of oscillations $\hat{v}$ in various layers of the fluid.

Solving (1.20) with homogeneous boundary conditions means finding such values α, β and ω at which there is a non-trivial solution (solvability condition). Such implicit relation between α, β and ω can be characterized by the so-called characteristic (dispersion) relation written for a particular Reynolds number in general implicit form as

$$\mathcal{F}(\alpha, \beta, \omega) = 0. \tag{1.28}$$

Spectral formulation of stability. The concept of stability in time in the spectral space is defined as follows. If there are complex eigenvalues $\omega = \omega_r + \mathrm{i}\omega_i$ of the characteristic equation (1.28) such that $\omega_i > 0$ at certain real α and β, the basic state is (linearly) unstable. If $\omega_i < 0$ for all α and β, it is stable to small disturbances. In the case when $\omega_i = 0$ at certain α and β, it is said that the system is *neutrally (indefinitely) stable*. The neutral disturbance can be of two types: if $\omega_r = 0$, it is stationary; if $\omega_r \neq 0$, it is periodic in time and represents a travelling wave.

The equation $\omega_i(\alpha, \beta, \text{Re}) = 0$ parametrically determines a surface in space $(\alpha, \beta, \text{Re})$ separating the region where the flow is stable from an instability region and is called *a boundary or surface of neutral stability* (Fig. 1.6). The value Re_{L} minimized over all frequencies at which the instability occurs is called the (first) critical Reynolds number of the spectral problem. As the neutral surface and the critical Reynolds number bound the parameter regions, at which the basic laminar flow is stable, or unstable to small disturbances, their determination is the primary problem of the stability theory.

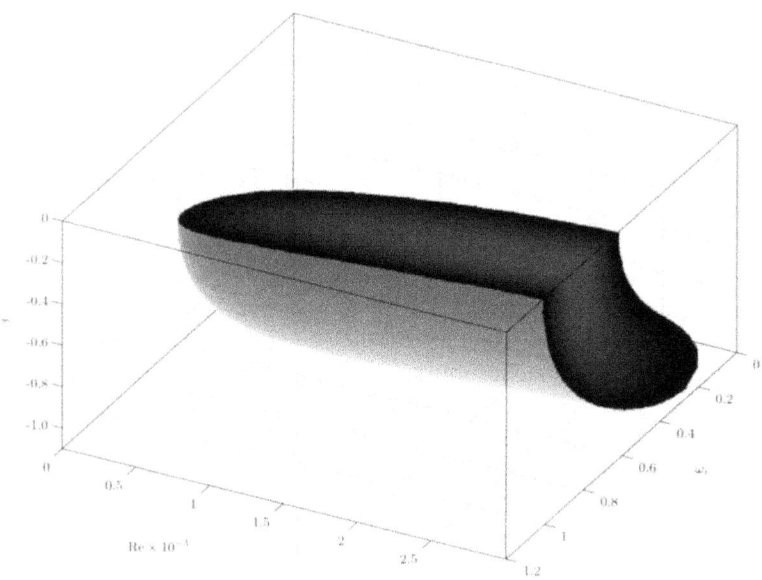

Fig. 1.6. A neutral surface cut with interior isocaps for Poiseuille flow (calculated by A.V. Boiko)

Of course, it is possible to formulate other problems, for example to find the parameters of the dispersion relation at which the growth rate is maximal.

For three-dimensional disturbances, it is necessary to analyze the normal vorticity equation (1.27) as well. It is non-homogeneous (with non-zero right-hand side) linear second-order differential equation, whose solution can be expressed as the sum of a linear combination of the solutions of the corresponding homogeneous problem and a particular solution of the non-homogeneous problem. The particular solution expresses the normal vorticity component of the corresponding Orr–Sommerfeld mode.

The solution of the homogeneous equation results from consideration of the corresponding eigenvalue problem and gives a set of eigenvalues and eigenfunctions of the normal vorticity waves for the flow of interest. However, these modes are always stable: multiplication of the homogeneous normal vorticity equation by its complex-conjugate solution $\hat{\eta}^*$, integration over y and subsequent separation of imaginary and real parts yield

$$\int_{y_1}^{y_2} \left[|\hat{\eta}'|^2 + \left(\kappa^2 + \omega_i \mathrm{Re}\right) |\hat{\eta}|^2 \right] \mathrm{d}y = 0, \tag{1.29}$$

$$\int_{y_1}^{y_2} \left(U - \frac{\omega_r}{\alpha} \right) |\hat{\eta}|^2 \,\mathrm{d}y = 0. \tag{1.30}$$

The first of these equations can be satisfied, only if the waves decay, i.e. $\omega_i < 0$. From the second equation it follows directly that the propagation velocity of a vorticity wave lies between the minimum and maximum mean flow velocities. Therefore, it is possible to study the linear stability of the flow by solving only the Orr–Sommerfeld equation. Of course, the normal vorticity equation is still significant in considering the initial-value problem of the disturbance propagation. Note that under certain conditions, e.g., in the presence of additives in fluid, the problem of linear stability is reduced to the standard one, but effectively with a complex mean velocity profile (Saffman 1962). In such cases the above proof becomes invalid and it is impossible in such a way to exclude the appearance of unstable modes of normal vorticity.

Squire theorem. If $\beta = 0$, the Orr–Sommerfeld operator for three-dimensional disturbances (1.24) is reduced to its two-dimensional analog. Written in the form

$$\mathcal{L}_{\mathrm{OS2D}} = \left(U_2 - \frac{\omega_2}{\alpha_2}\right)\left(\frac{\mathrm{d}^2}{\mathrm{d}y^2} - \alpha_2^2\right) - U_2'' + \frac{\mathrm{i}}{\alpha_2 \mathrm{Re}_2}\left(\frac{\mathrm{d}^2}{\mathrm{d}y^2} - \alpha_2^2\right)^2 \quad (1.31)$$

it is clearly shows that both operators are equivalent if

$$U_2 = U + W\beta/\alpha, \quad (1.32)$$

$$\omega_2/\alpha_2 = \omega/\alpha, \quad (1.33)$$

$$\alpha_2^2 = \alpha^2 + \beta^2, \quad (1.34)$$

$$\mathrm{Re}_2 = \mathrm{Re}\alpha/\alpha_2 = \mathrm{Re}\alpha/\kappa. \quad (1.35)$$

It is seen that in the case of the real wave numbers α and β, the plane oblique waves (i.e. waves with $\beta \neq 0$, also loosely called three-dimensional waves) have effective Reynolds numbers that are smaller compared to the corresponding two-dimensional waves. Hence, at $W = 0$ the problem of determining Re_L is reduced to the two-dimensional problem *for the same basic velocity profile*, but at a smaller Reynolds number Re_2. This statement constitutes *the Squire theorem* (Squire 1933). According to this, a two-dimensional boundary layer loses linear stability to the two-dimensional normal modes with real streamwise wave numbers α_2 at smaller Reynolds numbers, than in the three-dimensional case with the same length of wave vector $\boldsymbol{\kappa}$. Moreover, if the relation $\omega_i(\alpha_2, \beta = 0, \mathrm{Re}_2) = 0$ is known for a given basic velocity profile $U(y)$, then the relation $\omega_i(\alpha, \beta, \mathrm{Re}) = 0$ is determined directly from (1.32)–(1.35). Since the two-dimensional disturbances start to grow at smaller Reynolds numbers in the given approach, to find Re_L it is possible to consider only two-dimensional waves, and to build the *neutral curve* formed by values $\omega_i(\alpha, \beta = 0, \mathrm{Re}) = 0$ instead of the surface of neutral stability.

Typical results of the linear stability calculation on the basis of the two-dimensional Orr–Sommerfeld equation are shown in Fig. 1.7 for two-dimensional plane channel flow and a flat plate boundary layer. The outer lines represent the neutral curve. Branch I, shown in Fig. 1.7a, is called the lower

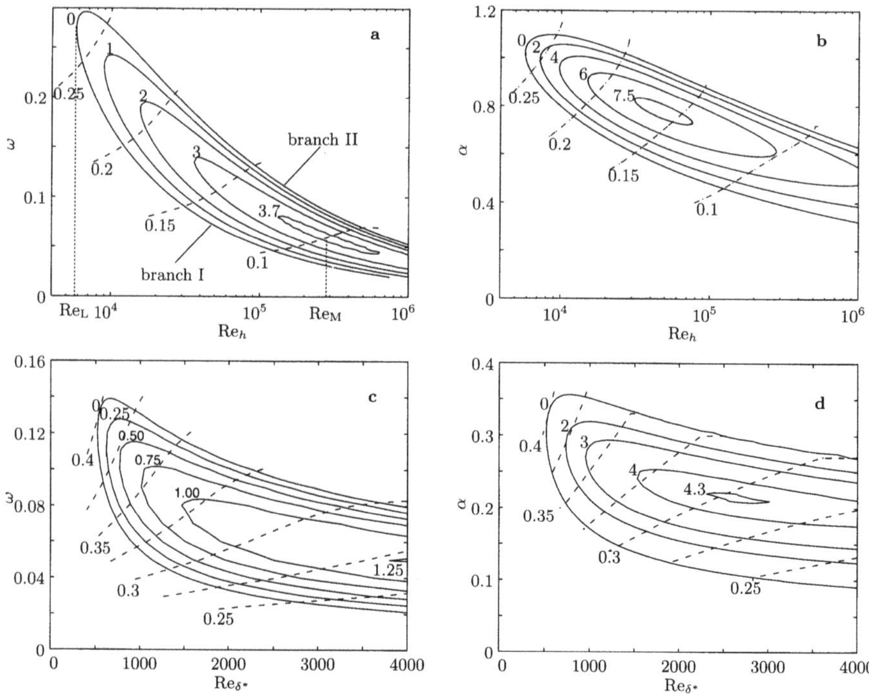

Fig. 1.7. Neutral stability curves and isolines of growth rates (*solid lines*) for: **a** spatial stability of plane Poiseuille flow $-\alpha_i \times 10^2$; **b** temporal stability of plane Poiseuille flow $\omega_i \times 10^3$; **c** spatial stability of Blasius boundary layer $-\alpha_i \times 10^2$; **d** temporal stability of Blasius boundary layer $\omega_i \times 10^3$. *Dashed lined,* isolines of the real part of phase velocities $c = \omega/\alpha$; $\mathrm{Re}_{\delta^*} = 1.72\sqrt{\mathrm{Re}_x}$ (calculated by A.V. Boiko)

branch of the neutral curve. Branch II is called the upper branch. The same definitions are valid for the other neutral curves in Fig. 1.7.

The neutral curves are perhaps the most important – but not unique – parameters describing the instability in the (Re–α)- or (Re–ω)-planes. In particular, the area inside the curves corresponds to unstable disturbances. The isolines or values with other (non-zero) growth rates, and points of the maximum amplification Re_M shown in Fig. 1.7a, can also be of interest.

In the case of a two-dimensional instability wave, the spanwise velocity component is absent and it is convenient to introduce a stream function of the perturbed motion $\psi = \hat{\psi}\mathrm{e}^{\mathrm{i}(\alpha x - \omega t)}$ so that

$$v = -\partial\psi/\partial x \qquad \text{and} \qquad u = \partial\psi/\partial y. \tag{1.36}$$

Then the problem related to disturbances is reduced to the analysis of only the Orr–Sommerfeld equation for the amplitude of the stream function $\hat{\psi}$.

Frequently the Squire theorem is misinterpreted. First of all, it is valid only for the stability in time (and as an obvious exclusion, for the neutral waves developing in space). Cases when the oblique waves amplify faster than their two-dimensional counterparts are not excluded, and under certain conditions can serve as the main source for the transition to turbulence as, for example, in scenarios by Berlin et al. (1994), and Elofsson and Lundbladh (1994). Secondly, it relates to the lower limit of the linear instability $\mathrm{Re_L}$. Meanwhile, situations in the boundary layer (when the Reynolds number grows downstream) are possible when a two-dimensional wave has already left the instability region, but some corresponding oblique waves still amplify. As the neutral curve for the waves with $\beta \neq 0$ represents the same curves displaced to larger Re, the three-dimensional waves have larger growth rates behind the the point of maximum amplification of the two-dimensional waves at (i.e. $\mathrm{Re} > \mathrm{Re_M}$); see Fig. 1.7a.

At last, for the solution of the initial-value problem, e.g., in studying the development of wave packets (see Sect. 2.1.3), the solution of the linear problem is only a starting point, since it is necessary to find not only the eigenvalues but also the eigenfunctions $\hat{v}$, which for the cases of straight waves and oblique waves are different. Then, to describe the development in time of an initial disturbance under homogeneous boundary conditions, it is necessary to find a complete set of the waves constituting it. Moreover, in this case an attempt to reduce the initial-value problem *only* to search for the eigenvalues and eigenfunctions $\hat{v}$ is not physically justified, as the Orr–Sommerfeld operator is not usually self-adjoint, and its eigenfunctions are not orthogonal (see Sect. 1.5).

1.2.4 Inviscid linear stability problem

Theoretical difficulties with the analysis of the Orr–Sommerfeld equation (1.20) associated with the presence of the small parameter $(\alpha\mathrm{Re})^{-1}$ at the highest derivative can be reduced by assuming that a flow loses stability at large Reynolds numbers. Then it seems natural to drop the viscous terms. This results in the Rayleigh equation of the inviscid instability (Schlichting and Gersten 2000):

$$\left(\hat{v}'' - \kappa^2\hat{v}\right) - \frac{\overline{U}''}{\overline{U} - c}\hat{v} = 0, \tag{1.37}$$

where $c = \omega/\alpha$ is a kind of phase velocity, cf. (1.17). Since it is of second order, it is possible to satisfy only two of the four boundary conditions of the viscous problem – keeping those of zero normal velocity at the walls and/or infinity and dropping no-slip conditions. As practice shows, such simplification is justified, in particular, for free or separated shear flows, when solid boundaries are absent or far from the shear layer of interest.

The instability waves connected to the solutions of the Rayleigh equation (1.37) are called Rayleigh waves. Considering the inviscid stability problem *in time* (but not in space) for two-dimensional waves ($\beta = 0$), Rayleigh (1880) proved some important general theorems. The first theorem (or *the inflection point criterium*) states that the necessary condition for inviscid instability is the presence of an inflection point in the mean velocity profile $\overline{U}(y)$; i.e. if $c_i > 0$, then $\overline{U}''$ changes sign between flow boundaries. The inflection point is a point y_s such that $\overline{U}''(y_s) = 0$.

This result is obtained by formal multiplication of the Rayleigh equation by its complex-conjugate solution $\hat{v}^*$ and integration by parts over y. The imaginary part of the resulting equation is

$$c_i \int_{y_0}^{y_1} \frac{\overline{U}''|\hat{v}|^2}{|\overline{U} - c|^2} \mathrm{d}y = 0. \tag{1.38}$$

It is seen that for the equality to be valid, $\overline{U}''$ has to change sign between the limits. Based on the inflection point criterium nothing can be said, about the linear stability of plane Couette flow with $\overline{U}(y) = y$, where $\overline{U}'' = 0$ at all y. An additional necessary condition for the instability of monotonous mean velocity profiles, the *criterium of the maximum vorticity* $\mathrm{d}\overline{U}/\mathrm{d}y$, was found later by Fjørtoft (1950), through a consideration of the real part of the equation leading to (1.38). Simple manipulation yields

$$\int_{y_1}^{y_2} \frac{\overline{U}''\left(\overline{U} - \overline{U}_s\right)}{|\overline{U} - c|^2} |\hat{v}|^2 \mathrm{d}y < 0,$$

where $\overline{U}_s$ is the velocity at the inflection point. Thus, only the velocity profiles with the inflection points associated with the maximum shear are unstable; i.e. the inequality $\overline{U}'(\overline{U} - \overline{U}_s) < 0$ should be satisfied over a certain range of y. Taking into account the first Rayleigh theorem, this statement is equivalent to the requirement of a relative maximum of the absolute value of the vorticity at y_s for the instability to occur (Fig. 1.8). It follows in particular from the criterium of maximum vorticity that the Couette flow is stable in the inviscid approach.

Howard (1961) proved the so-called *semicircle theorem* for the bound of the growth rates for unstable Rayleigh waves. It follows from multiplication of the Rayleigh equation by the complex conjugate of the auxiliary variable Z introduced above and integration of the obtained equation over y taking into account the homogeneous boundary conditions. After some transformations of the real and imaginary parts, it appears that

$$\left[c_r - \frac{1}{2}\left(\overline{U}_{\max} + \overline{U}_{\min}\right)\right]^2 + c_i^2 \leqslant \left[\frac{1}{2}\left(\overline{U}_{\max} - \overline{U}_{\min}\right)\right]^2,$$

where $\overline{U}_{\min}$ and $\overline{U}_{\max}$ are the minimum and maximum velocity in the flow, respectively (Fig. 1.9). All unstable modes lie inside a semicircle.

The *second Rayleigh theorem* states that the velocity of propagation of neutral disturbances in a shear layer is between the minimum and maximum flow velocity, i.e. $\overline{U}_{\min} < c_r < \overline{U}_{\max}$. If an auxiliary variable $Z = \hat{v}/(\overline{U} - c)$ is introduced in (1.37), and the equation is multiplied by its complex-conjugate

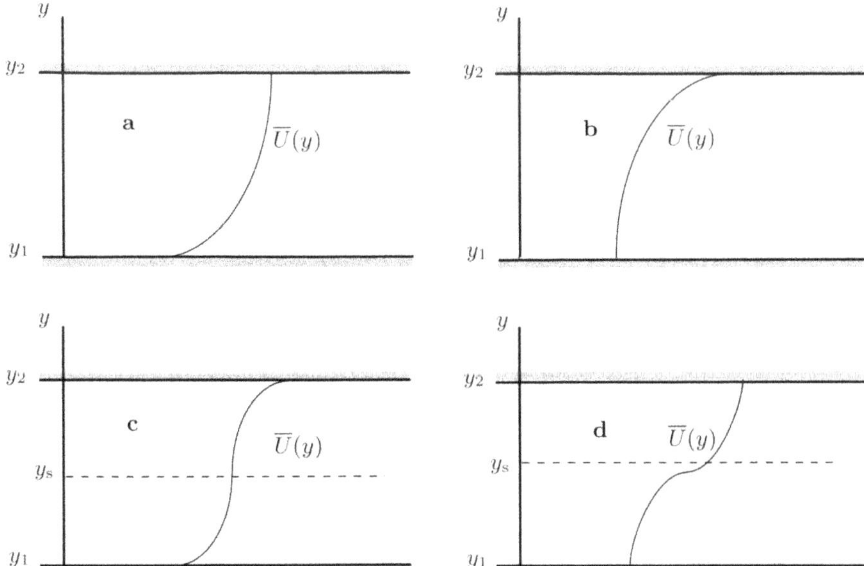

Fig. 1.8. Analysis of stability of flows through stability criteria: **a** stable, $\overline{U}'' < 0$; **b** stable, $\overline{U}'' > 0$; **c** stable, $\overline{U}''_s = 0$ in y_s, but $\overline{U}''(\overline{U} - \overline{U}_s) \geqslant 0$; **d** probably unstable, $\overline{U}''_s = 0$ and $\overline{U}''(\overline{U} - \overline{U}_s) \leqslant 0$

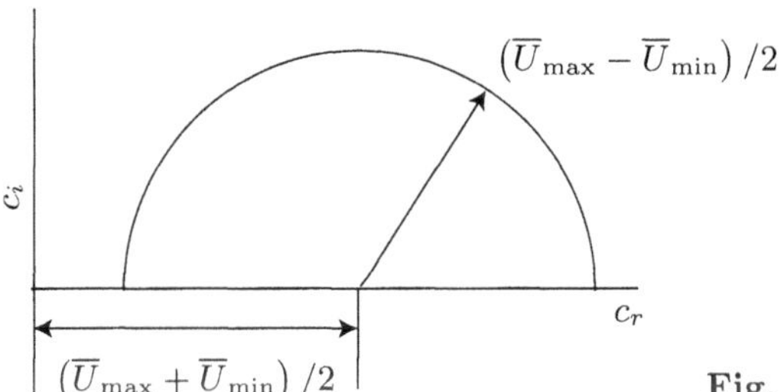

Fig. 1.9. Howard's semicircle

Z^* and integrated, the result of the theorem follows from the consideration of the real part of the final expression.

It means that there is a layer inside an inviscidly unstable shear flow where the local mean velocity is equal to the velocity of the disturbance propagation. The distance $y = y_c$, where $\overline{U} = c_r$, is called the *critical layer* of the mean flow. This layer frequently corresponds to a singular point in the Rayleigh equation for a neutral disturbance. The normal velocity component $\hat{v}''$ of the neutral disturbance tends to infinity at the critical layer, unless $\overline{U}''(y_c) = \overline{U}''(y_s) = 0$ or unless the critical layer does not coincide with one of the boundaries. In the latter case the singular solution is overlapped by the boundary condition $\hat{v} = 0$. The regular neutral modes are known to exist for some flows (Maslowe 1985). Otherwise, the presence of the singularity indicates that it is impossible to completely neglect an effect of viscosity in the vicinity of the critical layer, at least for the neutral disturbances.

Obviously, the considered stability problem formulation with precisely zero viscosity is not physically quite correct. It results in the disappearance of some solutions (because of the reduction of the order of the Orr–Sommerfeld equation) and the appearance of new ones (due to the possible singularity in the critical layer and weakened boundary conditions). An interesting conclusion follows if to take into account that the eigenvalues of the Rayleigh equation appears in complex-conjugate pairs. Combined with the inflection point criterion, this leads to the so-called Lin (1955) paradox of the selection of the solutions implemented in practice for profiles without the inflection. In such a case the Rayleigh equation can have no solutions – either stable or unstable — satisfying given boundary conditions (Maslowe 1985). This can be easily seen for plane Couette flow with $U = y$. The Rayleigh equation is then reduced to

$$v'' - \kappa^2 v = 0,$$

which has no solution equal to zero at both boundaries. In such a case the usual mathematical tool to describe the evolution of an initial disturbance is through an introduction of a set of singular modes of a continuous spectrum corresponding to the solutions of the equation

$$\left(\hat{v}'' - \kappa^2 \hat{v}\right) - \frac{U''}{U - c}\hat{v} = \delta(U - c)$$

at the same homogeneous boundary conditions (here δ is the delta function). Then $\hat{v}$ are generalized (Green) functions with singularity of the first derivative at $U = c$, which is smoothed in solving the initial-value problem (Case 1960; Lin 1961). It could be shown that for the particular case of the Couette flow, the inviscid solution consists of non-dispersive disturbances propagating with the local mean velocity $U(y)$. For flows with more complex non-inflectional velocity profiles, the stable dispersive solutions may, however, exist if to allow y to be complex for integration purposes (Schmid and Henningson 2000). Obviously, in such a case the Rayleigh criterion does not operate.

The Lin's paradox also takes place for unstable flows. It appears that the unstable oscillations, as a rule, relate directly to the solutions of the Orr–Sommerfeld equation at $\mathrm{Re} \to \infty$. For decaying waves this is not always so. At least a part of them should be rejected as physically meaningless in the framework of linear theory.

Yet another more physical paradox is stipulated by viscosity, which under certain conditions can lead to a destabilizing action. Otherwise it is impossible to explain the reason of the instability in channel flow or in the Blasius boundary layer, where the inflection points in velocity profiles are absent. Such instability is called the *viscous instability*. Obviously, it can manifest

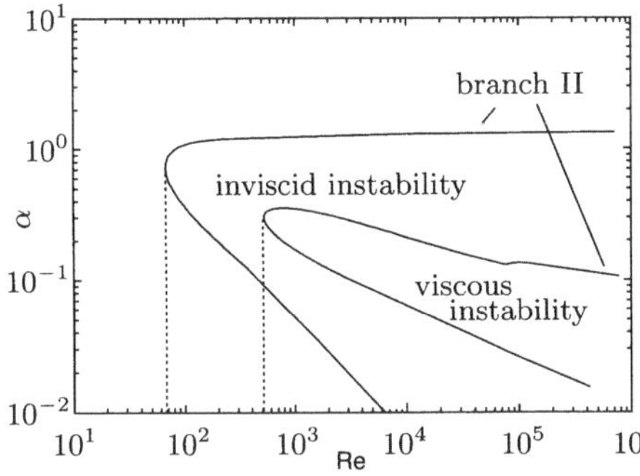

Fig. 1.10. Typical viscous – for Blasius boundary layer profile – and inviscid – for pre-separation Hartree profile – neutral stability curves (calculated by A.V. Boiko)

itself only at finite Reynolds numbers. Figure 1.10 illustrates that the asymptotic behaviour of the neutral stability branch II for the Blasius boundary layer, and for the profile close to separation, is different as Re approaches infinity. In the former case (no inflection point in the profile interior) it tends to zero; in the latter case (with the inflection), it becomes a constant value.

The physical mechanism of the viscous instability was described by Prandtl (1935) and Lin (1955). The two-dimensional counterpart of (1.7) is

$$\frac{\mathrm{d}E}{\mathrm{d}t} = \int \tau U' \mathrm{d}y - \frac{1}{\mathrm{Re}} \iint \left(\frac{\partial v}{\partial x} - \frac{\partial u}{\partial y} \right)^2 \mathrm{d}x \mathrm{d}y,$$

where τ is the Reynolds stress averaged in the streamwise direction x. Applying the averaging over one wavelength of the travelling wave gives $\tau = (\hat{v}\hat{u}^* - \hat{u}\hat{v}^*)/2$. As can be seen, the disturbance energy increases only if the phase shift between $\hat{u}$ and $\hat{v}$ changes the sign of τ. The Orr–Sommerfeld equation shows indeed that the viscosity (as a part of the imaginary multiplier containing Re at the higher derivatives) introduces a phase shift compared to the inviscid case. The phase shifts for $\hat{u}$ and $\hat{v}$ are different, since the viscous terms for these velocity components are different and $\hat{u}$ should satisfy additional no-slip boundary conditions. At certain parameters ω, α, β and Re, and velocity profile $U(y)$, this effect can appear to be destabilizing and exceed the effect of the energy dissipation by viscosity.

1.3 Instability in space

In the framework of the eigenvalue approach there is no means to guess a priori, which formulation – spatial or temporal – corresponds to a particular problem of interest in a given reference system. To identify how the disturbance develops, it is necessary to consider the impulse response of the system (Fig. 1.2). Such a general theoretical proof of convective character of instability for Poiseuille flow at Re $\gg$ 1 and for boundary layer velocity profiles without the inflection point was done first by Iordanskiy and Kulikovskiy

(1965). We consider how this problem is related to the wave packet development and can be attacked experimentally or numerically by the wave group velocity consideration in Sect. 2.1.3.

In any case, experiments show that disturbances propagate downstream in such open flows as the Blasius boundary boundary layer and plane channel flow even at quite small Reynolds numbers. Therefore, it is more appropriate for such flows to consider the waves with real frequencies ω and complex wave numbers α and β. However, spatial instability is a process which differs significantly from instability in time. The evolution problem for the normal velocity disturbance as $v(x,y,z,t) = \hat{v}(y,t)\mathrm{e}^{\mathrm{i}(\alpha x+\beta z)}$ obtained with the help of (1.12)–(1.15) is the quasi-linear parabolic partial differential equation:

$$\left[\left(\frac{\partial}{\partial t} + \mathrm{i}\alpha\overline{U}\right)\left(D^2 - \kappa^2\right) - \mathrm{i}\alpha\overline{U}'' \frac{1}{\mathrm{Re}}\left(D^2 - \kappa^2\right)^2\right]\hat{v}(y,t) = 0.$$

The spatial evolution problem for $v(x,y,z,t) = \hat{v}(x,y,z)\mathrm{e}^{-\mathrm{i}\omega t}$, on the contrary, is described by the elliptic equation:

$$\left[\left(-\mathrm{i}\omega + U\frac{\partial}{\partial x}\right)\nabla^2 - U''\frac{\partial}{\partial x} - \frac{1}{\mathrm{Re}}\nabla^4\right]\hat{v}(x,y,z) = 0.$$

The latter problem is, as a rule, ill-posed, as the solution depends on the conditions at the downstream boundary (e.g., at the trailing edge of the flat plate) and there are formal solutions propagating upstream. Note that the disturbance kinetic energy, which is almost exclusively used in the temporal stability approach as the measure of the disturbances (see Sect. 1.2) is, generally speaking, not appropriate in the spatial case (Henningson and Schmid 1994). At the absolute instability, the equality of the kinetic energy of a disturbance to zero means that it will be zero in all subsequent instances. At the convective instability, this is not always so. If, for example, the disturbance kinetic energy in the plane perpendicular to the flow direction is chosen as the measure, it cannot be appropriate for the mathematical description of instability which account standing waves whose periodic beatings ensure the absence of the energy at certain points in space.

One has to regularize the problem by imposing additional constrains to the initial data. Salwen and Grosch (1981) proposed to exclude all solutions propagating upstream. Zhigulev and Tumin (1987) gave a possible physical interpretation of this conjecture as a fast 'loss' of the effect of the downstream boundary conditions in the bulk of the channel or boundary layer flow due to a quick decay of the upstream propagating disturbances. With this constraint the stability criteria based on the energy of a disturbance, say, on the root-mean-square of the streamwise velocity component can still be used in physical experiments and numerical works.

In mathematical studies of the convective instability, the Reynolds number Re, the disturbance frequency ω, and one of the wave numbers, usually β, are given, while α is obtained from the solution of characteristic equation (1.28), which now parametrically determines the eigenvalues of the spatial problem. Compared to the temporal instability case, the problem is complicated, since one has to prescribe β, which is now generally complex. However, for given α and β it is always possible to rotate x and z axes so that in the new coordinates x' and z', the wave number α' along x' is complex and the wave number β' along z' is real.

For simple two-dimensional flows, the flow symmetry in respect to the z-coordinate ensures that these two coordinate systems coincide and all β are real. For general three-dimensional parallel flows, additional considerations have to be taken into account to choose the proper coordinate system (Nayfeh and Padhye 1979; Nayfeh 1980).

As in the temporal problem, it is possible to introduce a real wave vector $\boldsymbol{\kappa_r}$ such that its length and the angle between direction $\boldsymbol{\kappa_r}$ and axis x are

$$\kappa_r = \sqrt{\alpha_r^2 + \beta_r^2} \qquad \text{and} \qquad \gamma_r = \arctan\left(\frac{\beta_r}{\alpha_r}\right),$$

respectively. The phase velocity is then expressed by the formula

$$\boldsymbol{c} = \left(\frac{\boldsymbol{\kappa_r}}{\omega}\right)^{-1} = \boldsymbol{\kappa_r} \cdot \frac{\omega}{\kappa_r^2}.$$

However, in contrast to the temporal problem, it is possible to also define a vector $\boldsymbol{\kappa_i}$ through

$$\kappa_i = \sqrt{\alpha_i^2 + \beta_i^2} \qquad \text{and} \qquad \gamma_i = \arctan\left(\frac{\beta_i}{\alpha_i}\right).$$

If the direction along $\boldsymbol{\kappa_i}$ is denoted as $\tilde{\boldsymbol{r}}$, and the direction along $\boldsymbol{\kappa_r}$ as $\overline{\boldsymbol{r}}$, then the expression for the wave reads

$$\tilde{v} = \hat{v}(y)\mathrm{e}^{-\kappa_i \tilde{\boldsymbol{r}}}\mathrm{e}^{\mathrm{i}(\kappa_r \overline{\boldsymbol{r}} - \omega t)}.$$

As seen, $-\boldsymbol{\kappa_i}$ is the spatial amplification rate along the direction $\tilde{\boldsymbol{r}}$. In particular, if $\beta_i = 0$, the wave with $\beta_r \neq 0$ propagates at the angle $\gamma_r \neq 0$, but amplifies strictly downstream, $\gamma_i = 0$.

Assuming that β is real, the condition $\alpha_i = 0$ provides the neutral curves $\omega = \omega(\beta, \mathrm{Re})$ for different β constituting together the neutral surface separating the regions of flow stability at $\alpha_i > 0$ and flow instability at $\alpha_i < 0$ (Fig. 1.10). Due to differences in the mathematics of linear equations for eigenvalue problems, the general theorems of the inviscid temporal problem related to the real wave numbers and the complex frequency given in Sect. 1.2.4 are not applicable to the problem of spatial amplification. The Squire transformation (1.32)–(1.35) loses clear physical sense: the effective two dimensional mean velocity profiles and the Reynolds number become complex. However, since the equations for the neutral waves coincide in both formulations, the analytical conclusions about the spatial neutral modes are still valid.

1.4 Completeness of solutions for the Orr–Sommerfeld and Squire equations

As has already been mentioned in Sect. 1.2.3, to describe the development of a disturbance of an arbitrary shape in a particular flow, it is necessary to know the complete set of the waves constituting it. The completeness theorem, i.e. the existence of the complete set of discrete eigenvalues with corresponding eigenfunctions for the Orr–Sommerfeld equation was proved for plane Poiseuille flow by Schensted (1960), and later for *any* internal flow by Yudovich (1965) and DiPrima and Habetler (1969). For external flows, the general completeness theorem is absent. Usually to obtain a complete set of eigenfunctions it is necessary to supplement the normal mode analysis by the analysis of continous spectrum: to consider the so-called generalized eigenfunctions, i.e. the functions corresponding to the Orr–Sommerfeld equation which do not decay at infinity, but rather are only bounded.

For example, for a given finite Reynolds number and frequency of oscillations, there is only a finite number of discrete eigenvalues in an external flow such as the Blasius boundary layer (Fig. 1.11). However, as the Reynolds number increases, new discrete eigenvalues 'jump out' from the continuous spectrum (Murdock and Stewartson 1977; Grosch and Salwen 1978), so that at Re $\to \infty$ there is an infinite number of discrete eigenvalues, while at Re ≈ 1 the discrete spectrum is empty. Note, that at Re ≈ 1 the boundary layer approximation is not applicable and such a stability analysis is of a little practical importance.

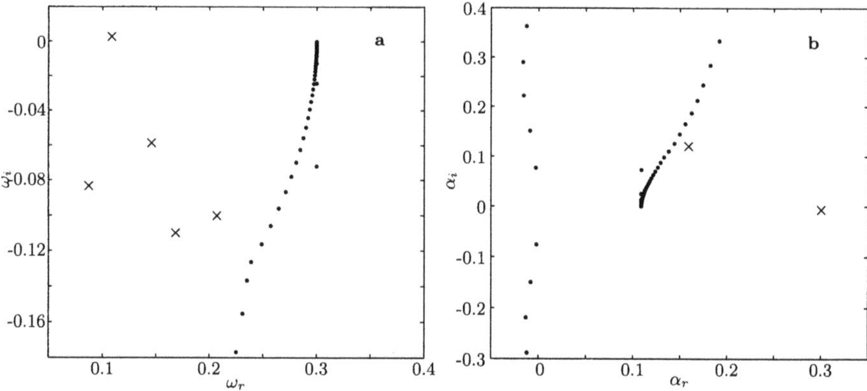

Fig. 1.11. Orr–Sommerfeld eigenvalues for Blasius boundary layer at $\beta = 0$, $\mathrm{Re}_{\delta^*} = 1000$: **a** temporal spectrum at $\alpha = 0.3$; **b** spatial spectrum at $\omega = 0.1087$. Discrete modes are shown by $\times$; other *points* are a 'numerical representation' of a part of poorly resolved continuous modes (calculated by A.V. Boiko)

The continuous spectrum of the Orr–Sommerfeld equation has a clear interpretation. Let us consider as a physical analogy a Schrödinger equation for an electron in the hydrogen potential. In this case there is an infinite set of discrete eigenvalues (levels) and corresponding eigenfunctions which, however, is not complete. The completeness requires the addition of generalized eigenfunctions, whose eigenvalues constitute the continuous spectrum. The discrete values correspond to the possible energy levels of the electron-bound states in the atom, whereas the continuous spectrum corresponds to the energy of a free electron scattered on the potential.

Similarly, in external flows far away from boundaries, nothing forbids the existence of disturbances generated, for example, by a remote source and 'scattered' at the wall. It is important to note that the nature of the continuous spectrum of the Orr–Sommerfeld equation differs completely from that of the continuous spectrum of the Rayleigh equation (see Sect. 1.2.4). In the presence of viscosity, the stability equations are regular and the continuous spectrum arises as a tool to describe disturbances in the external flow, whereas in the inviscid case it can arise even for flows limited by walls.

It was shown by Zhigulev et al. (1980) and Salwen and Grosch (1981) that the set of eigenvalues of the discrete and continuous spectra constitute the complete system both for temporal and spatial stability problems for flows of boundary-layer type. With no-slip boundary conditions at the wall and boundedness of disturbances in the free stream

$$\hat{v}(0) = 0, \qquad \hat{v}'(0) = 0, \qquad \hat{v}(y \to \infty) < \infty, \qquad \hat{v}'(y \to \infty) < \infty,$$

the solution contains the waves caused by the presence of background disturbances in the free stream.

The Orr–Sommerfeld equation has four linearly independent fundamental solutions. It is easy to establish their asymptotic behaviour at $y \to \infty$. For simplicity we assume that the axes are oriented so that β is real, $U \to 1$ and $W \to W_0$ beyond a narrow shear layer at the wall. The Orr–Sommerfeld equation is reduced at $y \to \infty$ to

$$(\alpha + \beta W_0 - \omega)\left(\hat{v}'' - \kappa^2 \hat{v}\right) = -\frac{\mathrm{i}}{\mathrm{Re}}\left(\hat{v}^{iv} - 2\kappa^2 \hat{v}'' + \kappa^4 \hat{v}\right), \tag{1.39}$$

and after the standard substitution $\hat{v} = \mathrm{e}^{qy}$ it becomes

$$(\alpha + \beta W_0 - \omega)\left(q^2 - \kappa^2\right) = -\frac{\mathrm{i}}{\mathrm{Re}}\left(q^2 - \kappa^2\right)^2 .$$

This equation has four roots, and the general solution of (1.39) is:

$$\hat{v} = C_1\varphi_1 + C_2\varphi_2 + C_3\varphi_3 + C_4\varphi_4,$$

where

$$\varphi_1 = \mathrm{e}^{\kappa y}, \qquad \varphi_2 = \mathrm{e}^{-\kappa y}, \qquad \varphi_3 = \mathrm{e}^{\xi y}, \qquad \varphi_4 = \mathrm{e}^{-\xi y}, \tag{1.40}$$

$$\xi = \sqrt{\kappa^2 + i\mathrm{Re}(\alpha + \beta W_0 - \omega)}, \tag{1.41}$$

and $\mathrm{Real}(\xi) < 0$.

It is seen that finite, but non-zero disturbances exist at $y \to \infty$, when either κ or ξ is purely imaginary. Let $\kappa = \pm ik$, $k > 0$. In this case α must be purely imaginary. The solutions limited at $y \to \infty$ take the form

$$\hat{v} = C_1\varphi_1 + C_2\varphi_2 + C_3\varphi_3.$$

The boundary conditions $\hat{v}(0) = \hat{v}'(0) = 0$ result in equations

$$C_1\varphi_1(0) + C_2\varphi_2(0) + C_3\varphi_3(0) = 0,$$
$$C_1\varphi_1'(0) + C_2\varphi_2'(0) + C_3\varphi_3'(0) = 0.$$

Based on them, two constants, e.g., C_2 and C_3, can be expressed through the third one. To be specific, let us choose $C_1 = 1$, then

$$\hat{v} = \varphi_1 + C_2\varphi_2 + C_3\varphi_3.$$

A particular form of C_2 and C_3 for the other coordinates y is determined by the mean velocity profile through the Orr–Sommerfeld equation and the boundary conditions. The solution exists at any k and consequently corresponds to the continuous spectrum. In the case under consideration

$$\lim_{y\to\infty} \hat{v} = \mathrm{e}^{iky} + C_2\mathrm{e}^{-iky}.$$

It is possible to show that at $y \to \infty$ the wave possesses a disturbance of vorticity equal to zero and a finite disturbance of pressure. Therefore, such waves are called *pressure waves.* Furthermore, these waves are standing, as their phase velocities $c = \pm\mathrm{i}\omega/k$ are purely imaginary and depending on the sign at k, the wave amplitudes decrease or grow in the direction of wave vector $\boldsymbol{\kappa}$.

Let now $\xi = \pm ik$ be purely imaginary. Then the equation for the wave numbers α is

$$\alpha^2 + \mathrm{i}\alpha\mathrm{Re} + k^2 + \beta^2 + \mathrm{i}(\beta W_0 - \omega)\mathrm{Re} = 0,$$

whence, supposing $\mathrm{Re} \gg 1$,

$$\alpha_1 \approx \omega - \beta W_0 + \mathrm{i}(k^2 + \beta^2)/\mathrm{Re} \qquad \text{and} \qquad \alpha_2 \approx \beta W_0 - \omega - \mathrm{iRe}.$$

If $W_0 = 0$ as in a two-dimensional case, the first wave has velocity of propagation of the free stream velocity ($c = \mathrm{Real}(\omega/\alpha) = 1$) and is damped weakly with x. The second wave propagates with the same velocity, but upstream ($c = -1$), and strongly decays in the direction of propagation. Their amplitude functions are

$$\hat{v} = C_2\varphi_2 + C_3\varphi_3 + C_4\varphi_4, \qquad \hat{v} = C_1\varphi_1 + C_3\varphi_3 + C_4\varphi_4.$$

As in the previous case, the two homogeneous boundary conditions at the wall determine two arbitrary constants through the third one. Thus, the considered waves exist at any k and consequently also constitute a part of the continuous spectrum. It could be shown that they possesses *vorticity disturbances.*

The characteristics of the continuous spectrum in flat plate flow has been the subject of a series of theoretical (Mack 1976; Murdock and Stewartson 1977; Grosch and Salwen 1978; Zhigulev et al. 1980; Salwen and Grosch 1981; Craik 1991) and experimental (Kachanov et al. 1982) studies. The structure of the pressure and vorticity waves can be understood by considering their amplitude functions in terms of wave scattering. The first term can be interpreted as the incident wave, the second as the reflected polarized (i.e. with complex amplitude) wave and the third one is an 'induced' near-wall wave, which is necessary to satisfy the boundary conditions at the wall.

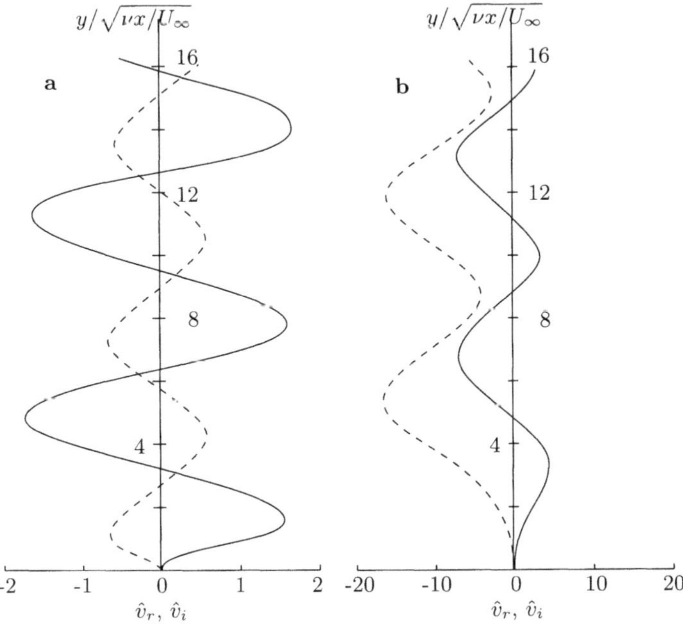

Fig. 1.12. Modes of the continuous spectrum for the Blasius boundary layer: **a** a pressure wave; **b** a vorticity wave. $\mathrm{Re}_x^{1/2} = 300$, $\omega = 0.006$, $\beta = 0$, $k = 1$ (Zhigulev and Tumin 1987)

Figure 1.12 shows the results of numerical calculations of the real $\hat{v}_r$ and imaginary $\hat{v}_i$ parts of the eigenfunctions for pressure and vorticity waves. As can be seen, when k increases, the disturbances of the continuous spectrum penetrate deeper inside the boundary layer; meanwhile, they decay faster.

Let us consider finally the case when κ and ξ are complex. Let $\mathrm{Real}(\kappa) > 0$, then the only allowable bounded solution exhibits $\hat{v} \to 0$ at $y \to \infty$ and

$$\hat{v} = C_2\varphi_1 + C_3\varphi_3.$$

The boundary conditions at $y = 0$ yield

$$C_2\varphi_2(0) + C_3\varphi_3(0) = 0,$$

$$C_2\varphi_2'(0) + C_3\varphi_3'(0) = 0.$$

For the existence of non-zero C_2 and C_3, it is necessary that the determinant of the system be zero:

$$\varphi_2(0)\varphi_3'(0) - \varphi_3(0)\varphi_2'(0) = 0. \tag{1.42}$$

Thus the spectrum of these waves is discrete; i.e. at a fixed ω, β, and Re there is a number of α_n ($n = 1, 2, 3, \ldots$) satisfying (1.42). They can both grow or decay in the direction of wave propagation as well as be neutral.

They are near-wall waves corresponding to that discussed in Sect. 1.2.3: their amplitudes approach zero at large y, i.e. they have a maximum (or maxima) near the wall (usually in the shear layer). Such elementary solutions of the eigenvalue problem of the Orr–Sommerfeld equation with $\omega \neq 0$, i.e. travelling eigenmodes of the shear layer, are called Tollmien–Schlichting waves.

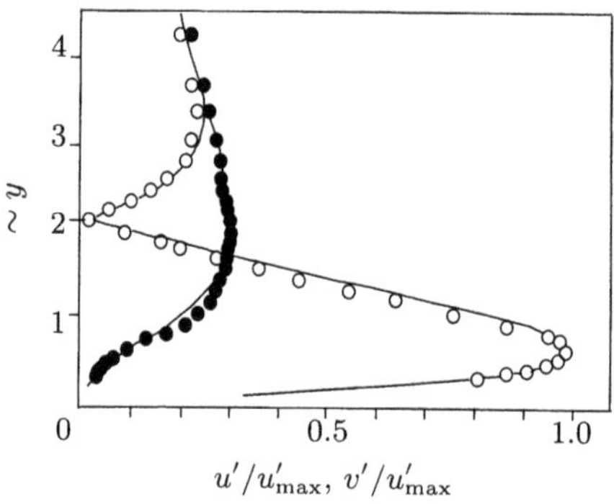

Fig. 1.13. A typical profile of two-dimensional Tollmien–Schlichting wave in the Blasius boundary layer. ∘, u'; •, v'. The velocities are normalized to the internal velocity maximum $u'_{\max}$. Lines shows corresponding calculation using the linear parallel theory (Westin 1997)

A typical distribution of the unstable Tollmien–Schlichting wave amplitude of the Blasius boundary layer measured as the root-mean-square of the periodic signal (in contrast to theoretical disturbance values denoted throughout the book by a prime) in an experiment by Westin (1997), together with the corresponding curves obtained with the parallel linear stability theory, is shown in Fig. 1.13. The wave has two characteristic maxima in the streamwise velocity component u', separated by a phase 'jump'. The phase jump is observed in accordance with (1.25) at the amplitude maximum of the normal velocity component v'.

At large angles of the wave inclination, when $\beta \gg \alpha$, the fundamental solutions φ_2 and φ_3 lose their linear independence so that at $\alpha = 0$, the Orr–Sommerfeld equation is reduced to $(D^2 - \beta^2 - \mathrm{i}\omega\mathrm{Re})(D^2 - \beta^2)\hat{v} = 0$ (D means here differentiation with respect to y), which has no discrete solutions at all (Hultgren and Gustavsson 1981).

The waves of the indicated classes constitute the complete set of solutions of the spatial Orr–Sommerfeld equation. In temporal stability formulation, the pressure waves are obviously absent, and the continuous spectrum consists only of the vorticity waves. Finally, the analysis of the homogeneous Squire equation shows that it also has a continuous spectrum of vorticity waves in external flows (Fig. 1.11), the discrete spectrum of the Squire modes being empty. The analysis of the spectrum for gradient or three-dimensional flows exhibits no methodological differences with that for Blasius boundary layer (Maslowe and Spiteri 2001). However, in the case of a three-dimensional velocity profile ($W \neq 0$), the discrete modes associated with its instability (crossflow modes) can appear (see Sect. 2.3.1).

The observation that parallel flows are always stable to the components of the continuous spectrum must be interpreted with caution. In particular, among these modes there are weakly decaying low-frequency waves. Though in a two-dimensional case they cannot penetrate deeply into the boundary layer, they can produce small quasi-stationary changes of the mean velocity profile close to the free-stream border (where U'' is small enough) resulting in quasi-stationary inflections in the mean velocity profile capable of affecting the flow stability. Besides, large-amplitude external disturbances and the boundary layer waves can interact non-linearly. As will be shown below, oblique disturbances can play a key role in the algebraic instability phenomenon in such external flows.

1.5 Lift-up effect

The classical analysis of linear stability treats the disturbances as separate modes of the Orr–Sommerfeld and Squire equations with exponential growth rates. However, the approach does not take into account the fact that these equations are not self–adjoint, i.e. the modes are not orthogonal.

Formally, for mathematically 'well-behaved' linear operators $\mathcal{H}$, such as the linearized Navier–Stokes equations (1.18), the *adjoint operator* $\mathcal{H}^\dagger$ is defined on a basis of the relation

$$(\mathcal{H}f, g) = (f, \mathcal{H}^\dagger g),$$

where the scalar product of functions $f(y)$ and $g(y)$ is given by expression:

$$(f, g) = \int_{y_1}^{y_2} f^* g \mathrm{d}y,$$

where values y_1 and y_2 limit the range of definition of f and g. In physics, the definition of the scalar products is connected to the notion of measure (a norm) of a process under consideration. In considering the Reynolds Orr equation, we have already used the scalar product in the definition of kinetic disturbance energy $E \sim (\boldsymbol{u}, \boldsymbol{u})$, i.e. it has a clear physical sense: in this notation the energy relates to the norm of velocity disturbances: $E \equiv \|\boldsymbol{u}\|^2$.

This definition of the adjoint relates also closely to the similar linear algebra definition of a matrix adjoint as its complex conjugate transpose. In such a way one can easily construct the adjoint of the linearized equations written in the matrix form (1.18), taking into account that the adjoint of the differentiation operator $\mathrm{d}/\mathrm{d}y$ is $-\mathrm{d}/\mathrm{d}y$, which follows directly from the differentiation-by-parts rule and assuming homogeneous boundary conditions both for the direct and adjoint problems.

The non-orthogonality of $\mathcal{L}_{\mathrm{OS}}$ and $\mathcal{L}_{SQ}$ is known to lead to dramatic consequences for most open flows. It appears that for the description of behaviour of the disturbances energy in finite intervals, it is necessary – in contrast to asymptotic statement of the Squire theorem (Sect. 1.2.3) – to supplement the analysis by considering three-dimensional disturbances. Such analysis can be especially important for the transition description at subcritical Reynolds numbers and in flows where the conventional mechanism of the linear stability manifests itself only at large Reynolds numbers or is completely absent.

From the point of view of the linear modal approach, subcritical growth can be observed when the amplitude functions of the non-orthogonal modes constituting a disturbance cancel each other initially, but then – due to the dispersion (difference of phase velocities and growth rates) – induce a large transient growth of the disturbance energy in the flow (cf. the earlier definition of conditional stability). As a simple geometrical analogy, it is possible to consider the behaviour of two non-orthogonal vectors on a plane. If the angle between these vectors decreases in time, the resulting vector can grow at first, even if the vector lengths monotonically decrease.

For the stability application, the mechanism is related to the selection of certain initial conditions or, more correctly, selection of such superposition of the non-orthogonal modes in the initial moment, which would trigger the processes of an establishment in the hydrodynamic system accompanied by a transient energy growth of the total disturbance (Henningson 1991; Reddy et al. 1993). In a flow subject to a stochastic effect, e.g., at a high level of disturbances in the free stream or at the channel entrance, such initial perturbations can arise naturally with enough energy to grow strongly for a limited time. In what follows we consider the application of these ideas to the case of temporally developing disturbances. Spatial amplification is considered in Chap. 5.

1.5.1 Inviscid algebraic instability

Ellingsen and Palm (1975) showed that the inviscid linear stability theory allows a linear temporal growth of kinetic energy of disturbances with $\alpha = 0$ in internal parallel shear flows. In this case the linearized momentum equation (1.12) in the absence of pressure gradients and spanwise mean velocity component W is reduced to

$$\frac{\partial u}{\partial t} + vU' = 0. \tag{1.43}$$

This expresses that most of the streamwise momentum in the presence of a shear is preserved when a fluid particle is displaced normal to the wall (i.e. normal to the shear). It gives the contribution to the streamwise disturbance velocity. Since the normal velocity v and the mean flow do not depend on time in this case, the disturbance streamwise velocity linearly increases in time. In physical space this effect manifests itself in elongation of the disturbance accompanied by a growth of its energy.

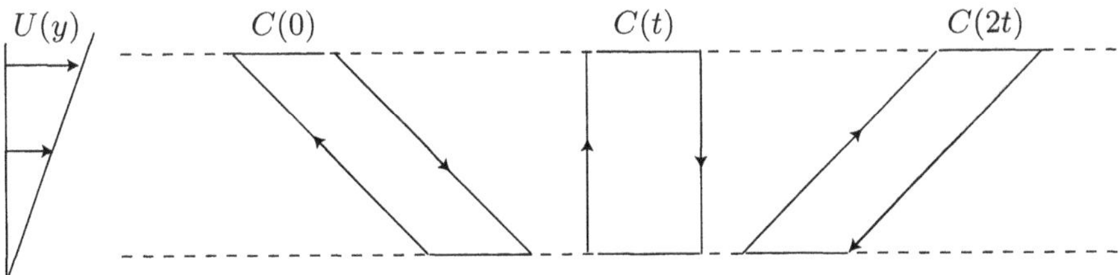

Fig. 1.14. Illustration of the Orr mechanism for Couette flow by consideration of the circulation $C(t)$ (Case 1960; Waleffe 1995a)

Such an inviscid mechanism of disturbance energy growth is called the algebraic instability (Landahl 1980), though Butler and Farrell (1992) note that it is more physical to characterize it as the lift-up effect. The growth mechanism of the algebraic instability is also somewhat similar to the hypothesis of the Prandtl's mixing length for the generation of the Reynolds stresses in turbulent shear flows.

For a two-dimensional disturbance ($\beta = 0$), a similar – the so-called Orr mechanism – also takes place. It is illustrated in Fig. 1.14. The example is based on considering the Couette flow in the inviscid limit, where only decaying disturbances convecting with the local mean velocity exist; see Sect. 1.2.4. If the initial disturbance inclined in the direction opposite to the mean shear, the latter leads to a continuous tilting of the disturbance. However, this mechanism provides a minor temporal growth in comparison to the three-dimensional disturbances, as shown by Butler and Farrell (1992), Reddy and Henningson (1993), Reddy et al. (1993) and others (see also Sect. 1.5.4).

Considering an isolated wave, it is possible to illustrate this process as follows. Let us transform the coordinate system so that axis x_1 is directed along the wave vector $\boldsymbol{\kappa}$, as in Fig. 1.15, and axis z_1 is perpendicular to it (Schmid and Henningson 2000). The mean flow in the new coordinate system has components:

$$U_1 = \frac{\alpha}{k}U, \qquad W_1 = -\frac{\beta}{k}U.$$

The components of the disturbance along the new coordinate axes are

$$\hat{u}_1 = \frac{\mathrm{i}}{k}\hat{v}', \qquad \hat{w}_1 = -\frac{\mathrm{i}}{k}\hat{\eta}.$$

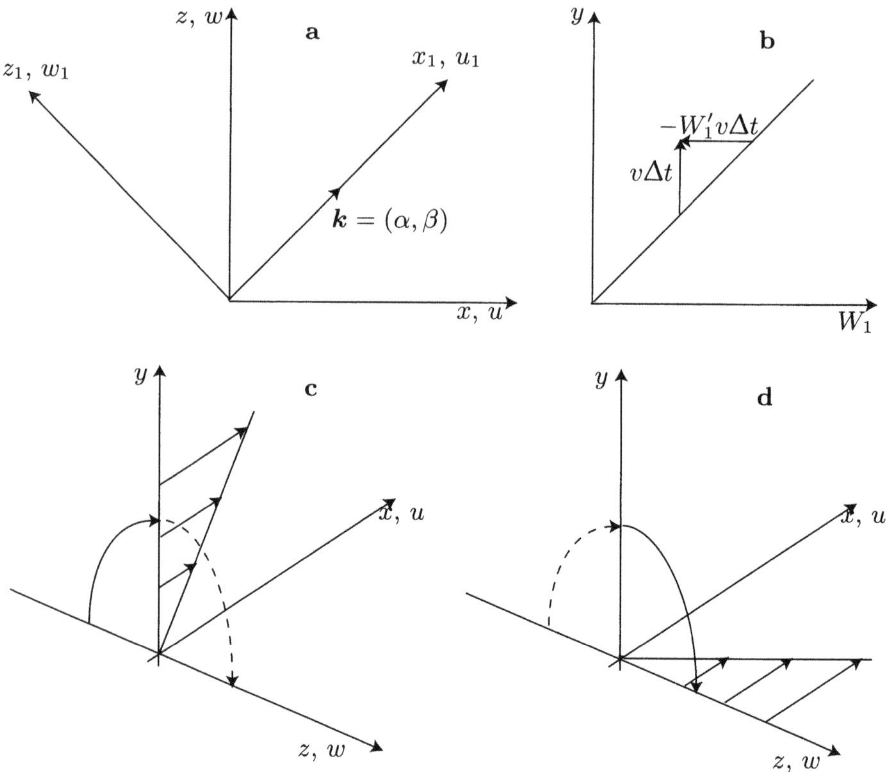

Fig. 1.15. Illustration of lift-up effect: **a** definition of the coordinate system, related to wave vector; **b** generation of horizontal velocity defect in direction z_1 by lift-up of the fluid element preserving horizontal momentum in this direction (Henningson et al. 1994). A simplified scheme of the lift-up mechanism: **c** initial state; **d** final state

Now the lift-up process can be imagined as a change of the velocity $\Delta\hat{w}_1$ during small time interval Δt. Applying the equations for $\eta(t)$, W_1 and $\hat{w}_1$ given above yields

$$\Delta\hat{w}_1 = -W_1'\hat{v}\Delta t,$$

where terms of $\mathcal{O}(\Delta t^2)$ are dropped and it is supposed that an observer moves together with the wave. The obtained expression for Δw_1 corresponds to an induced disturbance of horizontal velocity by the lift-up of a fluid particle by the normal velocity component.

More detailed analysis of the lift-up effect in the inviscid case for the Blasius boundary layer and the plane channel can be found in Gustavsson (1979), Hultgren and Gustavsson (1981), Henningson (1988).

1.5.2 Transient disturbance growth

As shown above, the energy growth due to the lift-up mechanism is an inviscid phenomenon. It can also occur, however, in the case when the viscosity is taken into account, though its effect eventually prevents the unbounded algebraic growth of the disturbance energy, i.e. it experiences a temporal (transient) growth (Henningson 1991).

Let us assume for simplicity that the flow is limited by solid walls at $y = \pm 1$ (the continuous spectrum is absent), and that an isolated l-mode of the normal velocity is initially excited. Then the solution is $\hat{v} = \hat{v}_l e^{-i\omega_l t}$. Substitution into the right-hand side of the normal vorticity equation (1.27) provides the general solution as a sum of solutions of $\hat{\eta}_h$ homogeneous equations and a particular solution of the equation with a non-homogeneous right-hand side:

$$\hat{\eta} = \hat{\eta}_h + \overline{\eta}_l^p e^{-i\omega_l t}.$$

To find $\hat{\eta}_h$ and $\overline{\eta}_l^p$, it is supposed that both solutions can be presented as a linear combination of eigenfunctions of the homogeneous part of the normal vorticity equation; i.e. can be expanded in the complete spectrum of the Squire modes. For the homogeneous part

$$\hat{\eta}_h = \sum_{j=1}^{J} B_j \overline{\eta}_j e^{-i\sigma_j t},$$

where B_j are the expansion coefficients and $\overline{\eta}_j$ are the Squire modes with eigenvalues σ_j. For the particular solution, a similar decomposition without time dependence is supposed. Their substitution in the normal vorticity equation gives (Gustavsson 1991)

$$\hat{\eta}_l = \sum_{j=1}^{J} \left[C_j \overline{\eta}_j e^{-i\sigma_j t} + D_{jl} \frac{e^{-i\omega_l t} - e^{-i\sigma_j t}}{\omega_l - \sigma_j} \right], \tag{1.44}$$

where

$$C_j = \int_{-1}^{1} \hat{\eta}_0 \overline{\eta}_j dy, \qquad D_{jl} = \beta \int_{-1}^{1} U' \overline{v}_l \overline{\eta}_j dy,$$

and $\hat{\eta}_0 = \hat{\eta}(t = 0)$. The more general form of this expansion for the wave packet case is given by Henningson (1991). The non-orthogonality of $\overline{\eta}_j$ and the resonance-like behaviour of the second right-hand side term in (1.44) ensure the presence of the transient phenomena in the system.

The possibility of algebraic instability and transient growth is found in calculations for most typical shear flows, for example in a round pipe (Schmid and Henningson 1994), free-shear flows (Farrell and Ioannou 1993) and two-phase flows (Schmid and Kytomaa 1994). Breuer and Kuraishi (1994) extended the analysis to a class of three-dimensional flows, and Hanifi et al.

(1994) to the case of compressible boundary layers. Gustavsson (1979), Hultgren and Gustavsson (1981) theoretically showed the necessity of accounting for modes of the continuous spectrum by consideration of an initial three-dimensional problem in the boundary layer. The analysis indicates that the eigenfunctions of the continuous spectrum at $\alpha \approx 0$ can strongly penetrate into the boundary layer, and stipulate a considerable temporal growth of the disturbance energy there (Fig. 1.16). In the case when the streamwise wave number $\alpha = 0$, only three-dimensional disturbances consisting exclusively of components of the continuous spectrum are allowed inside the boundary layer. This makes the modes of the continuous spectrum a necessary component of the disturbances.

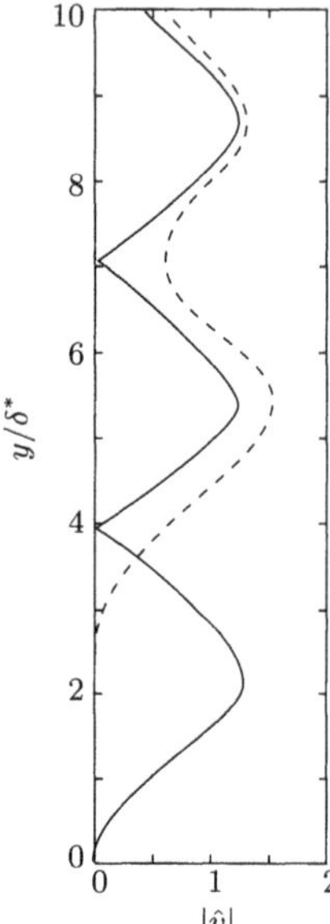

Fig. 1.16. Amplitude distribution of the wall-normal velocity of two modes of continuous spectrum, Re = 580: *solid curve*, $\alpha = 0$, $\beta = 1$; *dashed curve*, $\alpha = 1$, $\beta = 0$ (Hultgren and Gustavsson 1981)

1.5.3 Pseudospectra of the linearized Navier–Stokes operator

A general analysis of the non-orthogonal operators explains why the streamwise structures or streaks (the regions of local streamwise velocity excesses and defects elongated approximately along the streamwise direction) with very similar characteristics (see Fig. 1.17) are so widespread in various types of shear flows, though they are *not* eigenmodes of the linearized stability problem. To capture the general transient behaviour of non-normal operators, a generalized concept of spectrum can be used. Let us define the ε-pseudospectrum of the operator $\mathcal{L}$ as a set of spectra of all perturbed operators $\mathcal{L} + \mathcal{E}$, where $\|\mathcal{E}\| \leqslant \varepsilon$ for any number $\varepsilon \geqslant 0$ (Trefethen 1997).

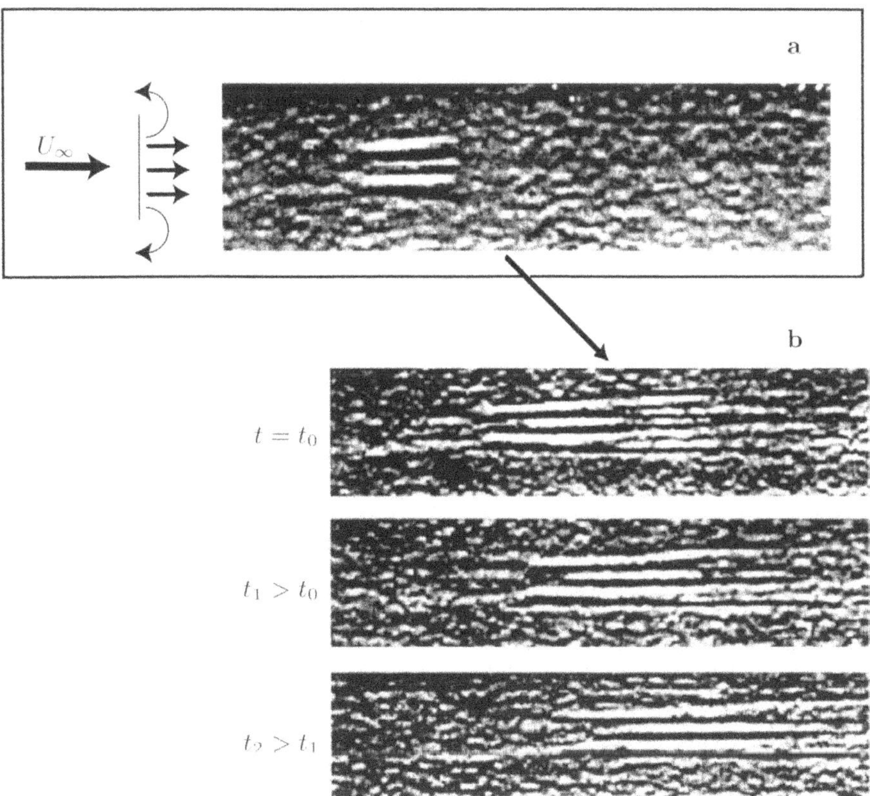

Fig. 1.17. Smoke visualization of the process of formation and development of streamwise structures stipulated by the lift-up mechanism: **a** scheme of the visualization – the disturbances excited by impulsive fluid blowing through a slot; **b** flow patterns at three consequent time instances (Alfredsson et al. 1996)

The pseudospectra $\Lambda_\varepsilon(\mathcal{L})$ form enclosed families of the sets on the complex ω-plane, and $\Lambda_0(\mathcal{L})$ corresponds to the true spectrum $\Lambda(\mathcal{L})$. For a normal operator $\mathcal{L}$, pseudospectrum $\Lambda_\varepsilon(\mathcal{L})$ is a set of all points at a distance less than ε from $\Lambda_0(\mathcal{L})$, while for a non-orthogonal case it can be much wider.

The boundaries of the pseudospectra can be interpreted as contours of resonance values. The presence of factor $1/(\omega_1 - \sigma_j)$ in (1.44) indicates that normal vorticity growth has a resonance character. However, the fulfilment (usually during very localized time) of the resonance conditions does not necessarily provide the greatest possible growth of disturbance energy, as the reason for the phenomenon lies primarily in the non-orthogonality of the Orr–Sommerfeld and Squire operators (Trefethen et al. 1993; Trefethen 1997).

Further mathematical analysis of pseudospectra shows that the possible transient growth can be estimated from the maximum values of imaginary part of a particular pseudospectrum γ_ε protruded into the unstable half-plane. The following inequality holds for the maximum possible energy growth optimized for each instance of time over all initial conditions (Trefethen 1997):

$$G_{\max} \leqslant \gamma_{\varepsilon}^{2}.$$

It appears that whereas the upper bound of the spectrum of the disturbances corresponds to the least stable two-dimensional wave, most unstable pseudo-modes, as a rule, are three-dimensional. However, there is a more quantitative description which makes it possible to determine the initial conditions which provide maximum possible amplification of energy $G(t) = E(t)/E(0)$ at a given time instance t. It is considered below.

1.5.4 Optimal disturbances

Let us outline the procedure of finding the condition for maximum allowed disturbance growth, the procedure being valid for a variety of stability approaches, rather then for the only linearized Navier–Stokes equations (Corbett and Bottaro 2000). We define formally a propagator $P(t)$ – the evolution operator which takes an initial condition $\boldsymbol{u}_0 = \boldsymbol{u}(t=0)$ to the state $\boldsymbol{u}_t = \boldsymbol{u}(t>0) = P(t)\boldsymbol{u}$. Then the expression for $G(t)$ in terms of the propagator is:

$$G(t) = \frac{E(t)}{E(0)} \equiv \frac{(\boldsymbol{u}_t, \boldsymbol{u}_t)}{(\boldsymbol{u}_0, \boldsymbol{u}_0)} = \frac{(P(t)\boldsymbol{u}_0, P(t)\boldsymbol{u}_0)}{(\boldsymbol{u}_0, \boldsymbol{u}_0)} = \frac{(\boldsymbol{u}_0, P(t)^{\dagger}P(t)\boldsymbol{u}_0)}{(\boldsymbol{u}_0, \boldsymbol{u}_0)}.$$

It is the Rayleigh quotient, cf. (1.8). It means that $G_{\max}(t) = \max_{\boldsymbol{v}_0 \neq 0} G(t)$ is positive and finite and is given by the largest eigenvalue of

$$P(t)^{\dagger}P(t)\boldsymbol{u}_0 = G(t)\boldsymbol{u}_0.$$

In numerical calculations, it is equivalent to finding the largest singular value of the propagator $P(t)$ approximated by a matrix P_N of a finite size $N \times N$. According to the singular value decomposition

$$P_N V = U\Sigma,$$

where V and U are unitary matrices and $\Sigma = \operatorname{diag}\{\sigma_1, \sigma_2, \ldots, \sigma_N\}$ is the matrix of singular values $\sigma_1 \geqslant \sigma_2 \geqslant \ldots \geqslant \sigma_N$; then

$$P_N \boldsymbol{v}_1 = \sigma_1 \boldsymbol{u}_1,$$

where $\boldsymbol{v}_1$ and $\boldsymbol{u}_1$ are right and left singular vectors for σ_1. This equation represents a mapping of the initial conditions given by $\boldsymbol{v}_1$ to the final state in time instance $t > 0$ given by $\boldsymbol{u}_1$ accompanied with the growth (mathematically, the stretching) factor σ_1, that is $G_{\max}(t) = \sigma_1$ (Schmid and Henningson 1994). Such initial conditions were called optimal perturbations by Butler and Farrell (1993).

Assuming that the singular values are well separated, the general way to find the optimal perturbations is through the power iterations (Corbett and Bottaro 2000; Luchini 2000):

$$v_0^{k+1} = P_N^\dagger P_N v_0^k.$$

For quasi-parallel flows yet another procedure is useful (Schmid and Henningson 2000). In this case the propagator is represented by the linearized Navier–Stokes equations (1.12–1.15) which are further reduced to the eigenvalue problem of the form:

$$\begin{pmatrix} \mathcal{L}_{OS}, & 0 \\ -\mathrm{i}k\overline{W}', & \mathcal{L}_{SQ} \end{pmatrix} \begin{pmatrix} \hat{v} \\ \hat{\eta} \end{pmatrix} = \mathrm{i}\omega \begin{pmatrix} k^2 - D^2, & 0 \\ & 0, 1 \end{pmatrix} \begin{pmatrix} \hat{v} \\ \hat{\eta} \end{pmatrix}, \tag{1.45}$$

with the energy norm in terms of normal velocity v and normal vorticity η given by

$$\begin{aligned} E &= \int_{y_1}^{y_2} (\hat{u}^*\hat{u} + \hat{v}^*\hat{v} + \hat{w}^*\hat{w})\mathrm{d}y \\ &= \frac{1}{k^2}\int_{y_1}^{y_2} \begin{pmatrix} \hat{v} \\ \hat{\eta} \end{pmatrix}^* \begin{pmatrix} \mathrm{i}\alpha\mathrm{d}/\mathrm{d}y, & -\mathrm{i}\beta \\ k^2, & 0 \\ \mathrm{i}\beta\mathrm{d}/\mathrm{d}y, & \mathrm{i}\alpha \end{pmatrix} \begin{pmatrix} \hat{v} \\ \hat{\eta} \end{pmatrix} \mathrm{d}y \\ &= \int_{y_1}^{y_2} \left[\hat{v}^*\hat{v} + \frac{1}{k^2}\left(\hat{\eta}^*\hat{\eta} + \frac{\mathrm{d}\hat{v}^*}{\mathrm{d}y}\frac{\mathrm{d}\hat{v}}{\mathrm{d}y} \right) \right] \mathrm{d}y. \end{aligned}$$

Formally this means a redefinition of the inner product in terms of weighted $\hat{v}$ and $\hat{\eta}$. A solution

$$q(t) = \begin{pmatrix} \hat{v} \\ \hat{\eta} \end{pmatrix}$$

of the linearized initial value problem which leads to (1.45), can be expanded in the basis of elementary disturbances $\hat{q}_i$, $i = 1, 2, \ldots$:

$$q(t) = \sum_{i=1}^{\infty} k_i(t)\hat{q}_i.$$

In this basis the initial-value problem takes very simple form which is useful for analysis of the optimal disturbances in the quasi-parallel approach:

$$\frac{\mathrm{d}\boldsymbol{k}}{\mathrm{d}t} = -\mathrm{i}\mathcal{P}\boldsymbol{k}, \qquad \boldsymbol{k} = \{k_1, k_2, \ldots\}, \qquad \mathcal{P} = \mathrm{diag}\{\omega_1, \omega_2, \ldots\}.$$

Operator $\mathcal{P}$ is the propagator $P(t)$ representation in this new basis.

Figure 1.18 illustrates different characteristics of the optimal disturbances found with the described procedure. A comparison of the growth of two-dimensional and three-dimensional optimal disturbances in the Blasius and Poiseuille flow at subcritical Reynolds numbers is given in Fig. 1.18a. This clearly demonstrates that the transient growth of three-dimensional disturbances is potentially much more dangerous to destroy the initial flow than the two-dimensional Orr mechanism. Corresponding initial and resulting velocity patterns (the streaks) in $(y$–$z)$-planes for the time instances of the maximum value of G are shown in the remaining part of the figure. Additionally, the profile of streamwise velocity amplitude for the optimal perturbation in the

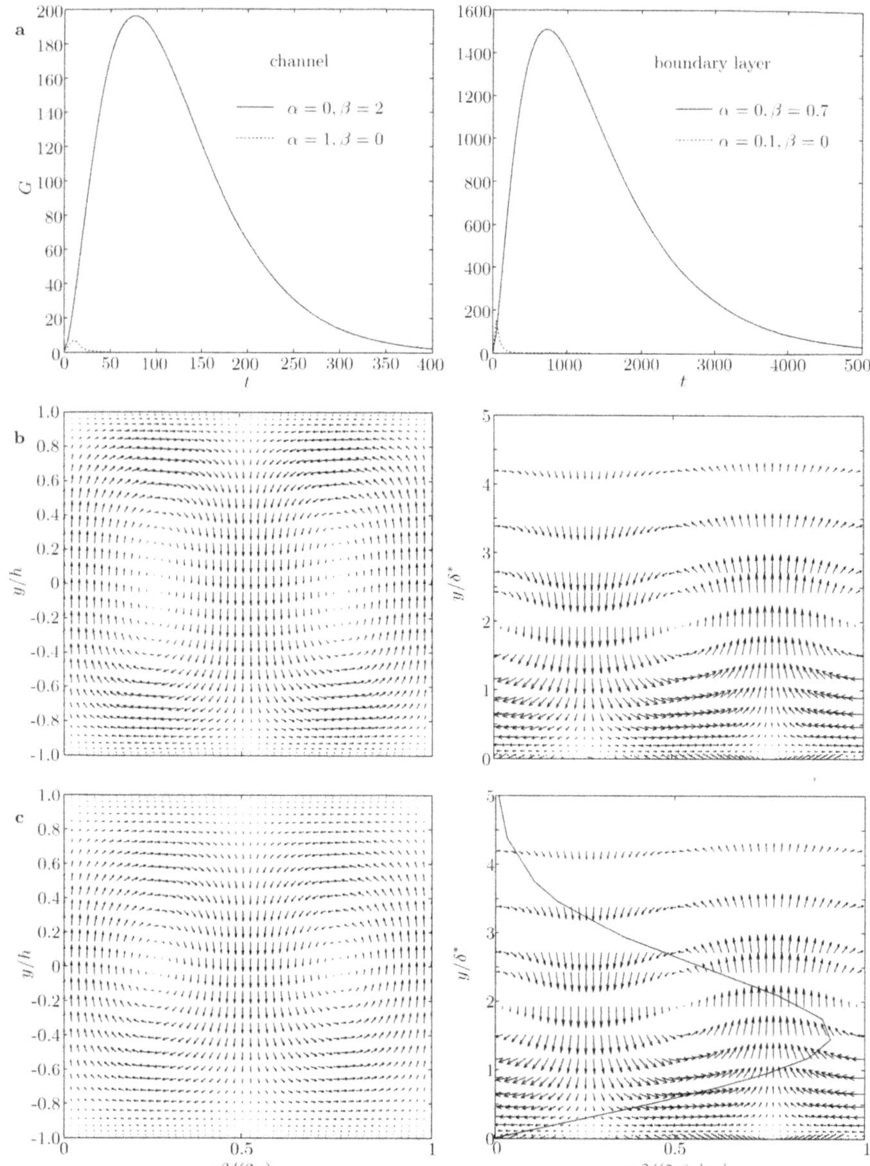

Fig. 1.18. Development of optimal disturbances for plane Poiseuille flow at $\mathrm{Re}_h = 1000$ and Blasius boundary layer at $\mathrm{Re}_{\delta^*} = 10000$. **a** amplification $G(t)$; *solid line*, behaviour of three-dimensional disturbance; *dotted line*, behaviour of two-dimensional disturbance; **b** initial spanwise waveform; **c** the 'optimal' waveform developed in the moment of maximum $G(t)$; *solid line*, streamwise disturbance amplitude profile $|u_1|$ (calculated by A.V. Boiko)

Blasius boundary layer is shown in Fig. 1.18c. This profile was proved to be one of the most typical characteristics of experimentally observed streaks caused by a wide variety of sources (see Chap. 5).

1.5.5 Numerical range

Let us recall that for incompressible Newtonian flows, any growth of disturbance energy is due to a linear process (see Sect. 1.2.2). It motivates to use the linearized formulation in solving the Reynolds–Orr equation (Schmid and Henningson 2000). With the above definition of the weighted inner product, it follows that

$$\frac{\mathrm{d}E}{\mathrm{d}t} = \frac{\mathrm{d}\|k\|^2}{\mathrm{d}t} = \left(\mathrm{i}(\mathcal{P}^\dagger - \mathcal{P})k, k\right) = 2\mathrm{Imag}\{(\mathcal{P}k, k)\}.$$

The set of values $z = (\mathcal{P}k, k)$ at $\|k\|^2 = 1$ is called the numerical range of $\mathcal{P}$. It follows, that the imaginary part of z determines $\mathrm{d}E/\mathrm{d}t$ at $t = 0$ exactly. Then a necessary and sufficient condition for the growth of energy density is that the numerical range protrudes into the upper complex half-plane; that is Re_E can be derived by finding conditions at which the numerical range has no positive imaginary part, the procedure of its determination being quite standard (Reddy and Henningson 1993; Henningson and Reddy 1994).

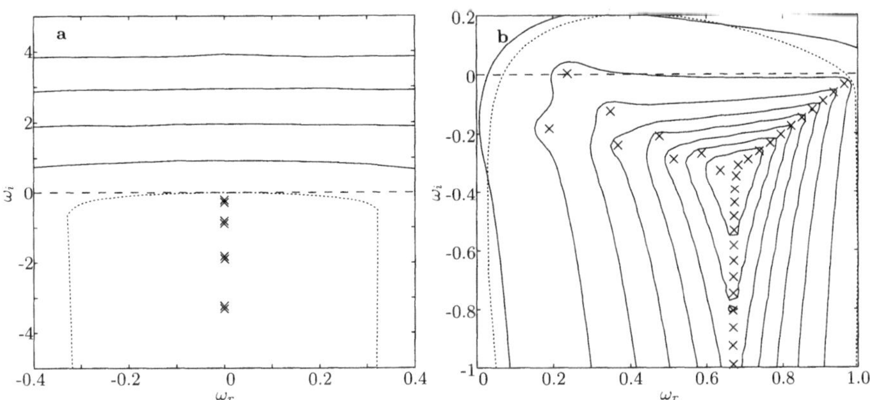

Fig. 1.19. Characteristics of linearized Navier–Stokes operator for plane Poiseuille flow at $\mathrm{Re}_h = 49, \alpha = 0, \beta = 2$ (**a**) and $\mathrm{Re}_h = 10000, \alpha = 1, \beta = 0$ (**b**): $\times$, spectrum; *solid lines*, boundaries of pseudospectra at $\varepsilon = 1, 2, 3, 4$ (*left*) and $10^{-8}, 10^{-7} \ldots, 10^{-1}$ (*right*); *dotted line*, numerical range (calculated by A.V. Boiko)

Examples of the spectrum, bounds of pseudospectra and the contours of numerical range for some characteristic Reynolds and wave numbers in Poiseuille flow are shown in Fig. 1.19. According to calculations, any disturbance below $\mathrm{Re}_{\mathrm{E},h} = 49$ monotonically decay in the flow. Comparison of the values of ω_i for the shown bounds of pseudospectra and corresponding values of ε indicates that the operator in this case is almost normal and the flow is globally stable. In contrast, at $\mathrm{Re}_h = 10000$ the flow is linearly unstable: one of the eigenvalues has positive imaginary part. However, the form of the bounds of pseudospectra and their strong penetration into the upper half-plane at very small ε testify that the flow is subject to a pronounced transient growth governed by a collective behaviour of the modes.

2 Development of linear disturbances in near-wall flows

In this chapter we consider the experimental and theoretical results devoted to the development of small-amplitude disturbances (both single waves and wave packets) in such prototypical flows as a flat plate boundary layer, plane channel flow, flows about a curved wall and swept wing.

2.1 The flat plate boundary layer

In this section the linear stability of the Blasius boundary layer is addressed. The reason to start with this flow is that being quite simple, but a quasi-parallel rather than a strictly parallel flow, it allows us to demonstrate the successful application of the approach developed in Chap. 1 as well as main difficulties in the investigation of small-amplitude disturbances from both experimental and theoretical points of view.

2.1.1 Historical issues

The Orr–Sommerfeld equation was derived independently by Orr (1907a, b) and Sommerfeld (1908). One of the first attempts to solve it belongs to Prandtl (1921, 1922), who studied stability of an idealized flow with a piece-wise linear velocity profile and came to the unexpected conclusion that the viscosity can destabilize the flow. Later Tollmien (1929), and then Schlichting (1933, 1935), for the first time completely calculated a curve of neutral stability for the Blasius boundary layer in the parallel approximation.

The least critical Reynolds number for the Blasius boundary layer, gained from the numerical solution of the linear two-dimensional parallel problem of stability, is

$$\mathrm{Re}_{\mathrm{L},\delta^*} = \frac{U_\infty \delta^*}{\nu} \approx 520,$$

where the displacement thickness, which is a measure of the boundary layer thickness, is defined as

$$\delta^* = \int_0^\infty (1 - U)/U_0 \mathrm{d}U,$$

where U_0 is the local external velocity. For the Blasius boundary layer $\delta^* = 1.72\sqrt{\nu x/U_\infty} = 1.72x/\sqrt{\mathrm{Re}_x}$, so that $\mathrm{Re}_{\delta^*} = 1.72\sqrt{\mathrm{Re}_x}$.

It is remarkable that the wavelengths of the unstable disturbances under consideration are quite large compared to the boundary layer thickness δ. The lowest wavelength is

$$\lambda_{\mathrm{min}} = \frac{2\pi}{0.36}\delta^* = 17.5\delta^* \approx 6\delta.$$

The boundary layer growth leads to quite specific behaviour of the instability waves during their propagation downstream. Let us consider a disturbance of a given frequency $\omega_0 = 2\pi f\delta^*/U_\infty$, where f is a corresponding dimensional frequency. Due to the continuous change of the characteristic boundary layer scale δ^* with x, ω_0 also experiences continuous change. This is the reason to introduce the so-called frequency parameter $F = 2\pi f\nu/U_\infty^2 = \omega_0/\mathrm{Re}_{\delta^*}$. The dimensionless variable F is independent of the local boundary layer thickness. Downstream propagation of the disturbance corresponds in the (Re–F)-plane to a continuous displacement to larger Re along a selected value of F. Thus the disturbance is at first damped, then, upon reaching branch I of the neutral curve, begins to grow. It grows to branch II, then decays again. Such behaviour is the first evidence of 'nonparallel effects', which we consider later.

The first experiments proving the correctness of the linear instability concept were carried out as recently as in 1942 by Schubauer and Skramstad (1948) using a unique-for-that-time experimental facility, setup and advanced techniques, which later became widely accepted. It is remarkable that the experiments were performed for the Blasius boundary layer which is not strictly parallel flow.

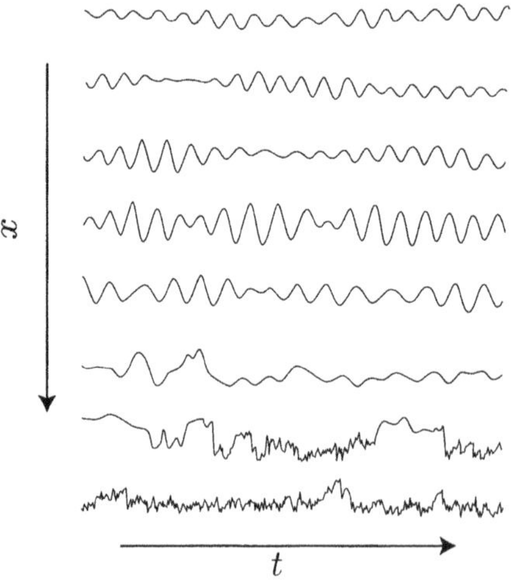

Fig. 2.1. Oscilloscope traces of the streamwise fluctuations of streamwise velocity in a flat plate boundary layer obtained at an arbitrary amplitude scale by Schubauer and Skramstad (1948)

Reducing the turbulence level in the test section of a subsonic wind tunnel down to Tu = 0.01–0.03%[1], Schubauer and Skramstad detected quasi-harmonic wave motions preceding the laminar–turbulent transition in the flat plate boundary layer on an oscilloscope screen using the hot-wire technique (Fig. 2.1). The data were obtained for various streamwise pressure gradients by placing the flat plate at different angles of attack to promote or attenuate the transition. The flow stability characteristics were studied with the help of small-amplitude periodic disturbances excited in a controlled manner in the laminar boundary layer. After several trials they chose a vibrating ribbon technique: an alternating current passing through a metal strip forced its vibration inside the boundary layer in a constant magnetic field caused by a permanent magnet located beneath the test surface. Using this method it appeared to be possible to excite two-dimensional eigenmodes of the boundary layer of a given frequency, and study their characteristics.

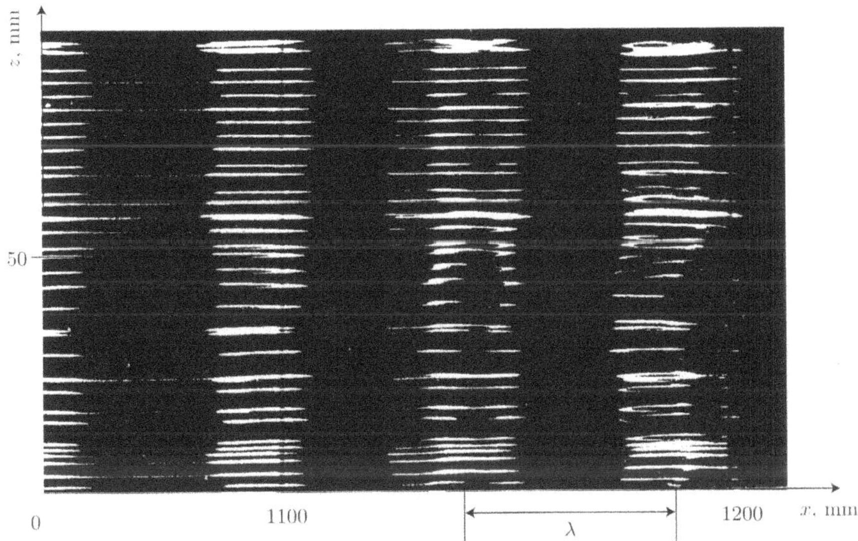

Fig. 2.2. Smoke visualization of a ribbon response in a flat plate boundary layer in the region of Tollmien–Schlichting waves development; λ, wavelength

In the linear normal mode approach, the ribbon response at a certain distance from the source – where the parallel flow approximation is applicable – can be expanded into the set of Orr–Sommerfeld modes consisting mostly of the discrete waves. In the Blasius boundary layer, as well as in many other subsonic flows, all modes of the discrete spectrum strongly decay with the streamwise coordinate x, except for the least stable eigenvalue. Therefore,

[1] The free-stream turbulence level is determined here and below as the value Tu = u'/U_∞, as measured near the leading edge of a body.

at a distance from the source comparable to the characteristic wavelength of the induced oscillations, only one Tollmien–Schlichting mode can 'survive' (Fig. 2.2).[2]

Table 2.1. Characteristics of the eigenvalues at $\sqrt{\mathrm{Re}_x} = 580$, $\alpha = 0.179$ (Mack 1976)

Mode	$\Delta E/E\|_\delta$	$E(T)/E(0)$
1	+0.0462	1.32
2	−0.827	6.11×10^{-6}
3	−0.574	6.74×10^{-3}
4	−0.756	2.58×10^{-4}
5	−0.646	2.33×10^{-3}
6	−0.692	1.01×10^{-3}
7	−0.634	2.79×10^{-3}

An example of the characteristic changes of the kinetic energy for the modes of the discrete spectrum is presented in Table 2.1. The variations on the scale of the Blasius boundary layer thickness δ and the wave period T are given.

Schubauer and Skramstad (1948) were able to localize both branches of the neutral stability curve by measuring the root-mean-square amplitude of the streamwise disturbance velocity component at a fixed dimensional distance from the wall (Fig. 2.3). They also measured the disturbance amplitude distribution in the wall-normal direction (velocity profile) and showed that it has two maxima: the larger one close to the wall and the other at the outer border of the boundary layer. Furthermore, a phase jump associated with the amplitude minimum was found. Comparison of the obtained data with the theoretical results of Schlichting (1935) shows qualitative agreement. The data confirmed excellently the validity of the linear theory of hydrodynamic stability and formed a classical basis for the majority of further experimental and theoretical studies of the laminar–turbulent transition in various other quasi-parallel flows, as channels, local separations, jets and wakes.

Wortmann (1953, 1964), and later Kozlov and Babenko (1978), investigated boundary layer stability in water by visualization of the flow using tellurium. Short pulses of electric current were passed through a thin tellurium wire placed in the stream. The wire served as a cathode, a metal plate installed downstream in the flow served as an anode, and water as the electrolyte. The electric current produced a stream of tellurium ions, whose

[2] Opposite examples are served by flat plate supersonic flow at Mach more than 3.5 and subsonic near-wall jet: two unstable modes appear in both cases.

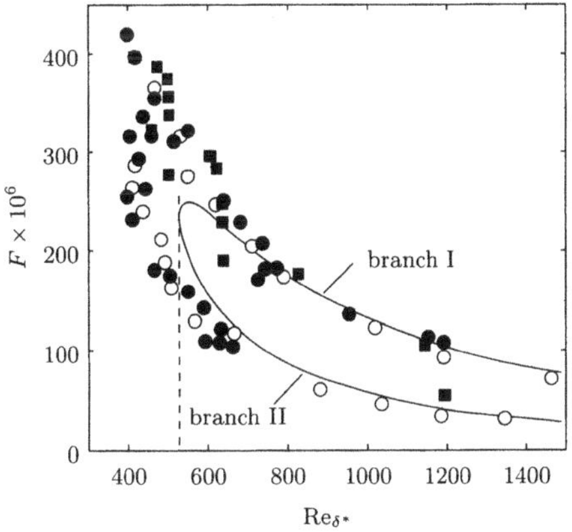

Fig. 2.3. Neutral stability curves for Blasius boundary layer: ■, Schubauer and Skramstad (1948); ∘, (Ross et al. 1970); •, (Kachanov et al. 1979a); *solid line*, numerical calculation using linear stability theory; *dashed line*, location of the theoretical critical Reynolds number

small colloidal particles moved with the flow. The stroboscopic visualization of the position of the particles in the boundary layer enabled the determination of the amplitudes of the velocity oscillations at various frequencies, and reconstruction of a part of the neutral stability curve. The results of the study coincided qualitatively with those obtained by Schubauer and Skramstad (1948).

Ross et al. (1970) published experimental results which included detail quantitative data on both the position of the neutral curve and the amplitude growth, as well as on the amplitude profiles of the waves at several Reynolds numbers. This study, in essence, was a revision of the experiments of Schubauer and Skramstad (1948) using improved experimental technique. The necessity of the revision, was apparently, driven by two factors: scattering of the experimental points in the previous study, due to limitations of the registration technique; and the appearance of the first precise calculations of Blasius boundary layer stability by Jordisson (1970). These calculations improved the position of the neutral stability curve obtained using asymptotic methods by Lin (1955) which was accepted earlier as the reference data.

Nevertheless, qualitative agreement with the theory was obtained only at low frequencies. The oscillations grew up to $F = 400 \times 10^{-6}$, though the theoretical calculations indicate that all disturbances above $F \approx 250 \times 10^{-6}$ should decay. A number of issues were left open concerning the correspondence between the temporal and spatial instability approaches, the non-parallel effects, the behaviour of naturally occurring wave packets and the influence of free-stream turbulence upon transition.

We address most of them below in this chapter. The final stated above point is related tightly to the receptivity problem and the so-called bypass transition is treated in detail in Chaps. 3 and 5.

2.1.2 Non-parallel effects

The discrepancy of results from linear stability theory and experiments in the region of small Re was for a long time attributed to the non-parallel flow that drives a long-standing interest in the phenomenon. It later become clear that the main reason for the discrepancy is primarily experimental difficulties, but the flow non-parallelity is still of a great significance to the development of both oblique waves or wave packets in the Blasius boundary layer as well as Görtler and crossflow vortices.

Theoretical approaches to flow non-parallelity. The non-parallel effects can be divided into the parametric growth of the Reynolds number with the streamwise coordinate in the boundary layer resulting in downstream changes of the profiles, and wave numbers of the instability waves and the effect of the mean flow divergence, i.e. presence of both the wall-normal velocity $V(y)$ and the streamwise velocity gradient.

The first of these can be treated parametrically with the quasi-parallel flow approximation using the parallel linear stability results and the streamwise-velocity profile. A sequence of parallel-flow calculations with parametric variation of the Reynolds number gives the wave numbers $\alpha(x)$. Then the imaginary part, the growth rate α_i, is related locally to the disturbance amplitude $A(x)$ (e.g., the streamwise velocity) as $\mathrm{d}A/\mathrm{d}x = -\alpha_i(x)A$, so that:

$$A(x) = \exp\left(-\int_{x_0}^{x} \alpha_i(\xi)\mathrm{d}\xi\right).$$

The stability of the growing boundary layer has been studied with perturbation methods by Benney and Rosenblat (1964), Van Dyke (1964), Nayfeh (1973), Volodin (1973), Bouthier (1971, 1972, 1973), Nayfeh et al. (1974), Gaster (1974) and Saric and Nayfeh (1977). For a stream function ψ of the two-dimensional disturbance that slowly varies with x, one can write

$$\psi \sim S(y, \varepsilon x)\mathrm{e}^{\mathrm{i}[\int(\alpha_0+\varepsilon\alpha_1)\mathrm{d}x-\omega t]},$$

where both the disturbance wave number $\alpha = \alpha_0 + \varepsilon\alpha_1$ and the eigenfunction profile $S(y, \varepsilon x)$ vary slowly downstream; $\varepsilon = 1/\mathrm{Re}_\delta$ is a measure of the non-parallelity (Saric and Nayfeh 1977). As a variant, Volodin (1973) used the expansion in two small parameters, one of which accounts for the mean velocity variations and the other for the changes in the disturbance amplitude function. The application of the perturbation technique for derivation of the stability equations for weakly non-parallel flows results in a chain of linear equations such as

$$\mathcal{L}_0\hat{\psi}_0 = 0; \qquad \mathcal{L}_0\hat{\psi}_1 = \mathcal{L}_1\hat{\psi}_0, \ldots$$

with homogeneous boundary conditions. Operator $\mathcal{L}_0$ corresponds to the Orr–Sommerfeld operator; operator $\mathcal{L}_1$ takes into account the first-order effects of the non-parallelity.

The developed perturbations theories are, in essence, equivalent. At the same time the results depend – as far as the decomposition is correctly chosen – on the number of terms taken into account. The studies showed an essential influence of a weak flow non-parallelity on the development of small disturbances. In particular, a strong dependence of the growth rates on the wall-normal coordinate was found. This leads to a considerable dependence of the stability characteristics on y (Fig. 2.4b). In general, the results of these theories lead to the conclusion that the effects of non-parallelity destabilize the boundary layer both at low Reynolds numbers and for high-frequency disturbances.

Another theoretical approach is based on the asymptotic triple-deck theory which is more consistent in considering the viscosity effect both for the mean and the disturbed flows (Van Dyke 1964). An interested reader is referred to Smith (1979a) (1979a, b) and Healey (1995).

Methods for the numerical integration of the full Navier–Stokes equations have been developing intensively over the past 2–3 decades. In a direct numerical simulation (DNS), the flow non-parallelity is automatically taken into account. Earlier DNSs were carried out by Fasel (1976), Murdock (1977) and Kachanov et al. (1979c). Fasel (1976) was one of the first to calculate the spatial development of the disturbances prescribed at a certain initial section by solving the Orr–Sommerfeld equation. Later DNS results on two-dimensional Tollmien–Schlichting waves by Fasel and Konzelmann (1990), with a more realistic downstream boundary condition, confirmed the data of Gaster (1974) obtained with the perturbation technique, and showed much less deviation from the parallel stability theory than previous experimental results.

One of the main difficulties of using the Navier–Stokes equations for the analysis of hydrodynamic stability is the formulation of proper downstream boundary conditions. This resulted in the application of the parabolized stability equations (PSE) to the problem (Bertolotti et al. 1992). The approximation enables the removal of all elliptic components in the Navier–Stokes equations, so that the equations obtained can be solved computationally inexpensively by marching. PSE is a relatively simple approach which can be easily applied to more complicated problems such as the receptivity (Chap. 3), the non-linearity (Chap. 4) and to complex flow geometries (Chaps. 5 and 6). For the description of the Tollmien–Schlichting wave development, the PSE were used by Itoh (1986), Herbert and Bertolotti (1987) and Bertolotti et al. (1992). A detailed review of this approach can be found in Herbert (1997).

We outline the technique, following Crouch (1998) closely. Substituting $\boldsymbol{U} = \{U(x,y), V(x,y), 0\}$ into the linearized Navier–Stokes equations yields the stability equations for a non-parallel flow. The disturbance velocity is written in the form

$$\boldsymbol{u} = \tilde{\boldsymbol{u}}(\boldsymbol{y})\chi(x,t) \qquad \text{with} \qquad \chi(x,t) = \exp\left(\mathrm{i}\int_{x_0}^{x} \alpha(\xi)\mathrm{d}\xi - \mathrm{i}\omega t\right),$$

where α is complex. Since the boundary layer as well as the disturbance vary slowly with x, we neglect the terms with second derivatives in x and the products of the first derivatives. Substituting the obtained expression for $\boldsymbol{u}$ into the linearized Navier–Stokes equations yields a partial differential equation for $\tilde{\boldsymbol{u}}$ and α. Writing the velocity in terms of two functions of x introduces an arbitrariness that requires the imposition of additional constraints. The derived PSE with appropriate boundary, initial conditions and the constraints (called auxiliary conditions) are solved by streamwise integration. A more detailed discussion of the PSE can be found in Bertolotti et al. (1992) and Herbert (1997).

Bertolotti et al. (1992) used both a DNS and the PSE to calculate the linear and non-linear development of the Tollmien–Schlichting waves. The results of both methods correlated quite well with each other, and confirmed the weakness of the non-parallel effects found by Fasel and Konzelmann (1990). Thus, perturbation analysis, the PSE calculations, and DNSs cannot completely explain why the experiments show the instability region extending into small Re and high F.

Modern view on the place and the role of the non-parallelity. The difference between the theoretical and experimental results for the Blasius boundary layer is connected partly to features of experimental setups used for the linear stability investigations and partly to differences in amplitude measures used in the experiments and theoretical studies. The study of the development of two-dimensional waves in a boundary layer has many traps (Saric 1990, 1994a). One of the main problems is the quality of the wind tunnel stream; in particular, the level of free-stream turbulence. Though the external disturbances are extremely important for excitation of 'natural' Tollmien–Schlichting waves[3] (i.e. for the boundary layer *receptivity*), as well as for further non-linear stages of the disturbance growth, it is desirable to reduce them as much as possible when studying the development of *controlled* Tollmien–Schlichting waves.

In the boundary layer above an ideal flat plate, there is no streamwise pressure gradient. In an experiment, a plate has a finite thickness and a pressure gradient at the leading edge. Therefore, the boundary layer differs substantially from the Blasius one in that region, which can be quite extended downstream. The pressure gradient at the leading edge depends on the thickness of the plate, the edge shape and arrangement of the stagnation point. Thinner plates are preferable as they allow reduction of the Reynolds number near the beginning of the gradientless region. The position of the stagnation point can be controlled, for example, by a plate trailing edge flap.

[3] The terms *natural waves* and *natural transition* hereinafter denote the variety of ways of the transition obtained in various experiments without an artificial introduction of disturbances to a flow that are additional to the background disturbances.

The non-parallel effects require a selection of a stability criterion both in experiment and in theory (Fig. 2.4). Unfortunately, such a physical measure of the linear stability as energy integrated across the boundary layer is rarely used (Boiko et al. 1994; Boiko 2000). Schubauer and Skramstad (1948), in their pioneer study, used an elementary philosophy by measuring the flow traversing the hot-wire probe at a fixed distance from the wall. Ross et al. (1970) and Kachanov et al. (1979a) were the first to pay attention to a downstream displacement of the near-wall maximum in the streamwise disturbance velocity profile caused by the flow non-parallelity. They applied a modified measurement technique: the development of the disturbances was traced along the line of equal velocity (that is along a fixed self-similar boundary layer coordinate) close to the maximum of the streamwise velocity component of the velocity oscillations. Additionally, the influence of the distance to the wall on the position of the experimentally determined neutral stability curve and on the spatial amplification was studied by Kachanov et al. (1979a) (Fig. 2.4). Various ways of defining the disturbance measure were also discussed by Bouthier (1973), Gaster (1974) and Saric and Nayfeh (1977). However, despite these improvements in the methodology of experiment, the growth of the oscillations was once again found to occur up to frequency parameter $F = 400 \times 10^{-6}$ (Fig. 2.3).

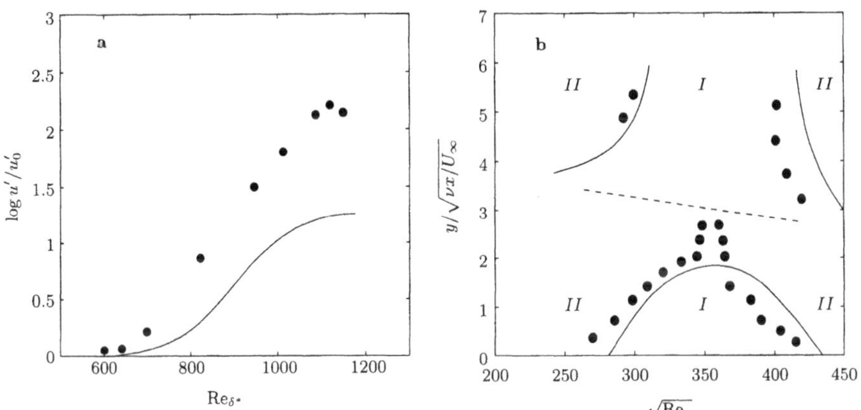

Fig. 2.4. a Tollmien–Schlichting wave amplification curves for $F = 110 \times 10^{-6}$: experiment (•) of Kachanov et al. (1979a) and linear stability theory (*solid line*) of Levchenko et al. (1975); **b** Location of neutral points for streamwise velocity disturbance in (Re_x–η)-plane for $F = 200 \times 10^{-6}$: experimental data (•) of Kachanov et al. (1979a); *solid lines*, calculations of Tumin and Shepelev (1980); *dashed line*, minimum $|\hat{u}|$ displacement; *I*, growth of disturbances; *II*, decay of disturbances

Experiments on flat plate boundary layer stability in which special care was undertaken to minimize the mean flow difference from the Blasius one were conducted by Klingmann et al. (1993). The plate used in that study had a leading edge whose shape was specially calculated with the help of two-dimensional potential flow theory to take into account the real design of the wind tunnel test section (see Fig. 5.8). The plate was supplied with a trailing edge flap to regulate the pressure distribution at the leading edge. Thus it became possible to remove the suction peak near the leading edge and to reduce substantially the size of the pressure gradient region. For excitation of the controlled instability waves, the vibrating ribbon technique was used.

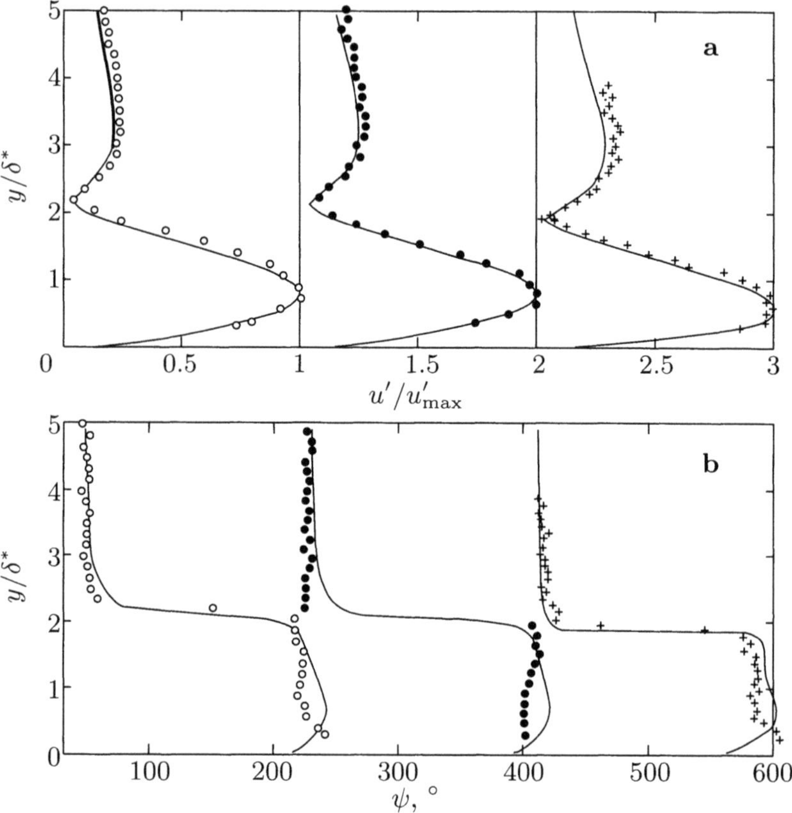

Fig. 2.5. Amplitude (**a**) and phase (**b**) distributions of Tollmien–Schlichting waves. $F = 250 \times 10^{-6}$; ○, $\mathrm{Re}_{\delta^*} = 343$; •, $\mathrm{Re}_{\delta^*} = 396$; +, $\mathrm{Re}_{\delta^*} = 574$. *solid lines*, corresponding results of the parallel linear theory (Klingmann et al. 1993)

The disturbance streamwise amplitude and phase profiles measured at three various distances from the leading edge are shown in Fig. 2.5. The Tollmien–Schlichting wave amplitude is normalized to its inner maximum. Characteristic features of the profiles that are typical for the Tollmien–Schlichting waves predicted by linear stability theory are the inner and the outer amplitude maxima as well as the phase 'jump' near the amplitude minimum.

Generally, the profile varies as the wave moves downstream. The phase jump and the amplitude maxima displace monotonically to smaller y/δ^*, indicating that the wave evolution is not self-similar in the boundary layer coordinates, as has been discussed earlier.

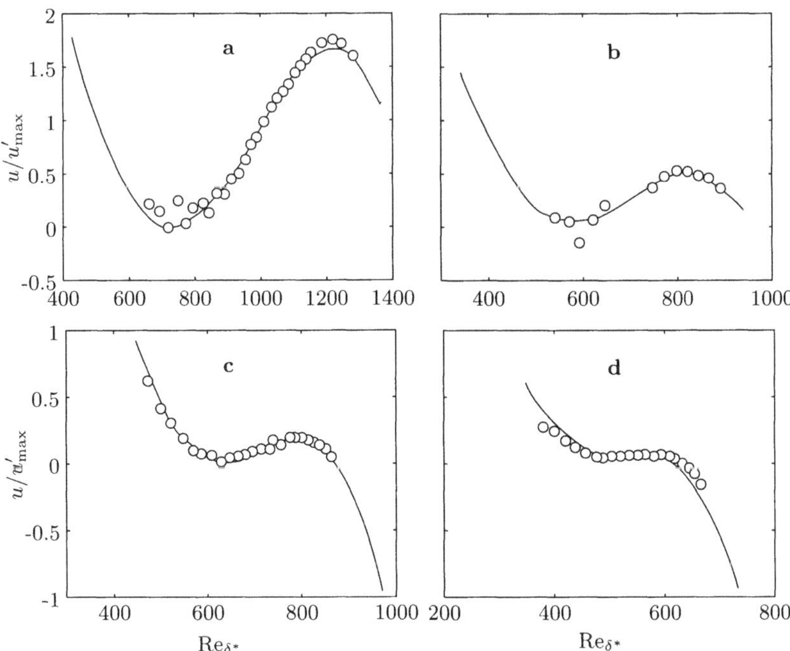

Fig. 2.6. Downstream behaviour of the Tollmien–Schlichting wave amplitudes: $F \times 10^6 = 100$ (**a**), 162 (**b**), 200 (**c**), 250 (**d**). ○, experimental data; *solid lines*, results of the linear parallel theory (Klingmann et al. 1993)

The development of the disturbance amplitude with Re measured at the inner Tollmien–Schlichting wave maximum is shown in Fig. 2.6 compared with the results of the parallel linear stability theory. The wave excited upstream of branch I of the neutral curve decays at first, then is amplified in the instability region and decays again behind branch II. At low frequencies, the amplitude evolution is predicted quite well by the theory; at high frequencies, the experimental amplitude curves experience a little deviation from that calculated by the theory, but much less than in the previous experimental studies. In particular, at $F = 250 \times 10^{-6}$, the experimental data show small amplification in the region about $Re_{\delta^*} = 520$, whereas the parallel linear stability theory predicts a monotonic decay of the Tollmien–Schlichting wave. The experimental and theoretical neutral curves are shown in Fig. 2.7. It can be seen that the experimental results are in good agreement with the Gaster (1974) non-parallel theory. For the low frequencies, the difference between the theoretical curves and experimental data is negligible.

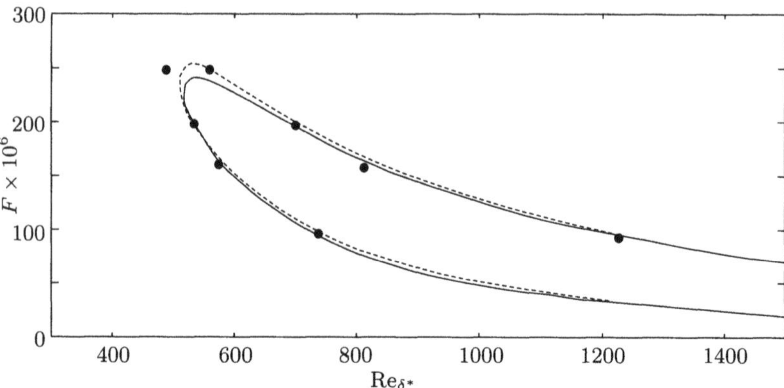

Fig. 2.7. Neutral stability curves: •, experimental data obtained at the inner amplitude maximum (Klingmann et al. 1993); *solid lines*, calculation for parallel flow; *dashed lines*, non-parallel theory (Gaster 1974)

The data of Klingmann et al. (1993) showed that the linear parallel stability theory predicts well the distributions of amplitudes, wave numbers, growth rates and neutral stability of the two-dimensional Tollmien–Schlichting waves. The assumption of boundary layer parallelity allows the acquisition of quite precise results for the disturbance growth rates, except for the upper most part of the neutral stability curve, where it is necessary to take into account the effects of flow non-parallelity, which are quite small. The discrepancy between the preceding experimental results and the predictions of the linear parallel stability theory arises mainly from inaccurate experimental simulation of the ideal Blasius boundary layer or, equivalently, from neglecting in the calculations the actual properties of the boundary layer. The results are most sensitive in the region of low Reynolds numbers near the leading edge of the plate where the pressure distribution is difficult to control with standard setups. The role of non-parallel effects for the development of three-dimensional waves is considered in more detail in the next section.

2.1.3 Wave packets

A disturbance consisting of a monochromatic wave represents an exclusive case. The capability tracing a single Tollmien–Schlichting wave in the experiments described earlier is due to the wavemaker (the ribbon) excites an unstable wave of a single frequency. In contrast, natural disturbances are mostly localized in space and time. In physics, for the description of such situations the wave packet concept is widely used. It appears that the concept is highly valuable also for the general analysis of flow instability properties.

In thc framework of the parallel linear theory, the localized disturbances can be expanded into frequency and wave number spectra of finite widths, the consideration of the disturbances in the physical and spectral spaces being equivalent. It appears that the spectral representation, as a rule, is more effective for the analysis of the linear stability of hydrodynamic systems, since it is possible to select a separate Tollmien–Schlichting wave and to study its behaviour independently of the other disturbances (see Chap. 1).

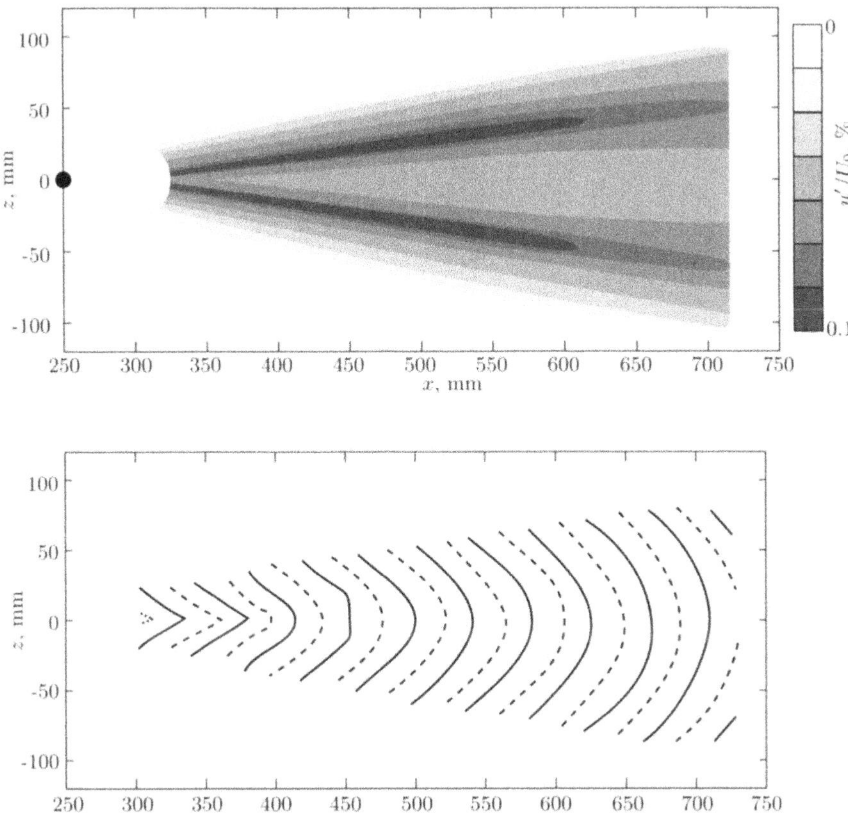

Fig. 2.8. Wave pattern of a typical wave train in a flat plate boundary layer: **a**, isolines of amplitude; **b**, isolines of phase with 180° shift, •, location of a point-like source of the periodic disturbances (Kozlov and Ryzhov 1990)

A non-periodic disturbance localized in space and time in a boundary layer represents a group of Tollmien–Schlichting waves with different frequencies and wave numbers, and also a set of modes of the continuous spectrum, *a wave packet*. When it is localized only in space, it becomes a degenerate purely spatial wave packet which is also called a *wave train* (Fig. 2.8). Such a packet can be created, for example, with the help of a point-like periodic disturbance source located inside the boundary layer. In turn, a degenerate case of the wave train is a periodic disturbance excited by, for example, a long vibrating ribbon. It consists of a packet of the modes of the discrete and continuous spectra with identical frequency ω and spanwise wave numbers β, but with variable streamwise wave numbers α (Gaster 1965; Ashpis and Reshotko 1990; Gaster and Gupta 1993), the least stable mode of which contribute usually to the disturbance energy far from the source. In the general case the 'shape' of the wave packet changes, since the discrete waves constituting it are *dispersive*, i.e. satisfy different dispersion relations (1.28) for different wave components.

The sets of isolated periodic waves and wave packets are represented by spectra of different types: line (amplitude) and power-density spectra according to Fig. 2.9. The difference between them relates to normalization, as the sets of isolated periodic waves are by definition spatially (temporally) infinitely extended, while the wave packets are localized. As the Fourier components of the wave packet do not interact, each element of the spectrum develops independently in time. This means that the spectrum at any moment is determined by its initial conditions, i.e. for the description of the behaviour of a wave packet it is necessary to solve the initial-value problem.

The initial-value problem from the viewpoint of disturbance excitation (the receptivity problem) is examined in Chap. 3. Below we will concentrate on the general description of the flow response to localized disturbances. In this section we consider the application of the wave packet concept to the description of the propagation of a localized disturbance. At the end we note the effect of flow non-parallelity on the development of both the packets and oblique waves. Further results of theoretical studies on wave packet development in the boundary layer can be found in Gaster (1975), Cohen et al. (1991) and Breuer et al. (1997).

Impulse response. Since the wave packets are localized, it is possible to determine the transportation of the disturbance energy in terms of velocity, direction and growth rate. Continuing the discussion of Sects. 1.1 and 1.3, we distinguish the wave packet development in time and/or space. Thus, if a growing localized disturbance is concentrated permanently at the source and covers it, absolute instability takes place. If the disturbance propagates at least along one direction from the source and leaves the position of source asymptotically with time, convective instability occurs.

The theory of dispersive waves was applied for instability in plasma jets by Briggs (1964) and later for hydrodynamic wave packets by Gaster (1968, 1975), Huerre and Monkewitz (1985), Brevdo (1995) and Oertel and Stank (1999). The theory states that the wave packet energy is transported with group velocity $\mathbf{c}_g = \partial\kappa/\partial\omega^{-1} = \nabla_\kappa\omega$ provided that the Reynolds number is 'frozen' at the differentiation.

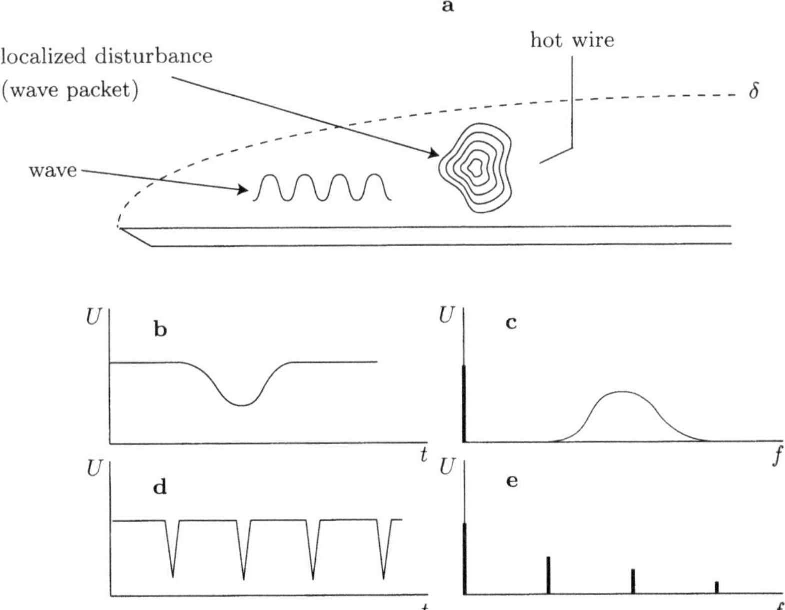

Fig. 2.9. To definition of a wave packet: a boundary layer sketch with a localized disturbance sensed by a hot wire (**a**); oscilloscope traces of a localized (**b**) and periodic (**d**) disturbances; the power spectrum of the localized disturbance (**c**); linear spectrum of the periodic signal (**e**)

In what follows we outline the general procedure for the analysis of the impulse response, closely following Schlichting and Gersten (2000). Let us consider the linearized Navier–Stokes operator with a non-homogeneous right-hand side in a general form, cf. (1.12–1.15)

$$\mathcal{L}_{\mathrm{NS}}\hat{\boldsymbol{v}} = \boldsymbol{G},$$

where $\boldsymbol{G}$ represents disturbance forces and sources. Using a Fourier transformation in x and z and a Laplace transformation in t, the equation is transformed to a system of ordinary differential equations in y:

$$\int_0^\infty \int_{-\infty}^\infty \int_{-\infty}^\infty \mathcal{L}_{\mathrm{NS}}\hat{\boldsymbol{v}}\mathrm{e}^{\mathrm{i}t\Psi}\mathrm{d}x\mathrm{d}z\mathrm{d}t = \hat{\boldsymbol{G}},$$

where $\hat{\boldsymbol{G}}(\alpha,\beta,\omega,y)$ is the Laplace–Fourier transformed forcing and $\Psi = \alpha x/t + \beta z/t - \omega$ is a phase. The solution in spectral space always takes the form $\hat{\boldsymbol{v}}(y) = \boldsymbol{Z}/\mathcal{F}$, where the numerator $\boldsymbol{Z}$ depends on the particular perturbation $\boldsymbol{G}$, while the denominator $\mathcal{F}$ is exactly the dispersion relation (1.28) which depends only on flow properties, and its roots give the eigenvalue spectrum of the corresponding homogeneous linearized problem. Both the frequency and the wave numbers can be complex, but since we expect that the least stable mode contributes most to the integral, we may exclude the rest of the spectrum from consideration.

The solution in physical space is obtained by inverse Fourier–Laplace transformation. The asymptotic behaviour of the solution in time for the waves with real or complex group velocity can be obtained by the steepest descent method or its variations (Schmid and Henningson 2000). The main idea of the method relies on the fact that at large t the phase Ψ provides rapid oscillations of the integrand over the most of the integration range, and only the part where $\nabla\Psi = 0$ contributes substantially to the integral. Therefore, we can concentrate on the solution for the wave vector $\boldsymbol{\kappa}_0 = \{\alpha_0, \beta_0\}$ (stationary point) which satisfies the equations

$$\frac{\partial\Psi}{\partial\alpha} = \frac{x}{t} - c_{gx} = 0, \qquad \frac{\partial\Psi}{\partial\beta} = \frac{z}{t} - c_{gz} = 0,$$

where $c_{gx} = \partial\omega/\partial\alpha$, $c_{gz} = \partial\omega/\partial\beta$. This means that an observer moving with the real part of the group velocity $\boldsymbol{c}_g = \{c_{gx}, c_{gz}\}$ of the packet follows the waves with the wave number vector equal to the real part of α_0 and β_0, and with frequency $\omega = \omega_0 - \boldsymbol{\kappa}_0\boldsymbol{c}_g$. Physically it means that in this frame of reference, the absolute instability is observed with a maximum possible growth rate. In the laboratory coordinate system, if the 'neutral curve' $\omega_i(\boldsymbol{c}_g) = 0$ in the (c_{gx}–c_{gz})-space surrounds the wave packet origin, the flow is absolutely unstable; otherwise it is convectively unstable.

If in a particular shear flow the dispersion relation varies significantly with spatial coordinates, the appearance of additional critical Reynolds numbers corresponding to the change of the convective instability to an absolute one, or vice versa, can occur. In this case the disturbance development will depend on the position of the disturbance source. It is necessary to note, however, that in such a case the transition from one instability to another is not stipulated by an appearance of a *qualitatively* new mechanism, but simply represents a *quantitative* shift in the balance between stabilizing and destabilizing forces (Huerre and Monkewitz 1990; Monkewitz 1990).

Transformations of Gaster's type. The results of Gaster (1962) provide a relation between the temporal and spatial growth rates for almost neutral modes. Because the spatial analysis is theoretically and numerically more difficult, the transformations are widely used to calculate the spatial growth rates from the temporal ones.

For brevity, let us consider the dispersion relation in the form

$$\mathcal{F}(\alpha, \omega) = 0,$$

i.e. $\beta = 0$, and apply it at two neighboring states (α, ω) and $(\alpha + \delta\alpha, \omega + \delta\omega)$. Expanding the relation at (α, ω) and leaving only linear terms we obtain $\delta\omega = c_g\delta\alpha$. To convert from the spatial to the temporal stability, we take $\delta\alpha = -\mathrm{i}\alpha_i$, which yields

$$\delta\omega = -\mathrm{i}c_g\alpha_i.$$

Considering the opposite transformation from the temporal to the spatial stability, we take $\delta\omega = \mathrm{i}\omega$ which gives

$$\delta\alpha = \mathrm{i}\omega/c_g.$$

Similar considerations can be applied to a general case of three-dimensional disturbances and three-dimensional flows (Nayfeh and Padhye 1979; Nayfeh 1980; Saric 1994a).

Global instability. In the case when the finiteness of the system – say, the flat plate length – is essential, the frequency spectrum of its eigenmodes is restricted by boundary conditions at the upstream and downstream ends. It consists of a large number of closely located discrete frequencies. Due to the boundaries, the difference between the convective and the absolute instability disappear; the flow becomes *globally unstable*, if at least one of the frequencies has a positive imaginary part. In this case a separation into independent normal modes is, strictly speaking, impossible. The general dispersion relations for this problem can be obtained for a system of finite, but large size (Kulikovskiy 1966), as it is reduced to a function only of properties of the fluid far from the boundaries using the method similar to the steepest descent technique outlined above. It appears that the presence of global instability is a sufficient condition for instability of an infinite system – related to it – at the removal of the boundaries to infinity. However, the opposite is true only for the absolutely – but not for the convectively – unstable fluid. The last case can become both unstable or stable following the 'installation' of the boundaries. Recent advances in global instability theory can be found in Huerre and Monkewitz (1990), Monkewitz (1990), Cossu and Chomaz (1997) and Huerre and Rossi (1998). Global instability phenomena, which one expects at the flow separation, are touched upon in Chap. 6.

Experimental investigations of the wave packets. There are at least two main reasons to consider the development of the wave packets experimentally. One of them is to simulate a naturally occurring disturbance and to study the stages of its transformation to turbulence as a transition scenario. The other is to consider only the initial (linear) stage of its development in order to extract wave characteristics of the waves constituting it, to verify some theoretical results and find their restrictions.

Vasudeva (1967) was the first to study the development of a three-dimensional wave packet in a boundary layer. He was not able to overcome certain technical difficulties, in particular the small signal-to-noise ratio (dynamic range) of hot-wire measurements. The experiments of Gaster and Grant (1975) were an essential step in the investigation of wave packets. Periodic pulses of the disturbances of quite small amplitude were excited by periodic sucking and blowing through a small hole in a flat plate. To reduce errors associated with the effects of non-parallelity, the measurement were performed in the region of the outer maximum of Tollmien–Schlichting waves, whose location weakly depends on the streamwise coordinate and the frequency of the disturbances. To increase the signal-to-noise ratio, synchronous detection with the help of a computer was applied. As a result, the pattern of downstream development of the excited wave train was obtained which corresponded to the result of calculations of the Tollmien–Schlichting wave superposition of the given frequency and different wave numbers.

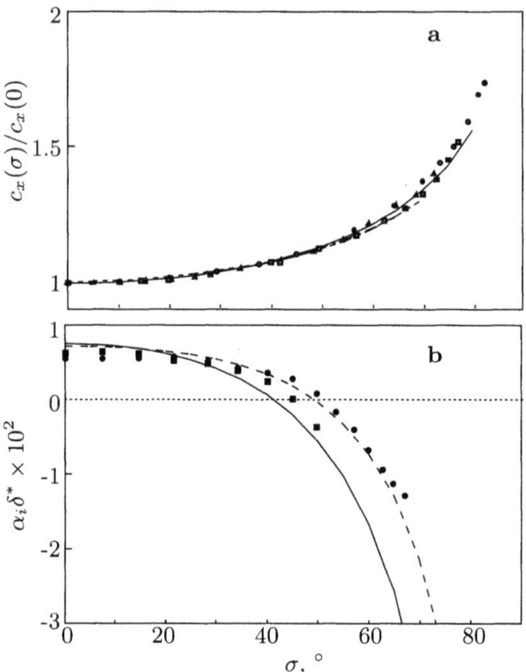

Fig. 2.10. Downstream phase velocity ratio as a function of wave angle (**a**): $\mathrm{Re}_{\delta^*} = 827$–$1226$, $F = (63.6$–$138.7) \times 10^{-6}$; *symbols*, experimental results; *lines*, corresponding theoretical curves. Normalized spatial wave growth rate as a function of wave angle (**b**): experimental data for $F = 94.1 \times 10^{-6}$ at $\mathrm{Re}_{\delta^*} = 1048$; ○, $y/\delta^* = 0.74$; □, $y = 4.5$ mm; theoretical curves at $\mathrm{Re}_{\delta^*} = 1089$ (*solid line*) and $\mathrm{Re}_{\delta^*} = 963$ (*dashed line*) (Kachanov 1985; Kachanov and Michalke 1994)

The experiments of Gaster and Grant (1975) were directed to the general analysis of the processes leading to the laminar–turbulent transition via the instability phenomena in the wave packet, while the studies of Kachanov (1985) and Kachanov and Michalke (1994) were performed to study the characteristics of the waves of a linear wave train. The main motivation of these experiments was the absence of detail measurements of the dispersion and stability characteristics for oblique Tollmien–Schlichting waves. In the latter study, an examination of the wave packet closely followed its theoretical description. The obtained periodic velocity distributions were subjected to spatial Fourier transformation to obtain the dispersion characteristics of the waves with different streamwise α and spanwise β wave numbers at a fixed frequency. As a result, the dispersion characteristics of the three-dimensional waves for a number of frequencies and angles of the wave inclinations $\gamma = \arctan(\beta/\alpha_r)$ up to 60–80° were obtained. Most of them appeared to be very close to the results predicted by the parallel linear stability theory. The general scape of the wave packet also agrees with the theoretical conclusions. However, the wave growth rates at large angles of inclination were larger than those obtained theoretically (Fig. 2.10). The last was attributed to an increased role of the non-parallel effects for the oblique waves.

Another possibility for the discrepancy are the transient phenomena described at the end of Chap. 1. Extensive measurements of the wave packet characteristics were carried out by Cohen et al. (1991), who created a perturbation without initial transient phenomena connected to the lift-up effect, by fluid injection through a perforated flat plate and tracing the disturbance behaviour before its transformation to a turbulent spot. We consider the transient phenomena and their role in the transition process in boundary layers in detail in Chap. 5.

2.2 Plane Poiseuille flow

At first sight, the instability of plane channel flow is much easier to describe than that of the Blasius boundary layer, since it is strictly parallel and bounded. This statement seems true for a theoretical description, but experimentally the answer is not so straightforward. The equations governing the flow instability are once again non-self-adjoint and strong transient phenomena take place. Moreover, in contrast to the Blasius boundary layer a subcritical bifurcation of the steady state is allowed (see Sect. 4.1.2).

2.2.1 Theoretical approach

The first theoretical results on the stability of plane Poiseuille flow were presented by Heisenberg (1924), who calculated the position of the second branch of the neutral stability curve using an asymptotic method. Lin (1955) advanced further the asymptotic analysis and confirmed the conclusions of Heisenberg. It was found that the instability occurs above a critical Reynolds number equal to about $\mathrm{Re}_{\mathrm{L},h} = U_\mathrm{a} h/2\nu \approx 5300$ at $\alpha \approx 1$. Since that time, $\alpha = 1$ has been widely used as a test value. The Reynolds number is based here on the maximum flow velocity in the channel U_a and a half distance between the walls[4] $h/2$ (Fig. 1.3). Thomas (1953) for the first time solved the problem using a numerical method on a computer and found $\mathrm{Re}_{\mathrm{L},h} = 5780$ at $\alpha = 1.02$. Orszag (1971) applied a very effective spectral method utilizing the fast convergence of Chebyshev polynomial approximation to the

[4] Sometimes $\overline{\mathrm{Re}} = 2/3\mathrm{Re}$, based on the mean flow velocity, is used.

solution of instability problem. He obtained $\mathrm{Re}_{\mathrm{L},h} = 5772.22$ at $\alpha = 1.02056$. The spectral approach appeared to be highly valuable for the study of the transient disturbance behaviour. It produces matrices of a moderate size, and provides an approximation of the set of least stable eigenmodes and corresponding eigenfunctions in one run with high accuracy which can be used for the following initial-value problem calculations (Fig. 2.11).

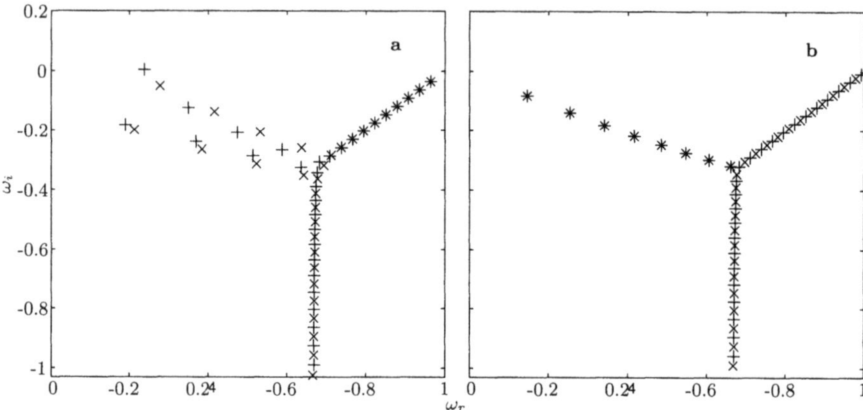

Fig. 2.11. Least stable temporal eigenvalues for Orr–Sommerfeld (**a**) and Squire (**b**) equations obtained by the pseudospectral method with 120 Chebyshev polynomial. First even (+) and odd (×) modes are shown. $\mathrm{Re}_h = 10000$, $\alpha = 1$, $\beta = 0$ (calculated by A.V. Boiko)

2.2.2 Experimental linear stability investigations

Despite the great variety of theoretical studies on the linear stability of Poiseuille flow, for a long time there was a lack of experimental results suitable for comparison. This situation is explained by technical difficulties connected with manufacturing the experimental facility for simulating the flow between two infinite plates. The length of the parallel plates in the experiment should exceed the extent of the input acceleration region, which increases with the Reynolds number (Davies and White 1928; Patel and Head 1969). According to calculations, the ratio of the length of the acceleration region to the channel height reaches approximately 400 for $\mathrm{Re}_{\mathrm{L},h} = 5772$ (Sparrow et al. 1964; Schlichting and Gersten 2000). The channel width should also be large enough. Only at large width-to-height ratios is it possible to obtain approximately two-dimensional flow in the central part of the channel. Disturbances and the turbulent wedges emanating from the angular areas reduce the effective width of the channel, destroying the two-dimensional velocity distribution. Such phenomena can be avoided by applying curved or adjustable walls, but their manufacture, as a rule, is not straightforward.

To achieve high channel length-to-width ratios, it is necessary to pick the least possible height which at the same time should be large enough to place the sensor of a hot-wire anemometer in the channel and to obtain a high spatial resolution. Besides, a small height increases the relative influence of technological inaccuracies of manufacturing the model: wall waviness and roughness.

Another difficulty is connected with the channel entrance, which should have a device ensuring smooth inflow of a fluid inside the channel to minimize the level of background turbulence and to prevent the subcritical transition stipulated by the algebraic instability or the disturbances of finite amplitude (Itoh 1974). As a rule, to reduce the background disturbances, contractors with smoothly varying contours are applied with entrance screens possessing small-size grids and honeycombs, but due to a strong contraction of the fluid, non-uniformity of the flow caused by the grids and the honeycombs can easily generate streamwise vortices in the channel that strongly affect the flow structure. In the majority of early studies, the inflow conditions, the mean flow uniformity and the turbulence level were insufficiently controlled (Davies and White 1928; Sherlin 1960; Narayanan and Narayana 1967; Patel and Head 1969), which produced large scattering of $\mathrm{Re_T}$ in the experimental studies; see Table 2.2. As a result, it was usually observed that the transition to turbulence begins with a 'sudden' appearance of turbulent spots (see Chap. 4) in the subcritical region.

Table 2.2. $\mathrm{Re_T}$ of early experiments on transition to turbulence in plane Poiseuille flow

Study	w/h	$\mathrm{Re}_{\mathrm{T},h}$
Davies and White (1928)	37–170	660–266
Sherlin (1960)	4	1265
Narayanan and Narayana (1967)	12	1425
Patel and Head (1969)	48	1035

Only the experiments of Nishioka et al. (1975) and Kozlov and Ramazanov (1982) were able to confirm definitively the applicability of linear stability theory for plane Poiseuille flow. In these studies, channels with width-to-height ratios of 27.4 and smooth elongated contractors at the channel entrance were used. Special care was taken to reduce the levels of turbulence inside the channels to 0.05% (Nishioka et al. 1975) and 0.1% of U_0 (Kozlov and Ramazanov 1982), and to keep the flow laminar to $\mathrm{Re}_{\mathrm{T},h} = 8000$ and 7000, respectively. The experiments allowed a detailed study of the spatial development of disturbances generated by a vibrating ribbon at various frequencies and over a large range of Reynolds numbers ($\mathrm{Re}_h = 3000$–8000) as well as obtaining points at the neutral stability curve. In Fig. 2.12a, the experimental and the-

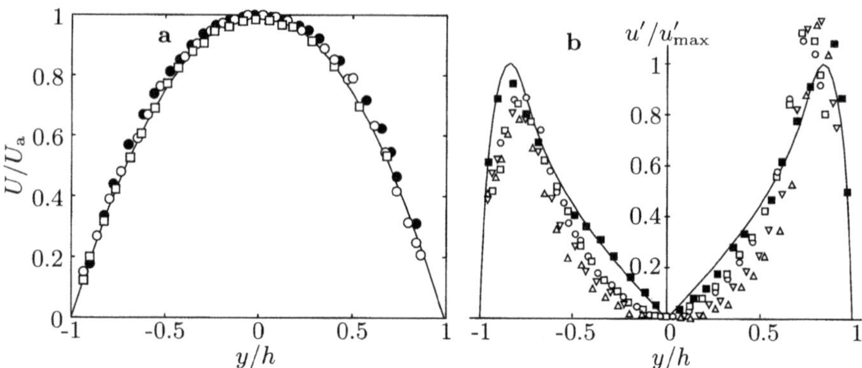

Fig. 2.12. Mean velocity profiles for plane Poiseuille flow (Kozlov and Ramazanov 1982) (**a**) and profiles of velocity oscillations $\omega = 0.27$ (**b**): $\mathrm{Re}_h = 3000$ (○), 4000 (□), 5000 (▽), 6000 (△) (Kozlov and Ramazanov 1982), 4000 (■) (Nishioka et al. 1975); *solid line*, calculations at Re = 4000 (Kozlov and Ramazanov 1982)

oretical mean velocity profiles are shown for $\mathrm{Re}_h = 3000$, 4000, and 5000 (Kozlov and Ramazanov 1982), ensuring that the flow in the channel is close to the plane Poiseuille one.

The profiles of small-amplitude disturbances excited by Nishioka et al. (1975) and Kozlov and Ramazanov (1982) are shown in Fig. 2.12b. The small asymmetry probably results from weak distortions introduced by the hot-wire probe and the ribbon. In the channel centre, a 180° phase jump occurs, which indicates the antisymmetric nature of the streamwise velocity disturbances.

Similarly, in Fig. 2.13a the amplification rate $-\alpha_i$ as a function of the frequency is given. Figure 2.13b shows the theoretical neutral stability curve. The symbols corresponding to the growing disturbances are filled; those of stable waves are empty. It can be seen that the disturbances decay at Re $\lesssim$ 6000; otherwise growing disturbances occur.

As a whole, the experiments of Nishioka et al. (1975) and Kozlov and Ramazanov (1982) showed that laminar plane Poiseuille flow with a proper control of the disturbances can be implemented up to Reynolds numbers $\mathrm{Re}_h = 7000$–8000. Excited small two-dimensional waves experience no growth up to the critical Reynolds number, as well-predicted by the linear stability theory.

2.2.3 Transient growth

The lift-up effect accounting for viscosity in plane channel flow was considered theoretically for the first time by Gustavsson (1991) and Henningson and Schmid (1994). The growth rates and 'the time of survival' appeared to be maximum for the waves with small streamwise wave numbers α (Fig. 2.14).

Klingmann (1991, 1992) studied the development of three-dimensional localized disturbances in channel flow experimentally, and conducted a com-

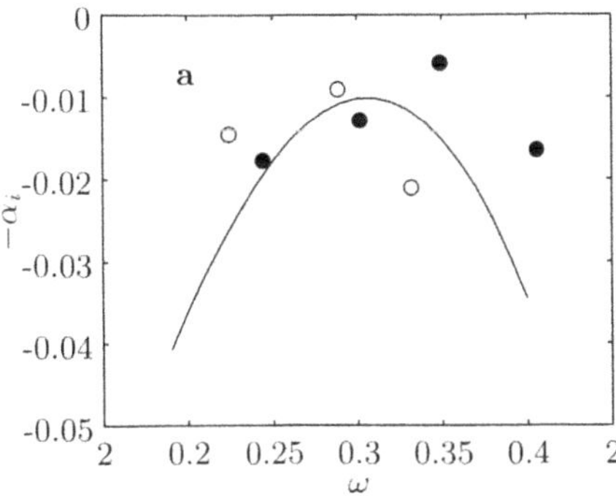

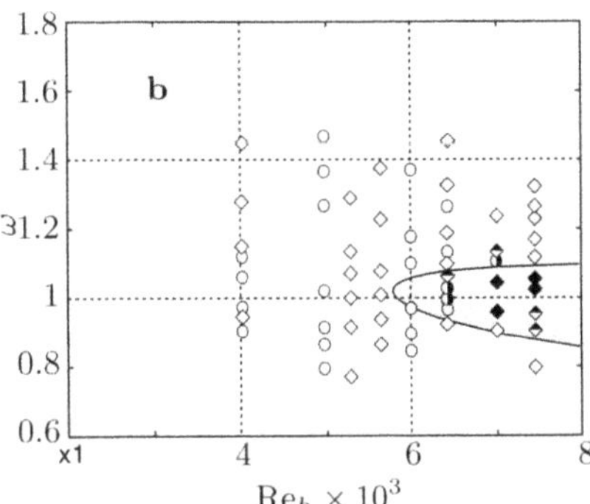

Fig. 2.13. Amplification rate as a function of frequency for plane Poiseuille flow at $\mathrm{Re}_h = 4000$ (**a**): ○, (Nishioka et al. 1975); •, (Kozlov and Ramazanov 1982); *solid line*, linear stability theory. Stability boundary of plane Poiseuille flow for small disturbances (**b**): ○, (Kozlov and Ramazanov 1982); ◇, (Nishioka et al. 1975); *filled symbols*, growing disturbances; *half-filled symbols*, nearly neutral disturbances; *empty symbols*, decaying disturbances; *solid line*, linear stability theory

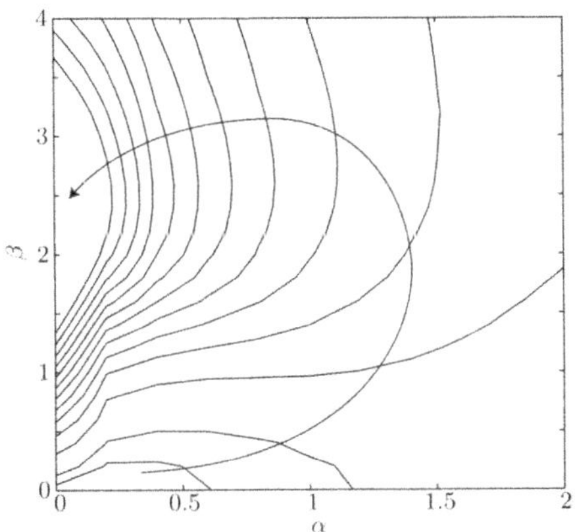

Fig. 2.14. Isolines of maximum growth of energy $G(t)$ in wave number (α–β)-plane for Poiseuille flow at $\mathrm{Re}_h = 3000$. The transition from one curve on another in a direction indicated by arrows corresponds to a change in $G(t) = 10, 25, 100, 200, \ldots, 1100$ (calculated by A.V. Boiko)

parison with the results of the viscous linear analysis of the corresponding initial-value problem. The disturbances were excited by an injection of fluid into the channel through a point-like hole in one of the channel walls. At the initial stage, despite the amplitude of the disturbances reaching 10–20% of U_a, a good correspondence with the theory was obtained: appearance of streamwise streaks of retarded and accelerated fluid with strong velocity shears in the spanwise direction which propagated with velocities between 0.5 and 0.9 of U_a; formation of new streaks at the downstream disturbance propagation; linear dependence of the energy growth of the disturbance in time; and proportionality of the amplification time to the Reynolds number. However, due to non-linear effects at late stages of the amplification, the disturbances grew longer than the theory predicts and downstream, depending on the initial intensity of excitation, they either decayed or led to turbulence through a stage of a non-linear development.

2.3 Three-dimensional boundary layers

The instability phenomena in three-dimensional flows are manifold. In this section we concentrate only at two prototypical instabilities: crossflow instability of swept-wing flow and Görtler instability at a concave wall.

2.3.1 Swept-wing flow

Swept-wing flow is of large practical interest (Kohama 1987; Zhigulev and Tumin 1987; Reed and Saric 1989; Arnal et al. 1990; Saric 1994a; Kachanov 1996; Bippes 1999). The growth of stationary vortices oriented approximately along an inviscid streamline were observed by Gray (1952) when visualizing the swept-wing flow of an aeroplane. This finding started the interest of investigators in the swept-wing stability problem. Later, such vortices were found also in flows on a rotating disc (Gregory et al. 1955), rotated cone (Kobayashi et al. 1983) and swept cylinder (Poll 1985).

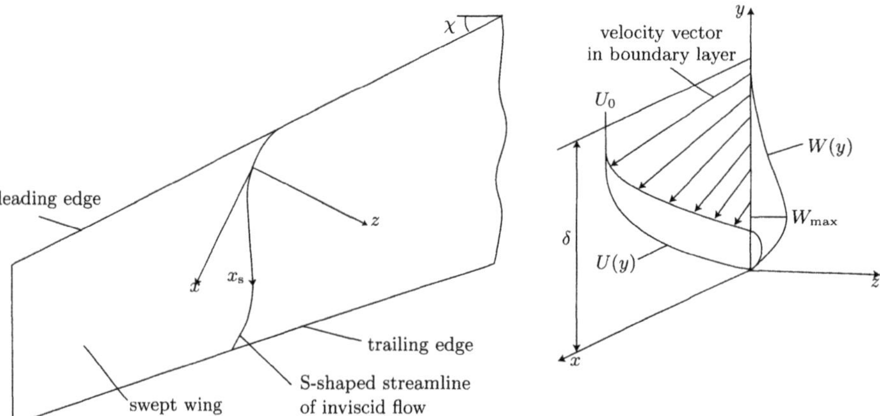

Fig. 2.15. Coordinate system and velocity components in a three-dimensional boundary layer with crossflow

In the flow around a swept wing, streamlines at the outer border of the boundary layer take an S-shaped form under the effect of the spanwise pressure gradient. Inside the boundary layer the pressure gradient forms the so-called crossflow. A traditional model for the swept-wing mean flow are two-parameter Falkner–Skan–Cooke equations (Cooke 1950). The flow is characterized by a pressure gradient Hartree parameter β_{H} and the angle of the free-stream streamline inclination to the wing chord direction ϕ_e. Supposing flow parallelity, its mean velocity can be represented as two components in the direction of the external streamline and that orthogonal to it:

$$\mathbf{U} = \{U(y), 0, W(y)\}. \tag{2.1}$$

The scheme of the velocity vector orientation in the boundary layer and the crossflow velocity profile $W(y)$ are given in Fig. 2.15.

Gregory et al. (1955) showed that the parallel linear stability approach is applicable to such flow. However, in contrast to the flat plate, when the flow independence in the spanwise direction presumes only real values of spanwise wave number β, the situation for *three-dimensional* flows is different. In this case, the dispersion relation reads $\alpha = \alpha(\omega, \beta_r, \beta_i)$. Thus it is necessary to take in some way into account additional parameter β_i; for example, to remove it by rotating the coordinate system, aligning x-axis with the direction of growth of the disturbances. In calculations this direction is determined taking into account some conjectures concerning the nature of perturbations and considering their group velocity Malik and Orszag (1980), Mack (1984).

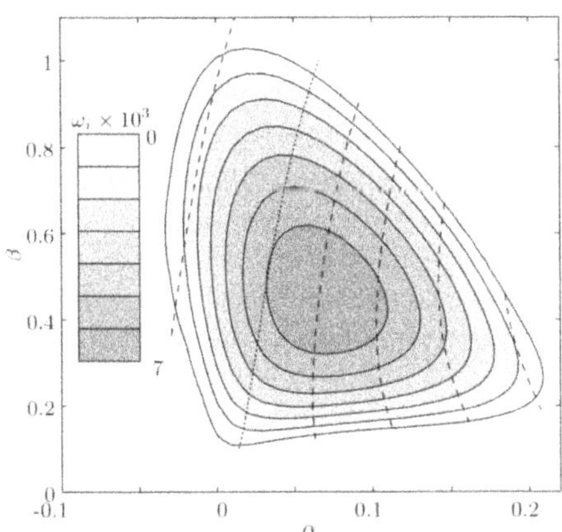

Fig. 2.16. A diagram of Falkner–Skan–Cooke boundary layer temporal stability for $\beta_H = 0.5$, $\phi_e = 45°$, $Re_{\delta_s^*} = 800$, δ_s^* is taken along the external flow streamline: *solid lines*, isolines of ω_i; *dotted line*, $\omega_r = 0$; *dashed lines*, isolines of ω_r with step 0.02 (calculated by A.V. Boiko)

Since the crossflow component $W(y) = 0$ at the wall and in the free stream, $W(y)$, is inflectional, the flow instability is determined by the inviscid mechanism and appears at low Reynolds numbers, with the transition occurring in the region subcritical to the Tollmien–Schlichting wave instability (Kohama 1987; Dagenhart et al. 1990; Kohama et al. 1991). Typical stability characteristics for a Falkner–Skan–Cooke profile arc shown in Fig. 2.16. It can be seen that stationary vortices ($\omega_r = 0$) as well as travelling waves ($\omega_r \neq 0$) can be unstable. The characteristic frequencies for the travelling disturbances are usually much lower than those for the Tollmien–Schlichting waves. The ratio of the wave numbers β/α for the most unstable travelling waves is quite high. This means that they propagate approximately in the crossflow direction. Interestingly enough, the amplitude distributions of the streamwise velocity component of the stationary vortices in the spanwise direction make them counter-rotating, and the widely observed co-rotation of the vortices in visualizations is connected with the mean flow superposition onto the vortex structure.

Due to the predominance of the stationary vortices, the majority of experimental investigations into crossflow instability were devoted to stationary disturbances (Michel et al. 1985b; Saric and Yeates 1985; Bippes 1999; Boiko 2000). The predominance of stationary vortices in the experiments is controlled by initial conditions in the form of natural surface roughness of microscopic size being nevertheless large enough compared to the boundary layer thickness at low Reynolds numbers. A development of a 'natural' wave packet originated from this roughness is characterized by vortex multiplication (Streett 1998), see also Sect. 3.6.1.

The early experiments on stationary modes under natural conditions showed qualitative agreement with the general properties of the crossflow vortices predicted by linear stability theory. However, the theoretical growth rates of the most unstable crossflow modes were found to be much higher than those observed in experiments (Radeztsky et al. 1994; Bippes 1999). In contrast, results obtained under controlled conditions demonstrate quite good agreement with the theory. The first quantitative experimental characteristics for spatially periodic crossflow vortices were obtained by Kachanov and Tararykin (1990) using the Fourier transformation technique, Boiko (2000), using a different experimental facility and excitation technique also found fairly good agreement between the experimental and theoretical characteristics of the most unstable stationary crossflow modes (see Fig. 3.15). Reibert et al. (1996) found under control conditions very good agreement of the vortex characteristics with results of PSE analysis.

Meanwhile, linear stability theory predicts that the travelling modes are the most unstable (Fig. 2.16). Consequently, to characterize the transition due to the crossflow instability over the practical range of conditions, both the stationary and the travelling modes are important. Usually the travelling modes are observed when the free-stream disturbance level is relatively high and the surface roughness is small (Nitschke-Kowsky and Bippes 1988; Dagenhart et al. 1990; Takagi and Itoh 1994; Deyhle and Bippes 1996). Most of these experiments were performed under natural conditions, which makes them very useful for the observation and documentation of typical properties of the travelling waves. However, under natural conditions it is only possible to obtain information about the development of the most intensive rather than most unstable disturbances, which is probably the main reason for discrepancies between observed experimental and calculated growth rates (Bippes 1999).

The experiments of Gaponenko et al. (1995), in which wave trains of the travelling modes were excited, confirmed the applicability of linear theory to travelling disturbances with relatively small amplification rates. The wave trains were decomposed to the normal modes by spatial Fourier transformation and stability characteristics of selected modes were investigated. The results provided a complete set of the stability characteristics of the travelling crossflow modes for a certain range of frequencies and spanwise wave numbers. Direct quantitative comparison of the experimental and theoretical data shows very good agreement for the stability characteristics for all modes considered (Fig. 2.17).

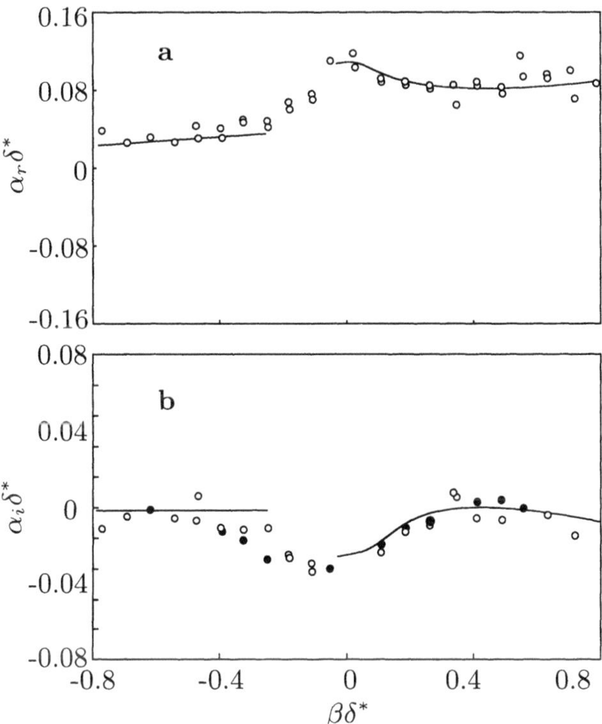

Fig. 2.17. Streamwise wave numbers (**a**) and amplification rates (**b**) of normal crossflow instability modes as a function of spanwise wave number in a local coordinate system, $F = 48.8 \times 10^{-6}$: *symbols*, experimental results; *lines*, corresponding theoretical curves (V.I. Borodulin, V.R. Gaponenko, A.V. Ivanov, Yu.S. Kachanov, J.D. Crouch, unpublished work, 2001)

When studying the crossflow instability its convective character is usually assumed, but the stability analyses of Lingwood (1995), Taylor and Peake (1998) and Ryzhov and Terent'ev (1998) showed that there is a range of parameters for which disturbances can grow both in time and space, so the absolute instability may supersede the convective one.

Besides growing disturbances produced by the development of crossflow instability modes, a transient amplification is also possible. Breuer and Kuraishi (1994) discovered the possibility of quite a pronounced mechanism for the algebraic growth in subcritical region leading to a more intensive formation of the crossflow vortices in the instability region. The relation between 'optimal' transient disturbances and stationary crossflow vortices is discussed in Boiko (2000), Corbett and Bottaro (2001). The behaviour of localized travelling disturbances in the swept-wing boundary layer will be considered in Chap. 5. Further details about the stability of three-dimensional boundary layers can be found in reviews of Reed and Saric (1989) and Bippes (1999).

2.3.2 Flow instability at a concave wall

In a boundary layer developing at a concave wall, the instability caused by an imbalance between pressure and centrifugal forces occurs, which promotes a formation of stationary Görtler vortices directed along the mean flow. These play an essential role during the transition at, for example, turbine blades and compressors, walls of supersonic nozzles and curved channels in heat exchangers. They can also be used as a prototype for the description of near-wall structures in turbulent boundary layers. It is necessary to note that the Görtler-like instability can occur not only in the boundary layers at a concave surface. It can arise also in the neighbourhood of a flow stagnation line, where the flow abruptly changes its direction of propagation and the streamlines are curved. Such conditions are observed at a flow around a wedge or near the point of a boundary layer separation.

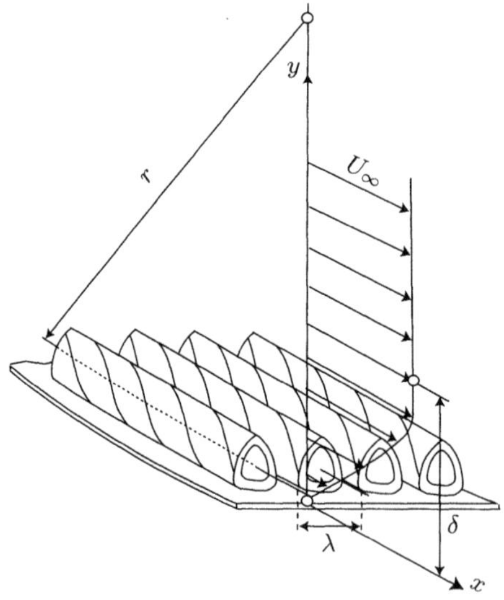

Fig. 2.18. Görtler vortices

This instability was studied for the first time by Görtler (1940), who assumed that the flow is parallel. He established that the flow is unstable in relation to counter-rotating vortex pairs directed along the mean flow (Fig. 2.18), when parameter

$$\mathrm{Gö}_{\vartheta} = \left(\mathrm{Re}_{\vartheta}\frac{\vartheta}{r}\right)^{1/2}$$

reaches a particular critical value (the momentum thickness $\vartheta = \int_0^{\infty} U(1 - U)/U_0^2 \mathrm{d}y$, U_0 is the local free-stream velocity, and r is the radius of curvature of the concave wall).

The existence of streamwise vortices at a concave wall was experimentally confirmed by Gregory and Walker (1956), Aihara (1962), Tani (1962) and Wortmann (1964). It was also revealed that the disturbances grow downstream and their wavelength depends primarily on the radius of the wall curvature r. An influence of external conditions on the development of the system of Görtler vortices was studied experimentally by Tani (1962) and Swearingen and Blackwelder (1983). It appears that the characteristic wave numbers of the excited vortices depend primarily on the nature of the field of free-stream disturbances.

Hämmerlin (1955) improved the Görtler results within the framework of the parallel approach. His calculations show that the instability occurs at first at zero wave number, but that the vortices protrude in this case far outside the boundary layer, which contradicts experimental observations. Only in recent decades has there been agreement between the results of various theoretical and experimental studies concerning the neutral stability curve for the Görtler vortices (Hall 1990). It appeared that the characteristic scale of the vortex growth is comparable with the scale of boundary layer development, and the parallelity approximation is evidently invalid except for the case of long wave number (short scale) vortices (Hall 1983; Floryan and Saric 1982; Bottaro and Luchini 1999). For the latter case Hall (1982) obtained the asymptotic form of the neutral stability curve. For vortices of a larger scale, the boundary layer approximation, PSE or multiple scale analysis are suitable to account for the effects of the flow non-parallelity (Hall 1988; Luchini and Bottaro 1998; Bottaro and Luchini 1999). In this case, the stability characteristics eventually depend on the initial conditions.

The Görtler instability does not play an essential role in three-dimensional flows. The inflectional crossflow velocity component destroys its delicate viscous mechanism (Floryan and Saric 1982). With the growth of crossflow velocity, the Görtler vortices transform to the crossflow ones (Collier and Malik 1987; Kobayashi et al. 1987). Nevertheless, the destabilizing influence of the centrifugal force takes place even in the presence of a strong crossflow, but its role is limited by a modification of the flow structures that arise from other phenomena.

The majority of investigators suppose that the Görtler instability is convective . Park and Huerre (1995) proved this for the case of a boundary layer of a constant thickness. Chomaz (1992) and Chomaz and Perrier (1991) confirmed this experimentally by a consideration of the wave packet development. However, Ruban (1991) and Savenkov (1991) showed theoretically that in the case of large Re and Gö numbers, the flow can becomes absolutely unstable due to vortices of large wavelengths.

Other details concerning the Görtler instability can be found in the book of Gaponov and Maslov (1980) and in reviews of Herbert (1976) and Floryan (1991).

3 Receptivity of laminar near-wall flows

The receptivity problem arises when one tries to provide a reason for the origin of disturbances in open near-wall flows. Turbulence in convectively unstable flows is usually a result of a development of certain disturbances which appear and start to grow far upstream of the onset of the final transition. Such flows are called 'noise amplifiers', in contrast to 'noise generators' governed by non-linear bifurcations (Huerre and Monkewitz 1985). The process of the transformation of the external disturbances into the shear flow ones is called receptivity and is the main concern in this chapter.

3.1 Formulation of receptivity problem

In the linear stability theory used to describe the development of waves in shear flows, the disturbance characteristics are found by the solution of an eigenvalue problem, which excludes the process of their excitation from consideration. As a consequence, the primary question in the course of the receptivity investigation is a search, selection, and experimental as well as theoretical modelling of those mechanisms that are most responsible for the formation of the near-wall disturbance field followed by the transition to turbulence in a particular flow (Morkovin 1968). A general sketch of the modern view for receptivity in different types of external disturbance fields and different physical mechanisms of the instability excitation is given in Fig. 3.1.

3.2 Localized and distributed generation

Two limit cases of excitation of the boundary layer structures can be selected. They are continuous or distributed generation along the majority of the boundary layer, and localized generation by a concentrated effect to the boundary layer. In practice, e.g. for Tollmien–Schlichting wave distributed generation, a source of the wave exists over an extended area of at least several wavelengths. The presence of such a source can lead to a change in the law of the wave development inside the region of generation. The localized generation occurs in regions of less than a wavelength, while downstream, the disturbances propagate as in the absence of the generator.

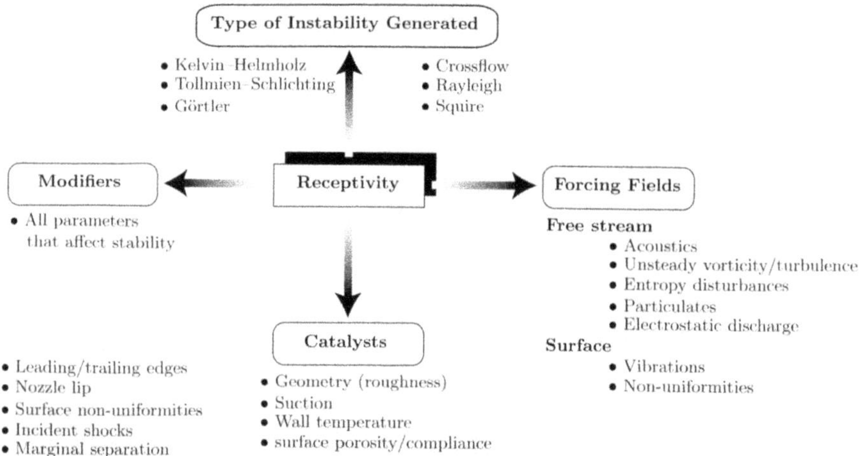

Fig. 3.1. Various aspects of the receptivity problem; based on the figure by Choudhari and Streett (1990)

For an effective excitation of the instability wave of a given frequency, forcing is usually needed which has not only the same frequency, but also comparable spatial scales. If the disturbances are introduced from inside the near-wall shear flow (e.g. by a vibrating ribbon, periodic heating of the surface or sucking and blowing), these scales are usually matched (Gaster and Gupta 1993; Hill 1995; Tumin 1996; Boiko et al. 1997a). The free-stream disturbances (acoustical and vortical) are controlled by inviscid dynamics and, consequently, the wavelengths of the disturbances there are, as a rule, quite different in subsonic flows. In particular, if we suppose that the Tollmien–Schlichting wave propagation velocity is approximately 0.35 of U_∞, the ratio between its wavelength λ_{TS} and the sound wavelength λ_{s} is $\lambda_{\mathrm{TS}}/\lambda_{\mathrm{s}} = 0.35\mathrm{Ma}$, where Ma is the Mach number.

Hence, for the effective transformation of some classes of external disturbances to boundary layer structures, there should be a mechanism of wavelength reduction. As the two-dimensional quasi-parallel stationary boundary layer has no short scales necessary for the effective process of the wavelength reduction, the transformation can be observed when the mean flow experiences 'fast' changes in the region of generation. This mechanism of wavelength reduction is called the 'scattering' of external disturbances at a boundary layer non-uniformity.

A special role during disturbance excitation in the boundary layer is played in most cases by stationary distortions of the flow caused by the presence of the leading edge (Kachanov et al. 1982; Sidorenko and Erofeev 1985; Zhigulev and Tumin 1987), pressure gradients (Nishioka and Morkovin 1986), sharp changes in the boundary conditions such as roughness (Aizin and Polyakov 1979), porosity (Heinrich et al. 1988), suction (Bodonyi and

Duck 1992), separation (Boiko et al. 1990) and flow non-parallelity (Crouch 1992b). Local *stationary* distortions of the surface geometry or of the pressure field result in the flow non-uniformity inside the boundary layer possessing a broad stationary wave spectrum. *Unsteady* external disturbances temporally modulate this stationary disturbance field resulting in a broad spectrum of travelling forced waves which resonantly excite the modes matched with the instability waves of interest (Goldstein and Hultgren 1989; Crouch 1992a). The asymptotic analysis of Goldstein (1985) and Nishioka and Morkovin (1986) shows that the local distortions of the boundary conditions of different types should lead to comparable results with efficiency dependent on the degree of the flow non-uniformity. Generally, more localized boundary layer variations cause an increase of the energy scattered to the short-wave part of the spectrum related to instability waves, which leads to enlarged amplitudes of the excited waves.

An opposite example of the flow, in which the distributed generation of the instability waves is effective, is served by a boundary layer with extended region of small-scale distortions, e.g. waviness (Zavolskiy et al. 1983; King and Breuer 2001), with the spacing of the wall waviness tuned to the wavelength of a particular instability mode. The receptivity to randomly and deterministically distributed surface imperfections were studied also by Choudhari (1993) and Wu (2001). The analysis of the distributed receptivity was extended to the case of three-dimensional disturbances based on the parabolized stability equations by Crouch and Bertolotti (1992), and to some three-dimensional flows by Choudhari and Streett (1990) and Crouch (1993b, 1994b).

3.3 Outline of theoretical approaches to receptivity

A major problems in studying the receptivity phenomenon is searching for the most important mechanisms of generation of the shear flow disturbances. The mechanisms can be characterized by a degree of response of the boundary layer to the disturbances imposed on it, i.e. by a receptivity function R. As the final aim, a theoretical or an experimental analysis must enable the determination of this function at least in the region or at points of interest in a parameter space of the boundary layer instability. For a localized modal excitation (e.g. for the Tollmien–Schlichting or crossflow instability), this problem is reduced to finding this receptivity function at fixed Re as a complex ratio of amplitudes of an external *harmonic* disturbance and an 'initial' amplitude of the instability wave, with the phase between them characterizing a 'delay' in the transformation.

Theoretical approaches to the analysis of receptivity at small-amplitude forcing on the basis of asymptotic methods and perturbation theory were used by Aizin and Polyakov (1979), Zavolskiy et al. (1983), Goldstein (1983), Ruban (1985), Goldstein et al. (1987), Kerschen (1991) and Crouch (1992a, b,

1993a, b, 1994a, b). Goldstein and Hultgren (1989) provided a detail description of the available theoretical approaches and their limitations. An extensive review of different theoretical approaches and experiments is given by Sarik et al. (1994). Kozlov and Ryzhov (1990) reviewed experiments on the receptivity and theoretical studies for a broad spectrum of disturbances: vibrations, periodic sucking and blowing, surface heating, sound and vortices.

One of the approaches suitable for receptivity analysis has already been outlined in Sect. 2.1.3 from the viewpoint of the system impulse response. The fully asymptotic receptivity approach using triple-deck theory and its variations was reviewed by Goldstein (1985), Nishioka and Morkovin (1986), Kozlov and Ryzhov (1990) and Rothmayer and Smith (1998), and is not considered here in more detail. The essence of some other theoretical approaches is considered in the next sections with emphasis on their physical relevance.

3.3.1 Perturbation analysis

Perturbation analysis is used successfully in a variety of physical applications (Landau and Lifshitz 1982). It was applied to the receptivity problem in a number of studies (e.g. Crouch 1993a; Crouch and Spalart 1995). The idea of the method can be illustrated by a simple mechanical model of a forced harmonic oscillator with a slowly varying frequency, as given by Schmid and Henningson (2000):

$$\frac{\mathrm{d}^2 u}{\mathrm{d}x^2} + \omega^2(x)u = \cos \Omega x$$

In the case of a constant frequency, the solution to this equation reads:

$$u(x) = A\cos(\omega x + \phi) + \frac{\cos \Omega x}{\omega^2 - \Omega^2},$$

where the first term represents the homogeneous solution with amplitude A and phase ϕ and the second term is the particular solution due to the forcing. In the case of slowly varying $\omega(x)$, A and ϕ also vary slowly with x. As in the receptivity approach, let us assume that the initial amplitude of the homogeneous solution $A(0) = 0$ and that there is certain mismatch between the frequencies $\omega(0) - \Omega \neq 0$. Then the functions $A(x)$ and $\phi(x)$ can be found iteratively starting at each step with a complete solution at $x \geqslant 0$ to compute an approximate solution at $x + \Delta x$, taking into account only the fast periodic terms (the essence of the perturbation technique).

As can be seen in Fig. 3.2, the behaviour of such a linear system is dominated initially by the particular solution and the system oscillates with frequency Ω. If at a certain x_0 the frequency $\omega(x_0) = \Omega$, a resonance occurs. At $x > x_0$, the amplitude of the particular solution decays to zero as $1/(\omega(x) - \Omega)$, but the amplitude A saturates slowly and the system starts to oscillate with the 'natural' frequency ω.

For the receptivity problem, the linearized Navier–Stokes equations play the role of spatio-temporal hydrodynamic resonator system with external

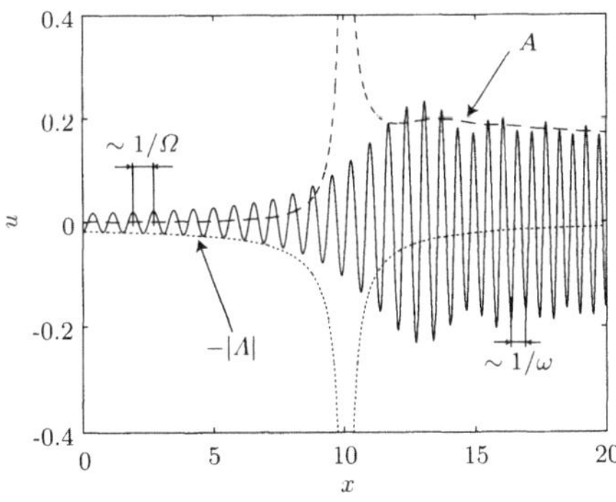

Fig. 3.2. Excitation of eigenmode of a forced quasi-harmonic oscillator: *solid line*, full solution; *dotted line*, negated amplitude of particular solution; *dashed line*, amplitude of eigensolution. System parameters: $w = \pi + x/2$; $\Omega = 2.6\pi$ (calculated by A.V. Boiko)

forces as the driving right-hand side term and the (usually locally) non-homogeneous boundary conditions.

3.3.2 Adjoint functions and the 'efficiency' of generation

One of the most straightforward and physically justified approach describing the generation of linear instability waves (as Tollmien–Schlichting and cross-flow ones) is based on a consideration of adjoint equations of a corresponding linear problem (see Chap. 1). The adjoint approach was also extended to consider the receptivity of spatially varying flow with Görtler vortices to wall roughness in terms of parabolized evolution equations (Luchini and Bottaro 1998).

Following to Marchuk (1977), we consider below the adjoint approach from a general physical background that is applicable not only to fluid dynamics. Let us consider a non-homogeneous linearized operator L describing a physical process of interest:

$$L\varphi(x) = q(x), \tag{3.1}$$

where $q(x)$ is a distribution of forcing sources in a physical medium, x is a set of all problem parameters (e.g., coordinates, time, energy, velocities, etc.) and, for simplicity (without lack of generality of the approach), suppose that φ and x are real. To proceed, we need to define a scalar product

$$(g, h) = \int g(x)h(x)\mathrm{d}x,$$

where the integration is over the whole region of definition of functions f and h. Any physical value p linearly dependent on $\varphi(x)$ and characterizing our physical process can be described through such a scalar product:

$$J_p[\varphi] = (\varphi, p).$$

Let us introduce operator $L^\dagger$ adjoint to L and, formally, the inhomogeneous adjoint equation by definition:

$$(g, Lh) = (h, L^{\dagger} g), \tag{3.2}$$

$$L^{\dagger} \varphi_p^* = p(x), \tag{3.3}$$

where g and h are arbitrary functions, φ_p^* is the adjoint function and p is so far an arbitrary function. Substituting φ, and φ_p^* into (3.2), instead of h and g, respectively, yields

$$(\varphi_p^*, L\varphi) = (\varphi, L^{\dagger} \varphi_p^*).$$

Using (3.1) and (3.3), this is reduced to $(\varphi_p^*, q) = (\varphi, p)$ or $J_q[\varphi_p^*] = J_p[\varphi]$. It means that if we need to find $J_p[\varphi]$ we can solve (3.1) or use

$$J_p[\varphi] = J_q[\varphi_p^*] = (\varphi_p^*, q).$$

Consequently, for any linear functional $J_p[\varphi]$, we can define function $\varphi_p^*(x)$ which satisfies (3.3) with right-hand side function $p(x)$ that characterizes the process.

Let the medium have a 'unit-power source' placed at point x_0 of the parameter space, i.e.

$$q(x) = \delta(x - x_0).$$

Since

$$(\varphi(x), \delta(x - x_0)) = \varphi(x_0),$$

then

$$J_p[\varphi] = J_{q=\delta(x-x_0)}[\varphi_p^*] = \varphi_p^*(x_0).$$

Consequently, the adjoint function $\varphi_p^*(x)$ describes the dependence of the functional $J_p[\varphi]$ on the location of the unit source in the parameter space. That is, the adjoint function is an efficiency function related to the functional $J_p[\varphi]$. For the receptivity application, the adjoint to the linearized Navier–Stokes operator can be used to filter a general disturbance field to identify the amplitude of the corresponding eigensolution.

Such a receptivity analysis was for the first time undertaken by Hill (1995). It provided a simple and direct means for solving a wide class of the receptivity problems suitable for modal analysis based on the linearized Navier–Stokes equations. In this context the adjoint eigensolutions to the equations define the efficiency with which a particular unit-power forcing excites the eigensolutions. For example, for momentum sources such as a ribbon vibrating parallel and normal to the wall, the corresponding adjoint velocities are essential, whereas for vorticity sources it is the adjoint stream function, for mass sources the adjoint pressure, etc.

An example illustrating the efficiency of Tollmien–Schlichting wave excitation is shown in Fig. 3.3 for the Blasius boundary layer. It can be seen that in this particular case the unit source of the wall-normal velocity $\hat{v}$ excites the wave most efficiently close to the critical layer at $U/U_0 \approx 0.28$. Out of this region, the efficiency of excitation decays rapidly.

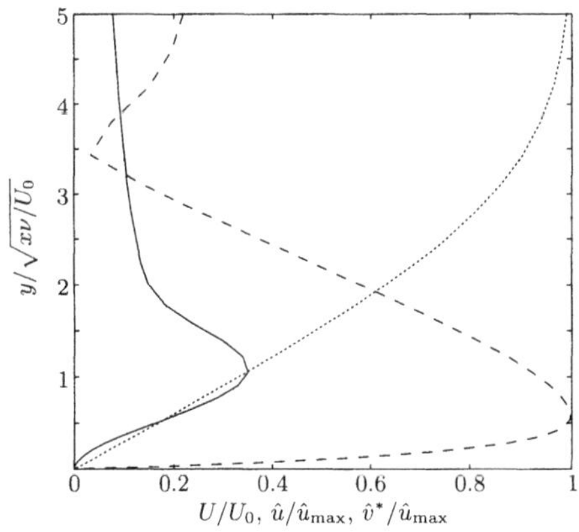

Fig. 3.3. Efficiency of Tollmien–Schlichting wave generation by normal velocity oscillations in the Blasius boundary layer at $\mathrm{Re}_{\delta*} = 1274$ and $\omega = 0.0439$: *dashed line*, absolute value of streamwise velocity of Tollmien–Schlichting wave $|\hat{u}|$; *solid line*, normal component of the adjoint $|\hat{v}^*|$; *dotted line*, mean velocity profile (calculated by A.V. Boiko)

3.3.3 Numerical simulation

Earlier calculations of instability wave generation by external disturbances were based on direct numerical integration of the two-dimensional Navier–Stokes equations with vortical or sound forcing terms (Rogler and Reshotko 1975; Lindors and Laine 1976; Murdock 1980; Tam 1981; Maksimov 1979). This technique extended for three-dimensional cases is now also the only, apart from laboratory experiments, to describe the receptivity in complex flows or at a high intensity of forcing (Gatski and Grosch 1987). This technique is heavily related to applied mathematical and numerical methods, which are not considered here. An interested reader is referred to Gatski and Grosch (1987), Crouch and Bertolotti (1992), Sarik et al. (1994) and Boiko et al. (1997a), and references therein.

3.4 Experimental approaches to the receptivity investigation

To achieve the goals of a receptivity experiment, i.e. to find, select and model a particular receptivity mechanism, it is reasonable to consider such situations which allow one to 'purify' an effect under consideration and reveal the underlying physical mechanisms which manage the process. This can be achieved through a choice of simplified flow geometries and eliminating possible 'side effects', e.g. by noise reduction, and subjecting the flow to the controlled influence of different factors, e.g. excitation of controllable disturbances.

The basic requirements to investigate a particular receptivity mechanism of interest are application of an appropriate forcing, quantitative characterization of the flow and forcing fields along the outer border of the boundary layer and the leading edge, and the possession of local information on the fluctuations in the boundary layer in the region under consideration (Nishioka and Morkovin 1986). This is why the idea of experimental simulation or modelling, i.e. performing an experiment under control conditions, appears to be a powerful tool for the investigation of receptivity. In contrast to the data obtained in a 'natural' environment, these results can be directly compared with calculations without any significant assumptions about the physical nature of the disturbances under investigation (Aizin and Polyakov 1979; Kozlov and Ryzhov 1990; Boiko 2000; Kachanov 2000).

For modal excitation, the receptivity functions are considered as the main results of such experiments. To obtain the receptivity functions in the case of a localized excitation, one needs to measure experimentally the initial forced field of disturbances and to measure or estimate the field of the excited eigenmodes at the position of the source, with both tasks being non-trivial. A classical example of such an experiment is provided by Aizin and Polyakov (1979); see Sect. 3.5.2.

Additional complexity in the case of the acoustic excitation comes from the presence of acoustic waves in the whole boundary layer far beyond the receptivity region. This necessitates the development of a technique to separate the fields measured by a conventional device such as a hot wire. Let us consider the possible approaches to their separation.

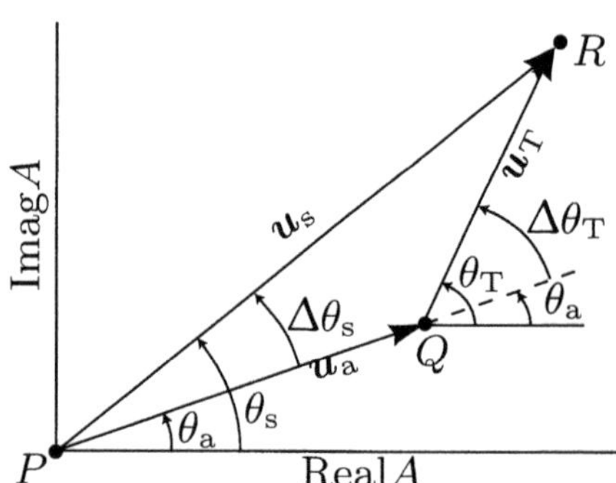

Fig. 3.4. Vector diagram of interference between the acoustic and instability waves

We take temporally periodic Tollmien–Schlichting $\boldsymbol{u}_{\mathrm{w}}$ and acoustic $\boldsymbol{u}_{\mathrm{a}}$ wave oscillations:

$$\boldsymbol{u}_{\mathrm{a}} = B_{\mathrm{a}}(y)\mathrm{e}^{\mathrm{i}(\theta_{\mathrm{a}}-\omega_t)}, \qquad \boldsymbol{u}_{\mathrm{w}} = B_{\mathrm{w}}(x,y)\mathrm{e}^{\mathrm{i}(\theta_{\mathrm{w}}-\omega_t)},$$

where

$$\theta_{\mathrm{w}} = \alpha_r x' + \theta_0, \qquad \theta_{\mathrm{a}} = \alpha_{\mathrm{a}} x', \qquad B_{\mathrm{w}} = A_{\mathrm{w}}\mathrm{e}^{-\alpha_i x'},$$

where $\alpha_{\mathrm{w}} = \alpha_r + \mathrm{i}\alpha_i$ is the Tollmien–Schlichting streamwise wave number and $x' = x - x_0$, x_0 is an approximate coordinate of the excitation and θ_0 is an initial phase at x_0.

The measured total signal is

$$\boldsymbol{u}_{\mathrm{t}} = \boldsymbol{u}_{\mathrm{a}} + \boldsymbol{u}_{\mathrm{w}} = \mathrm{e}^{\mathrm{i}(\theta_{\mathrm{a}} - \omega t)} \left[B_{\mathrm{a}}(y) + B_{\mathrm{w}}(x, y)\mathrm{e}^{\mathrm{i}(\theta_{\mathrm{w}} - \theta \mathrm{a})} \right].$$

In measuring at a fixed y, the dependence on y is eliminated and we have

$$\boldsymbol{u}_{\mathrm{t}} = \mathrm{e}^{\mathrm{i}(\theta_{\mathrm{a}} - \omega t)} \left[B_{\mathrm{a}} + B_{\mathrm{w}}(x)\mathrm{e}^{\mathrm{i}\Delta\theta_{\mathrm{w}}} \right],$$

where $\Delta\theta_{\mathrm{w}}(x) = \theta_{\mathrm{w}}(x) - \theta \mathrm{a}(x)$. Denoting $\Delta\theta_{\mathrm{t}}(x) = \theta_{\mathrm{t}}(x) - \theta \mathrm{a}(x)$ yields

$$\boldsymbol{u}_{\mathrm{t}} = B_{\mathrm{t}}\mathrm{e}^{\mathrm{i}(\Delta\theta_{\mathrm{t}} + \theta_{\mathrm{a}} - \omega t)},$$

so that

$$B_{\mathrm{t}}\mathrm{e}^{\mathrm{i}\Delta\theta_{\mathrm{t}}} = B_{\mathrm{a}} + B_{\mathrm{w}}\mathrm{e}^{\mathrm{i}\Delta\theta_{\mathrm{w}}}.$$

Equating real and imaginary parts we obtain formulas which connect B_{t}, $\Delta\theta_{\mathrm{t}}$ with B_{w}, $\Delta\theta_{\mathrm{w}}$ and B_{a}, $\Delta\theta_{\mathrm{a}}$:

$$B_{\mathrm{t}} \cos \Delta\theta_{\mathrm{t}} = B_{\mathrm{w}} \cos \Delta\theta_{\mathrm{w}} + B_{\mathrm{a}}, \tag{3.4}$$

$$B_{\mathrm{t}} \sin \Delta\theta_{\mathrm{t}} = B_{\mathrm{w}} \sin \Delta\theta_{\mathrm{w}}, \tag{3.5}$$

corresponding to the vector diagram in Fig. 3.4, provided $\overline{\mathrm{PQ}} = B_{\mathrm{a}}$, $\overline{\mathrm{QR}} = B_{\mathrm{w}}$, $\overline{\mathrm{PR}} = B_{\mathrm{t}}$.

Taking in account the fact that $\alpha_i = \alpha_i(x)$, $\alpha_r = \alpha_r(x)$, and supposing that α_{a} is a constant, we obtain finally:

$$B_{\mathrm{w}}(x) = A_{\mathrm{w}} \exp \int_{x_0}^{x} \alpha_i \mathrm{d}x, \tag{3.6}$$

$$\Delta\theta_{\mathrm{w}}(x) = \int_{x_0}^{x} \alpha_r \mathrm{d}x + \theta_0 - \alpha_{\mathrm{a}}(x - x_0). \tag{3.7}$$

The functions to find are usually the amplitude and the phase of the instability wave, i.e. $B_{\mathrm{w}}(x)$ and $\theta_{\mathrm{w}}(x)$. Then, differentiation of (3.6) and (3.7) with respect to x gives approximate values of α_r and α_i.

The amplitude and the phase of the total fluctuations $B_{\mathrm{t}}(x)$ and $\theta_{\mathrm{t}}(x)$ are found directly during the measurements. To determine $B_{\mathrm{w}}(x)$ and $\theta_{\mathrm{w}}(x)$, it is necessary to know additionally the amplitude and the phase of the sound $B_{\mathrm{a}}(x)$ and $\theta_{\mathrm{a}}(x)$, or more precisely its wave number α_{a}.

Let us assume that B_{a} as well as α_{a} are constants. This is valid at the boundary layer in the case of high Reynolds numbers and frequency parameters. Then B_{a} and α_{a} can be determined in the free stream. The acoustic wavelength can also be calculated, since its frequency and phase velocity are known.

In the case of small variations of $B_{\mathrm{a}}(x)$ and $B_{\mathrm{w}}(x)$, another simplified method can be used which allows finding $B_{\mathrm{a}}(x)$, $B_{\mathrm{w}}(x)$ and $\Delta\theta_{\mathrm{w}}(x)$ based on known $B_{\mathrm{t}}(x)$ and $\theta_{\mathrm{t}}(x)$.

It follows from (3.4) and the vector diagram that when the instability wave has either the same or half-a-period shifted phase with respect to the sound, the amplitude and the phase of the total signal have the values indicated in

Table 3.1. Wave parameters in minima and maxima of beatings

B_{w}	$\Delta\theta_{\mathrm{w}}$	$\Delta\theta_{\mathrm{t}}$	B_{t}	θ_{t}
$B_{\mathrm{w}} < B_{\mathrm{a}}$	$2\pi n$	0	$B_{\mathrm{a}} + B_{\mathrm{w}}$	$\alpha_{\mathrm{a}}(x_1 - x_0)$
	$2\pi(n+1)$	0	$B_{\mathrm{a}} - B_{\mathrm{w}}$	$\alpha_{\mathrm{a}}(x_2 - x_0)$
$B_{\mathrm{w}} > B_{\mathrm{a}}$	$2\pi n$	$2\pi n$	$B_{\mathrm{w}} + B_{\mathrm{a}}$	$\alpha_{\mathrm{a}}(x_1 - x_0) + 2\pi n$
	$2\pi(n+1)$	$2\pi(n+1)$	$B_{\mathrm{w}} - B_{\mathrm{a}}$	$\alpha_{\mathrm{a}}(x_2 - x_0) + 2\pi(n+1)$

Table 3.1. Values x_1 and x_2 are coordinates of the peak maxima and minima, respectively. At $B_{\mathrm{w}} < B_{\mathrm{a}}$ we have

$$B_{\mathrm{w}}(\xi_1) = [B_{\mathrm{t}}(x_1) - B_{\mathrm{t}}(x_2)]/2 - [B_{\mathrm{a}}(x_1) - B_{\mathrm{a}}(x_2)]/2,$$
$$B_{\mathrm{a}}(\xi_2) = [B_{\mathrm{t}}(x_1) + B_{\mathrm{t}}(x_2)]/2 - [B_{\mathrm{w}}(x_1) + B_{\mathrm{w}}(x_2)]/2,$$

where $B_{\mathrm{w}}(\xi_1) = [B_{\mathrm{w}}(x_1) - B_{\mathrm{w}}(x_2)]/2$, $B_{\mathrm{w}}(\xi_2) = [B_{\mathrm{w}}(x_1) + B_{\mathrm{w}}(x_2)]/2$ and $x_1 \leqslant \xi_1 \leqslant x_2$, $x_1 \leqslant \xi_2 \leqslant x_2$. If B_{w} and B_{a} depend slowly on x, then we may neglect terms $B_{\mathrm{a}}(x_1) - B_{\mathrm{a}}(x_2)$ and $B_{\mathrm{w}}(x_1) - B_{\mathrm{w}}(x_2)$. This yields

$$B_{\mathrm{w}}(\xi_1) \approx [B_{\mathrm{t}}(x_1) - B_{\mathrm{t}}(x_2)]/2, \tag{3.8}$$
$$B_{\mathrm{a}}(\xi_2) \approx [B_{\mathrm{t}}(x_1) + B_{\mathrm{t}}(x_2)]/2, \tag{3.9}$$

where $\xi_1 \approx \xi_2 \approx (x_1 + x_2)/2$. In the case of a constant acoustic amplitude B_{a}, (3.8) is an exact equality.

To conclude, function B_{t} has the following properties (see also Fig. 3.5). It experiences quasi-periodic oscillations from maximum values $B_{\mathrm{w}}(x_1)+B_{\mathrm{a}}(x_1)$ to minimum values $B_{\mathrm{w}}(x_2) - B_{\mathrm{a}}(x_2)$, with the amplitude and mean value equal to $B_{\mathrm{w}}(x_2)$ and $B_{\mathrm{a}}(x_2)$ at $B_{\mathrm{w}} < B_{\mathrm{a}}$, respectively, and vice versa at $B_{\mathrm{w}} > B_{\mathrm{a}}$. The oscillation period is $\Delta x = 2\pi/\Delta\alpha = 2\pi/(\alpha_r - \alpha_{\mathrm{a}})$.

When α_{a} can be found by another method, the value of Δx unequivocally determines the instability wavelength λ_{w} and consequently its phase. In particular, at small subsonic velocities (Ma $\ll 1$), $\alpha_{\mathrm{a}}/\alpha_r \ll 1$ and

$$\Delta x = 2\pi/\alpha_r = \lambda_{\mathrm{w}}.$$

In the case of combined vibro-acoustic forcing, another technique for the signal separation has been proposed (Ivanov et al. 1994). The excitation was investigated there at combination frequencies $f_{\mathrm{a}} \pm f_{\mathrm{v}}$, where f_{a} is the acoustic and $f_{\mathrm{v}} \ll f_{\mathrm{a}}$ is the vibrational frequencies. The quasi-stationarity of vibrations with respect to f_{a} and virtually linear dependence of the receptivity function on f_{v} allowed the authors to extrapolate their results for the case $f_{\mathrm{v}} = 0$, i.e. to consider the sound scattering at a stationary wall nonuniformity.

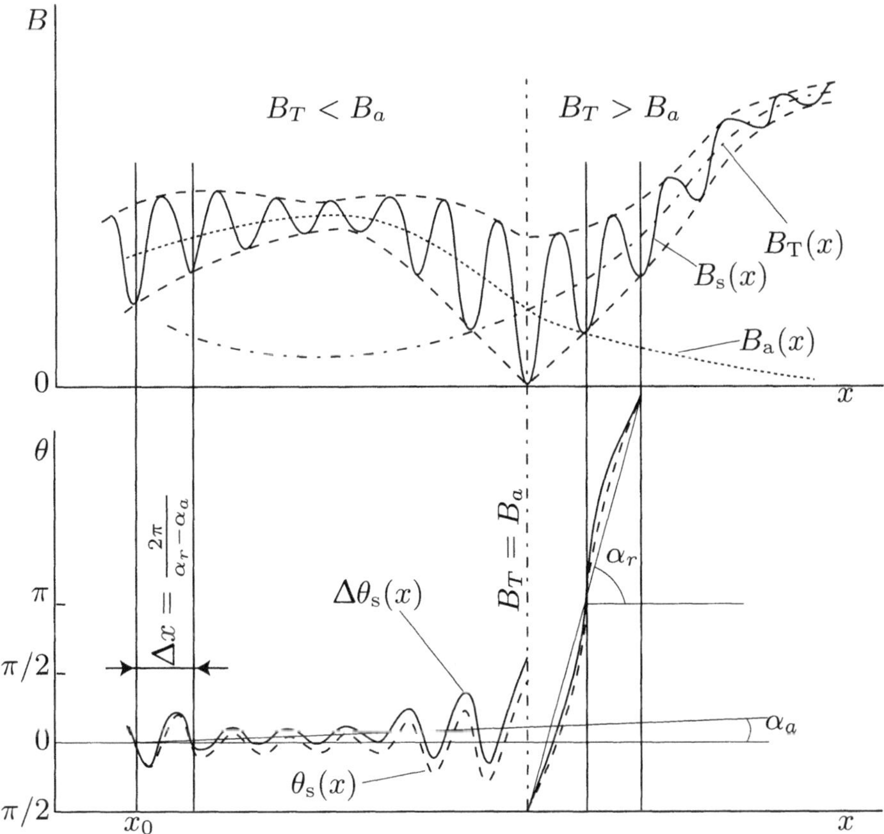

Fig. 3.5. Typical amplitude and phase of the total signal, and the approximate separation of the acoustic and instability waves

3.5 Receptivity of two-dimensional boundary layers

From a practical viewpoint several kinds of boundary layer structures are important. They are Tollmien–Schlichting waves, crossflow vortices and travelling waves as well as quasi-stationary vortices and streaks in some three-dimensional flows. The Tollmien–Schlichting wave receptivity is the most significant for two-dimensional boundary layers, the crossflow instability is dominant in three-dimensional flows, while the streaks can be effectively excited by non-modal receptivity. Below we outline some basic receptivity mechanisms that are responsible for the generation of the first three that have been investigated to date. The receptivity for streaks in the context of the bypass transition is considered later in Chap. 5. The receptivity of local separation bubbles is treated in Chap. 6. Discussion of the receptivity mechanisms for free-shear flows, secondary instability and non-aerodynamic disturbances can be found elsewhere (Bushnell 1989; Ustinov 1995; Choudhari 1998).

The external acoustics, surface vibrations and vortical disturbances in the form of localized flow modulations or free-stream turbulence are those which most frequently contribute to the boundary layer receptivity as reviewed by Nishioka and Morkovin (1986), Kozlov and Ryzhov (1990), Saric (1990) and Bippes (1999). Because of the viscous nature of the linear instability of the flow about a flat plate, the localized mechanism of scale reduction of the wavelengths of inviscid free-stream disturbances is dominant (Goldstein 1985; Ruban 1985). In particular, the effective transformation or scattering of the free-stream disturbances to the Tollmien–Schlichting waves occurs preferably alongside sudden changes of the mean flow (e.g. over the leading edge, separation region or suction slits). Below we consider some of these mechanisms.

3.5.1 Leading-edge receptivity

The role of the leading edge in the mechanism of transformation of external disturbances to instability waves was first indicated by Kachanov et al. (1975, 1979a, b), Polyakov (1979) and others. In the study of Kachanov et al. (1975), Tollmien–Schlichting waves generated by sound were identified. Polyakov (1975) studied 'natural' transition in the flat plate boundary layer and also observed the acoustic excitation of the Tollmien–Schlichting waves. Kachanov et al. (1979c) and Murdock (1980) carried out an analysis of the excitation of the instability waves using two-dimensional numerical simulation.

One of the first experimental confirmations of this mechanism for acoustic oscillations were the observations of Shapiro (1977) and Leehey and Shapiro (1980). It was found that a plane acoustic wave propagating in the streamwise direction can excite a Tollmien–Schlichting wave of the same frequency. Thomas and Lekoudis (1978) compared the results of Shapiro (1977) with a model in which the flow field consisted of a simple superposition of the Tollmien–Schlichting waves and the sound oscillations. The comparison showed the absence of interaction between the disturbances, and it was supposed that it occurs earlier, near the leading edge.

The excitation of flat plate boundary layer oscillations near the leading edge by two-dimensional free-stream vortical disturbances was studied by Kachanov et al. (1979b). The results of the measurement of the mean flow field are presented in Fig. 3.6. The regions of flow deceleration and acceleration about the leading edge as well as the formation of the boundary layer are clearly visible. The data on the fields of disturbance amplitude and phase are shown in Fig. 3.7. For convenience of comparison, the phase contours are presented mirrored with respect to x. Amplitudes are normalized to the disturbance maximum. The distance between the leading edge and the disturbance minimum in the boundary layer (point A in Fig. 3.7) characterizes the size of the region of generation. Downstream of this point, a travelling wave

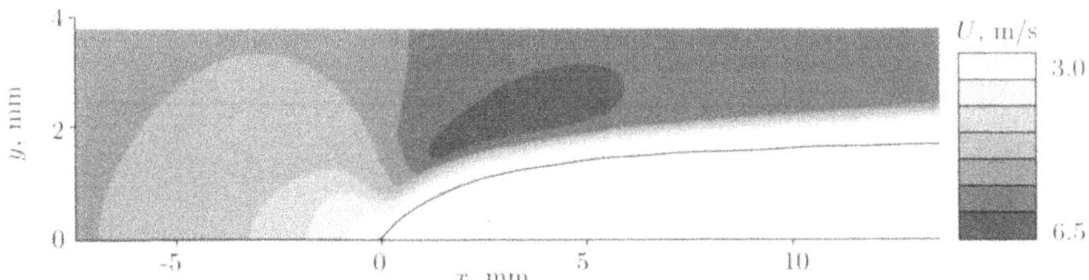

Fig. 3.6. Isolines of mean flow velocity in the leading-edge region of a plate (Kachanov et al. 1979b)

with growing amplitude propagates. Starting from point B, the amplitude of the wave decays; to this point the boundary layer has already been quite well formed. The influence of the region of generation decreases with distance from the leading edge, and the velocity of wave propagation stabilizes. The value of the jump of the disturbance amplitude, i.e. the ratio of the streamwise velocity amplitudes in the leading-edge region and in the free stream, characterizes the receptivity of the Tollmien–Schlichting wave generation in this case. It was found that the ratio does not depend on the amplitude of the instability wave (at least, up to about 0.25% of U_∞); i.e. the receptivity process is linear. The efficiency of the receptivity, however, changes with the frequency of the excitation and the mean flow velocity.

The experiments of Kachanov et al. (1979b) showed the important role of the wall-normal disturbance velocity v' for Tollmien–Schlichting wave excitation by free-stream vortices. A vibrating ribbon was placed in the free stream along the wall-normal coordinate y. The vortex street behind it near the leading edge contained quite large u' and w' velocity components, but a v' component was virtually absent. Generation of the Tollmien–Schlichting wave or a wave packet was not observed in this case.

The primary role of v' in process of the free-stream disturbance transformation at the leading edge and a similarity between the mechanisms of the Tollmien–Schlichting wave generation by external disturbances of various types were shown, also, in the numerical experiments of Yang and Voke (1991), where the velocity components were separately artificially suppressed in the free stream, which revealed that external turbulence can efficiently excite the Tollmien–Schlichting waves only by v'-velocity disturbances, while u' and w' velocity disturbances do not influence the process. Following Kachanov et al. (1979b), the transformation of the external disturbances can be qualitatively presented as a result of a convergence of the perturbed motion streamlines in front of the leading edge, the increase of the wall-normal component and the sharp change of direction of the streamlines. As a result, a 'transformation' of the wall-normal to the streamwise disturbance velocity occurs, resulting in a formation of the localized vortex, see in Fig. 3.7. Hence, the role of the streamwise velocity component of the external disturbance is rather insignificant.

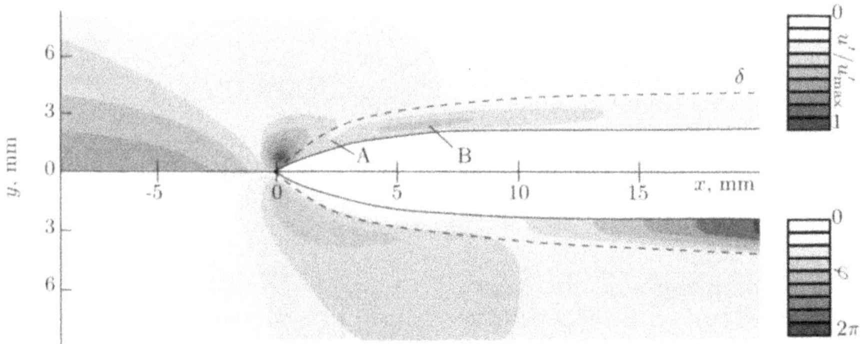

Fig. 3.7. Amplitude and phase isolines of streamwise disturbance velocity about the leading edge (Kachanov et al. 1979b)

In the experiments of Kachanov et al. (1975), where the flat plate boundary layer was forced by acoustic oscillations, periodic wall-normal vibrations of the whole plate were found to generate Tollmien–Schlichting waves of the same frequency in the leading-edge region, even at a low sound intensity. This observation motivated Kachanov et al. (1975) to investigate the isolated vibrational receptivity mechanism. For this purpose, a vibrator was fastened to the plate and the corresponding receptivity mechanism was studied.

In the experiments of Kachanov et al. (1979b), the possibility of distributed excitation of the flat plate instability waves by external two-dimensional vortical disturbances was studied. The experiment considered the interaction of the boundary layer with a vortex street behind a vibrating ribbon located in the free stream in front and a little above the plate, which meant that the core of the vortex wake passed the leading edges at a distance from it (Fig. 3.8). It appeared that the disturbance strongly decayed inside the boundary layer and the Tollmien–Schlichting wave was not excited. The distributed generation of the Tollmien–Schlichting waves due to boundary layer non-parallelity was not found. The results of this experiment qualitatively agree with the data of the theoretical study of Rogler and Reshotko (1975), in which rapid energy attenuation of free-stream fluctuations inside the boundary layer occurred, and the generation of instability waves by free-stream turbulence was not revealed.

3.5.2 Receptivity caused by surface geometry variations

Flow non-uniformity in a developed boundary layer is an important agent of receptivity, not only because of its general occurrence, but also due to

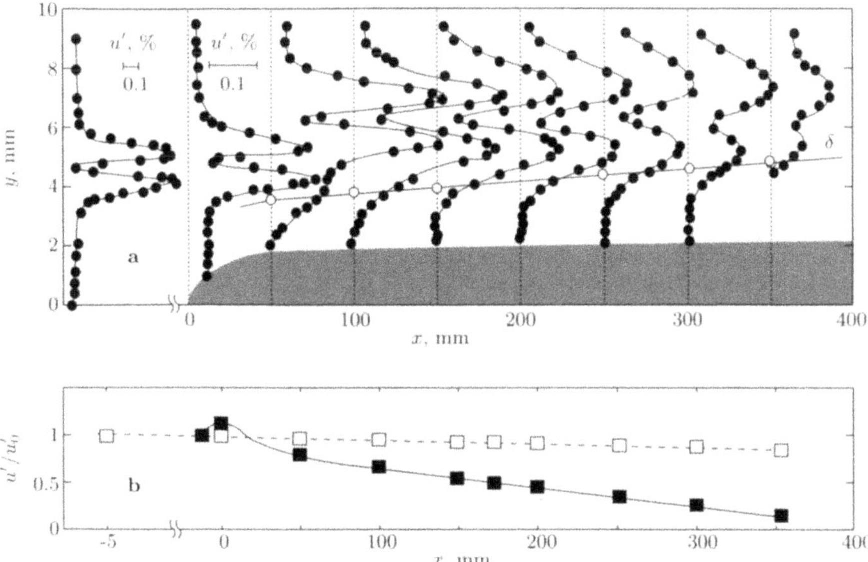

Fig. 3.8. Development of a vortical wave (**a**) and change of its intensity (**b**) in the free stream (■) and at the contact with the boundary layer (□); estimated boundary layer thickness δ marked by ∘ (Kachanov et al. 1982)

its possible proximity to the instability region. In contrast to the Tollmien–Schlichting waves generated near the leading edge, which are subject to a strong decay before they start to grow, the waves generated by roughness close to the critical Reynolds number can exhibit immediate downstream growth. The possibility of Tollmien–Schlichting wave generation at small local surface roughness by streamwise acoustic oscillations was shown in the experiments of Aizin and Polyakov (1979). With micron-size non-uniformities and a boundary layer thickness of several millimeters, the coefficient of the transformation of the acoustic disturbances into the Tollmien–Schlichting waves was about 1%. An asymptotic analysis of the excitation mechanism was later performed by Goldstein (1985) and Ruban (1985). The Tollmien–Schlichting wave generation occurs as a result of the imposition of short scale vorticity modulation – caused by the surface non-uniformity – on the boundary layer Stokes wave induced by the sound or vibrations at a given frequency. The Stokes wave small modulations provide a wide spectrum of modes including that of the Tollmien–Schlichting waves.

In Fig. 3.9, the data of Aizin and Polyakov (1979) illustrating localized boundary layer receptivity at a two-dimensional obstacle on a flat plate are shown. Characteristic beatings of the amplification curve of the streamwise disturbance velocity obtained by traversing a hot-wire probe along the boundary layer are caused by an interference of the acoustic and the excited Tollmien–Schlichting waves. Based on the discussion in Sect. 3.4, it is seen that the Tollmien–Schlichting wave is generated at a short distance (about one wavelength) behind the obstacle, i.e. the mechanism of the excitation is localized.

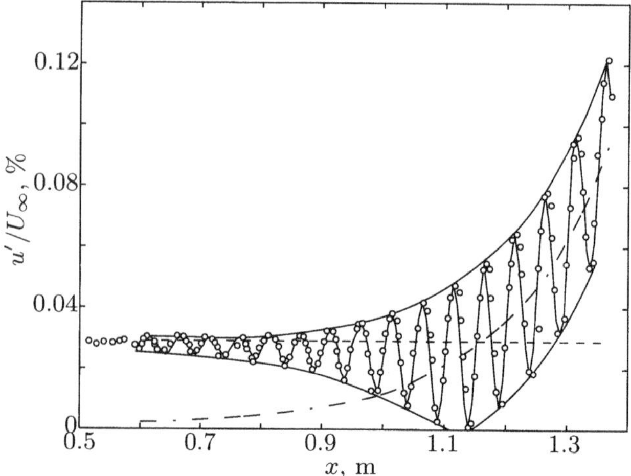

Fig. 3.9. Interference of acoustic and Tollmien–Schlichting waves on a flat plate behind a plain plastic strip: *dash-dotted line*, Tollmien–Schlichting wave amplitude; *dotted line*, sound amplitude; o, total hot-wire anemometer signal (Aizin and Polyakov 1979)

Investigating experimentally boundary layer acoustic receptivity at roughness elements, Saric et al. (1991) found that the receptivity coefficient exhibits linear growth with the roughness height, until it exceeds a certain threshold value (related to the thickness of the viscous near-wall area of the Stokes wave localization), when a pronounced deviation from the linear receptivity mechanism occurs. The effect of the roughness shape and size on the flat plate boundary layer receptivity was investigated later by Boiko et al. (1990).

The two-dimensional mechanism of instability wave generation in the Blasius boundary layer by means of *unsteady* surface non-uniformities was studied experimentally by Gilev and Kozlov (1984) and theoretically by Gaster (1965). The results of the experiment were compared with calculations by Terent'ev (1984). The theoretical analysis was advanced by Fedorov (1984), who studied the boundary layer receptivity to arbitrary periodic surface perturbations (vibrations, suction, blowing and heating). A combined experimental and theoretical study for the case of three-dimensional vibrations was performed by Ivanov et al. (1998) and reviewed by Kachanov (2000).

The process of localized receptivity to a 'gust' (a vortical two-dimensional disturbance) was surveyed by Kerschen (1991). Such a disturbance can serve as a simplified two-dimensional model of external turbulence. Far from the leading-edge region, the free-stream disturbance penetrates into the boundary layer only by diffusion. At large Reynolds numbers, the intensity of this penetration is small, and the non-stationary motion occurs predominantly at the outer border of the boundary layer without generation of a Tollmien–Schlichting wave similar to the experimental results of Kachanov et al. (1982). However, in this case local surface irregularities also lead to a local pressure fluctuation, which interacts with the gust causing wave excitation. Though the physical receptivity mechanisms for the gust and sound wave are quite different, asymptotic analyses produce receptivity coefficients of the same order of magnitude.

3.5.3 Excitation of instability waves by spatially localized disturbances

When the boundary layer is subject to an external localized disturbance, its broad wave spectrum can contain scales that are comparable with characteristic instability wavelengths. Thus, in this case a mechanism of wavelength reduction at local non-uniformities of the mean flow is not necessarily required for the generation of eigenmodes in a developed boundary layer.

Tam (1978) studied the possibility of instability wave excitation in a shear layer by external acoustic waves oriented at various angles to the mean flow direction. The results of his calculations showed that the efficiency of the transformation of the acoustic waves depends significantly on the width of the sound beam with respect to the boundary layer thickness: the narrower the beam, the better the conditions for the excitation of instability waves. By the concentrated effect on the boundary layer, it is possible to explain the 'saw-tooth' flow response to an acoustic forcing found by Vlasov et al. (1977) as an effect of non-uniformity of the acoustic field in the wind tunnel test section associated with loudspeaker non-ideality, reflections of the sound waves from the test section walls, etc.

Tollmien–Schlichting wave generation in a developed boundary layer by locally non-uniform external vortical disturbances was studied by Fasel (1977). Integrating the full Navier–Stokes equations about the flat plate for a spatially localized and temporally periodic boundary layer forcing, he observed Tollmien–Schlichting wave generation in the boundary layer.

An experimental study related to the case under consideration was carried out by Dovgal et al. (1980), whose conclusions agree with the data of the calculations conducted by Fasel (1977). Periodic disturbances were excited in the flow by a vibrating ribbon located outside of the developed boundary layer above the flat plate at the distance equal approximately to double boundary layer thickness. The ribbon produced a decaying vortex street behind it. Thus the boundary layer was affected strongly by a non-uniform field of disturbances along the streamwise coordinate in the vicinity of the ribbon.

It was found that the free-stream disturbances efficiently excite the typical Tollmien–Schlichting waves in the boundary layer, with the receptivity being localized in the vicinity of the vibrating ribbon in the free stream. In

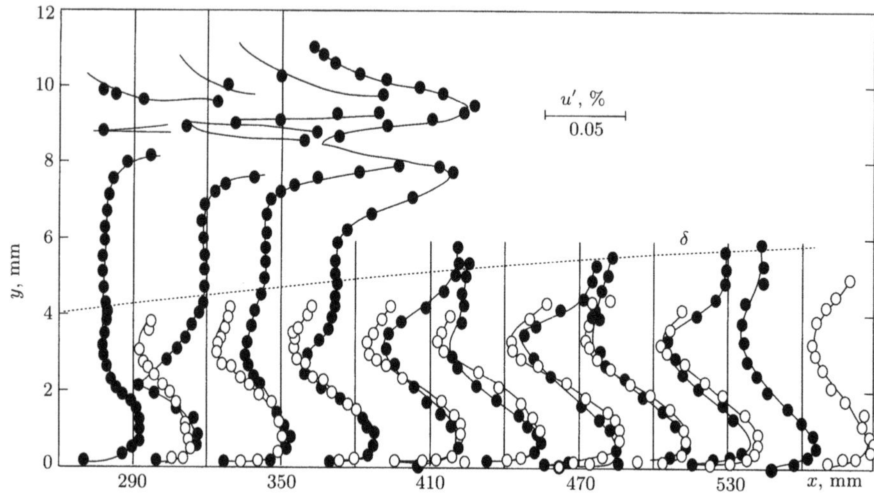

Fig. 3.10. Development of the disturbance amplitude in the boundary layer and in the free stream behind the ribbon for a disturbance entering from outside (•) and inside (∘) of boundary layer (Dovgal et al. 1980)

Fig. 3.10, the amplitude profiles of the streamwise velocity oscillations u' excited from inside and outside the boundary layer are presented. In the first sections, the amplitude distribution of the disturbance excited in the boundary layer differs qualitatively from the Tollmien–Schlichting wave profile. In this area a considerable distortion is introduced by the peripheral part of the vortex street. It decays whilst propagating downstream, most significantly near the wall, so that downstream the Tollmien–Schlichting wave develops. At the same time, at the outer border of the boundary layer a superposition of the external disturbance and the Tollmien–Schlichting wave occurs. The disturbance created by the ribbon outside the boundary layer propagates with the free-stream velocity. The phase velocity and the growth rate of the disturbances excited in the boundary layer coincide with those of the Tollmien–Schlichting wave. A similar receptivity mechanism was later considered experimentally and numerically by Berlin and Henningson (1994) and Elofsson and Alfredsson (2000) for a pair of symmetrical oblique vibrating ribbons in free-stream excited transient structures, and Tollmien–Schlichting waves in the flat plate boundary layer.

The three-dimensional receptivity of the flat plate flows with zero and negative pressure gradients to localized surface vibrations was considered by Ivanov et al. (1998) and S. Bake, A.V. Ivanov, H.H. Fernholz, K. Neemann and Y.S. Kachanov, unpublished work, (2000). For boundary layer excitation, a specially designed surface vibrator was used. The receptivity functions, expressed as the ratio of the initial wave number spectrum of the excited Tollmien–Schlichting waves to the corresponding resonant spectrum of the surface vibrations were determined as outlined in Fig. 3.11. Only those wave

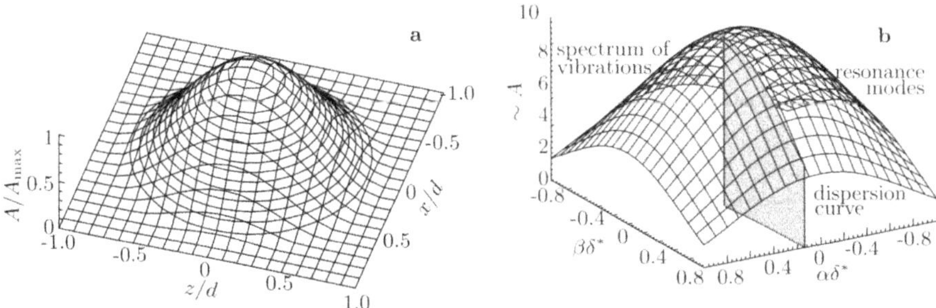

Fig. 3.11. Spatial distribution of the membrane oscillation amplitude normalized to the amplitude in its center (**a**), and amplitude of the wave number spectrum of the receptivity source vibrations together with the Tollmien–Schlichting wave dispersion curve and the selected resonant modes – the projection of dispersion curve onto the spectrum of vibrations (**b**) (S. Bake, A.V. Ivanov, H.H. Fernholz, K. Neemann and Y.S. Kachanov, unpublished work, 2000)

components excited by the vibrator contribute to the receptivity, which has the same dispersion characteristics as the corresponding Tollmien–Schlichting waves.

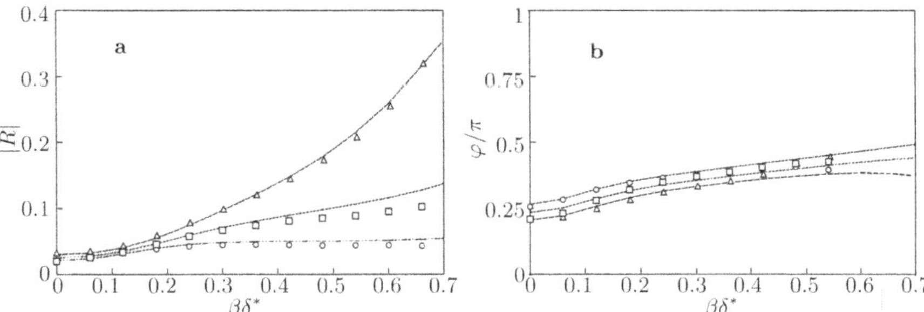

Fig. 3.12. Amplitude (**a**) and phase (**b**) of vibrational receptivity functions for the Blasius boundary layer: $f = 37$ (○); 55 (□); and 75 Hz (△); *lines*, calculations (Ivanov et al. 1998)

The receptivity functions R were found for each fixed frequency as a function of the spanwise wave number (Fig. 3.12). The initial spectra of the excited perturbations were obtained both by linear stability calculations and by means of a second reference source mounted upstream, and they provided the amplification curves in the vicinity of the vibrator. The numerical results showed good qualitative agreement with the experiments. It was concluded that the favourable pressure gradient increases the boundary layer receptivity to the surface vibrations.

3.6 Receptivity of a three-dimensional boundary layer

Three-dimensional boundary layers are the most common type of near-wall flows. However, their receptivity has only been investigated recently (for reviews see e.g. Saric 1990; Crouch 1993b; Bippes 1999; Kachanov 2000). Studies of the spatial boundary layer receptivity are complicated by the possibility of several instability mechanisms which are capable of causing the transition to turbulence in various practical situations. Of most interest are the crossflow, Tollmien–Schlichting, separation bubble and Görtler instabilities. Because in aerodynamic applications the transition on swept wings is caused most of all by crossflow instability, below we consider the receptivity mechanisms that excite crossflow vortices. Since curved walls frequently occur in engineering applications and three-dimensional flows, the receptivity to Görtler vortices is also considered. The swept wing receptivity to the streaks and receptivity of three-dimensional separation bubbles are treated in Chaps. 5 and 6, respectively.

3.6.1 Swept-wing receptivity

The flow about a swept wing is a typical example of a three-dimensional boundary layer which is subject to several types of instabilities, each of which can lead – under certain conditions – to the transition to turbulence. Usually the two-dimensional mechanisms dominate at low sweep angles. However, starting from sweep angle of 30–40° and at rather small Reynolds numbers, a row of streamwise vortices is usually formed and travelling waves start to grow. The disturbances are caused by inviscid instability of inflectional velocity profiles which appear due to the presence of spanwise pressure gradient and crossflow (Sect. 2.3.1).

At Tu $\lesssim 0.1\%$ the stationary crossflow vortices usually dominate the flow structure due to relatively large amplitude of the initial surface roughness 'seeds', as was shown by Müller and Bippes (1988), Deyhle and Bippes (1996) and Kozlov et al. (1999) and observed in other experiments; for references see Bippes (1999). However, with the growth of the free-stream turbulence level, their saturation amplitudes become smaller which lead to a domination of the travelling crossflow waves. This implies that the velocity fluctuations in the free stream contribute significantly to their development. It appears that the travelling modes are generated mostly by the vortical rather than by the acoustic perturbations (Bippes 1999; Kachanov 2000).

Receptivity to acoustics and mechanical vibrations. The receptivity of the swept-wing boundary layer to a local boundary layer non-uniformity in the form of roughness or suction has been considered in a series of recent studies by Crouch and Spalart (1995), Collis and Lele (1999), Ng and Crouch (1999) and Bertolotti (2000) using different theoretical methods. Choudhari and Streett (1992) have also shown theoretically that free-stream noise coupled with roughness elements on the wing surface is an effective generator of stationary disturbances.

Since the instability of crossflow modes is governed by inviscid mechanisms, their receptivity characteristics are quite different from that for the Tollmien–Schlichting waves. Moreover, stationary boundary or flow nonuniformity have the scales necessary for the effective excitation of the stationary crossflow vortices, and an additional mechanism of scale reduction is not needed, which is in contrast to the excitation of the travelling waves. The sensitivity of the three-dimensional boundary layer to excitation by stationary disturbances is shown in the experiments of Müller and Bippes (1988) and Takagi et al. (1991). Generation of the stationary mode in the Falkner–Skan–Cooke boundary layer was studied theoretically by Fedorov (1988) and Crouch (1993b). Herbert and Lin (1993) analyzed the receptivity under local boundary layer suction with the help of parabolized stability equations. A direct numerical simulation of the wave excitation for localized and distributed suction was carried out by Spalart (1993). Analytically, the crossflow receptivity on a rotated disc was considered by Choudhari and Streett (1990) and Balakumar et al. (1990).

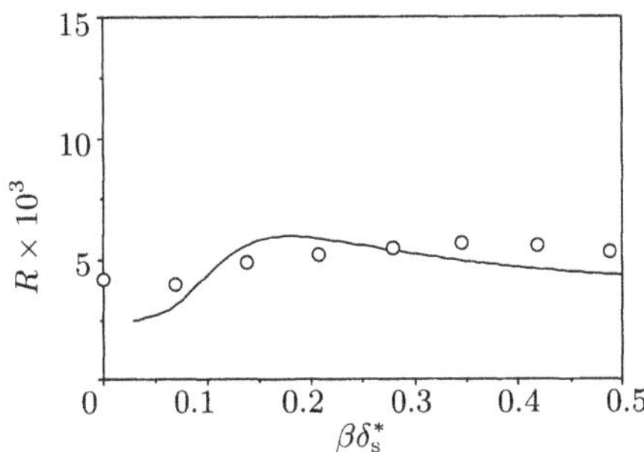

Fig. 3.13. Experimental (○) and theoretical (*solid line*) amplitudes of vibrational receptivity functions for a swept-wing boundary layer (Crouch et al. 1997, 1998)

Gaponenko et al. (1995, 1996), Kachanov (1996) and Kachanov et al. (1997), performed experimental studies of the swept-wing boundary layer receptivity to mechanical vibrations and acoustic excitation. The results of these studies are reviewed in detail by Kachanov (2000). All of the basic receptivity characteristics over a broad range of oscillation frequencies correlated well with theoretical results. An example of such a comparison for the vibrational receptivity is outlined in Fig. 3.13. In the case of combined vibro-acoustic forcing, an extrapolation of the experimental data to a vibration frequency of zero allowed estimation of a factor of receptivity for the stationary crossflow vortices with a stationary roughness, which is difficult to estimate using other experimental techniques.

Receptivity to localized vortical disturbances. The effect of free-stream vorticity on the generation of swept-wing boundary layer structures was carried out experimentally by Boiko (2000). A model free-stream vortex originated at the tip of a micro-wing positioned above the wing model and in front of the leading edge. The forcing led to the formation of pronounced stationary disturbances in the boundary layer. The distribution of the streamwise velocity defects is presented in Fig. 3.14. In contrast to the streaks in two-dimensional flat plate boundary layer, the presence of the crossflow leads to a multiplication of the excited velocity defects downstream. Another characteristic feature of the defect development is an initial decay of the disturbance intensity followed by the disturbance growth with a smaller spacing between the disturbance defects and excesses.

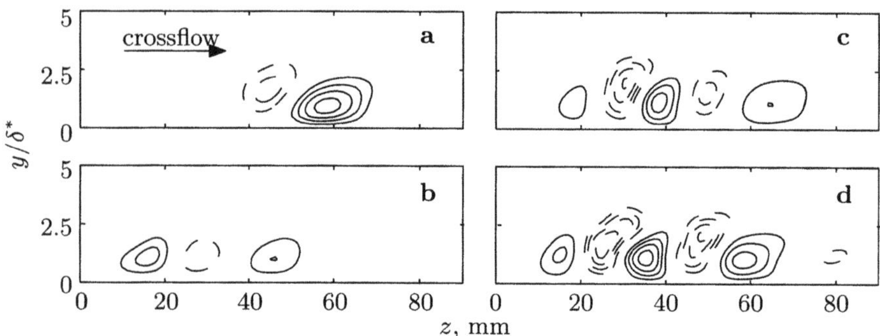

Fig. 3.14. Isolines of streamwise stationary velocity disturbances in a swept-wing boundary layer induced by a free-stream vortex: $x/l = 0.38$ (**a**), 0.48 (**b**), 0.70 (**c**) and 0.83 (**d**); wing chord $l = 500$ mm (Boiko 2000)

The multiplication of stationary vortices is a known feature that is specific for the development of a localized wave packet of stationary crossflow modes generated from a small-scale boundary non-uniformity, as was shown by numerical simulations in a series of studies by Joslin and Streett (1994), Joslin (1995) and Huai et al. (1997). In particular, it was shown by Streett (1998) that a small variation of the wave angle and the growth direction across the range of the crossflow instability (see Sect. 2.3.1) is the main reason for the vortex multiplication, which in those cases is basically a linear phenomenon. As can be seen, the disturbances with shorter spanwise wavelengths appear on one side of the wave packet, whereas those of longer wavelengths propagate on the other side. This suggests that one of the components of the developing packet arose due to the crossflow instability and was excited by a localized receptivity to the free-stream vortical disturbance close to the leading edge.

After Fourier transformation along the spanwise coordinate z, integration across the boundary layer and normalization to the front-most measured amplitudes, the data are presented in Fig. 3.15. Additionally, the corresponding results of the linear stability analysis in a parallel flow approximation of the Falkner–Skan–Cooke profiles are shown by solid lines. It can be seen that in

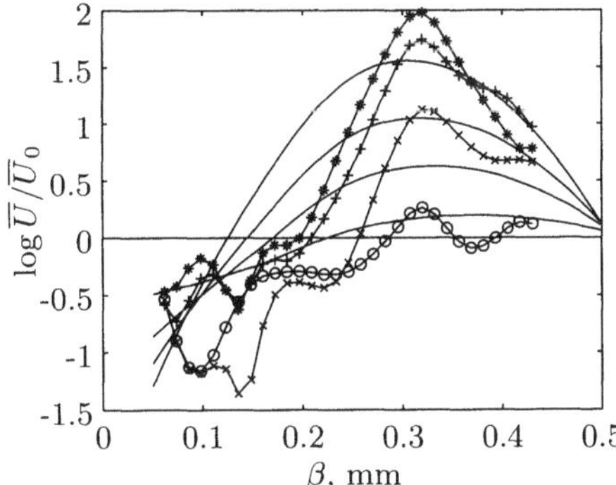

Fig. 3.15. Averaged spectral disturbance growth normalized to 'initial' amplitude: x/l = 0.38 (○), 0.48 (×) 0.70 (+), 0.83 (∗); *solid lines*, corresponding growth calculated by linear stability theory, l, the chord length (Boiko 2000)

this case the theory correctly predicts the wavelength of the most-amplified stationary mode. Meanwhile, the parallel linear stability theory does not seem to be ideal for describing the development of the small-amplitude crossflow vortices for the wing model used in the experiment (Bippes 1999). A strong deviation from the theory is also found in the present experiments, especially for the case of low wave numbers. While the theory predicts neutral disturbances at $\beta \approx 0.1$, the experimentally observed neutral point is close to 0.2.

Moreover, the theory cannot explain the appearance and development of amplitude maxima at $\beta \approx 0.1$–0.2, below the neutral spanwise number. Their origin can be attributed to the distributed receptivity due to the presence of the tip vortex along the whole model chord.

3.6.2 Excitation of Görtler vortices

As described in Chap. 2, the Görtler vortices at a concave wall – except for the short scale vortices – evolve over the same streamwise length scale as that of the undisturbed boundary layer, i.e. the effect of flow non-parallelity is large (see Sect. 2.3.2). Consequently, a local linear stability analysis based on the parallel flow assumption is invalid in this case (Hall 1983). To solve the problem, one has to consider, for example, parabolized stability equations for the flow which contain terms that are dependent on both wall-normal and streamwise coordinates. Being essentially an initial-value problem, the instability analysis in this case must always be accompanied by consideration of the receptivity.

It was found experimentally that the Görtler vortices are excited effectively by both surface roughness and free-stream disturbances (Gregory and Walker 1956; Swearingen and Blackwelder 1987). Hall (1990) analyzed the interaction of a concave plate flow with steady spanwise free-stream variations. He showed that the interaction is effective close to the leading edge and the intensity of the excited vortices is comparable with that of the external disturbances. His results concerning the following development of the Görtler vortices are in a quite good agreement with the available experimental data.

Denier et al. (1991) and Bassom and Hall (1994) showed that a streamwise localized roughness promotes the vortex growth at higher Görtler numbers in respect to the free-stream disturbances, while a distributed roughness (with streamwise extent comparable with the evolution scale of the boundary layer) can excite the vortices at arbitrary small values of Gö. In the case of the short scale vortices, their development can be approximated by local linear stability analysis, while their receptivity can be analyzed by inhomogeneous linearized Navier–Stokes equations. Since these modes are localized close to the wall, they can be more effectively excited by roughness than by free-stream vortical disturbances (Bassom and Seddougui 1995).

4 Late stages of transition

The region of non-linear disturbance activity is most important for understanding the physical processes which are responsible for the final transition to turbulence: in this area the perturbed flow becomes completely turbulent. Below we consider the prototypical mechanisms active both at the exponential growth of a single travelling two-dimensional linear instability wave, when it dominates in the flow, and at the predominance of stationary streamwise vortices generated by, for example, a surface roughness or crossflow instability. Mechanisms to explain the transition at high free-stream turbulence levels, with early formation of streaks are discussed in Chap. 5 and in separation bubbles in Chap. 6.

4.1 Onset of non-linearity

Since two-dimensional Tollmien–Schlichting waves become unstable first in two-dimensional flows, it can be assumed that at a relatively homogeneous level of initial disturbances, they have better chances to trigger the final transition. A typical sequence of stages of the laminar–turbulent transition initiated in a boundary layer by such a two-dimensional instability wave is shown in Fig. 4.1.

In the beginning, when a linear instability wave reaches a large-enough amplitude, it enters the region of its essentially non-linear, but still two-dimensional development. Usually the wave amplitude saturates in this region that is reminiscent of the formation of new quasi-steady state. Possible tendencies of the Tollmien–Schlichting wave development at this stage can be described on the basis of a model of non-linear critical layer. Another mechanism for a flow to reach a non-linear regime is through the mechanism of exchange of stability leading to a new steady state as for a flow modulated by vortical structures. In some cases this mechanism can be specified based on weakly non-linear theories.

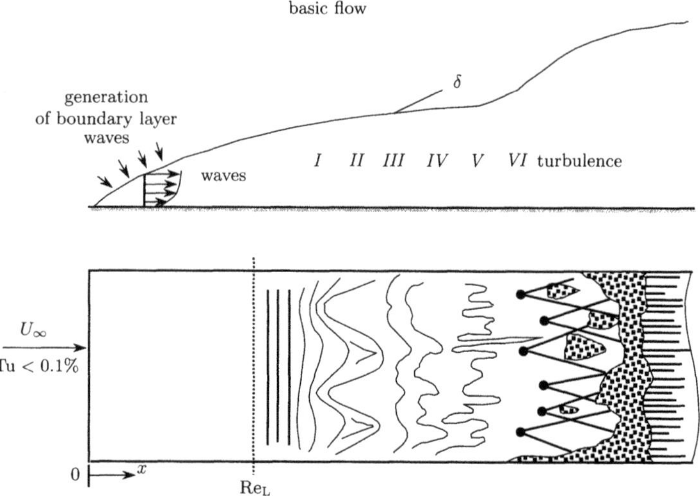

Fig. 4.1. Typical stages of the transition process at low free-stream turbulence (Schlichting and Gersten 2000): I, linear instability stage of the Tollmien–Schlichting waves; II, development of three-dimensional Λ-structures; III, formations of the streamwise vortical structures; IV, appearance of strong shear layers; V, region of the turbulent spots origination; VI interaction and merging of turbulent spots

4.1.1 Amplitude criteria for appearance of non-linearity stimulated by Tollmien–Schlichting wave growth

The non-linear critical layer approach is built on a conjecture that non-linear terms in the Navier–Stokes equations first become large in the critical layer close to the maximum of the disturbance wall-normal velocity component.

By transforming to a coordinate system moving with the wave, it is possible to introduce a stream function Ψ as

$$\Psi = \int (U - c)\mathrm{d}y + \varepsilon\psi$$

and to consider an equation for the disturbance stream function ψ of an arbitrary amplitude ε, derived from the Navier–Stokes equations (Zhigulev and Tumin 1987):

$$\underbrace{\frac{1}{\varepsilon}\frac{\partial}{\partial t}(\varepsilon\nabla^2\psi)}_{\mathrm{D}} + \underbrace{(U-c)\nabla^2\psi_x - \nabla^2 U\psi_x}_{\mathrm{A}} + \underbrace{\varepsilon(\psi_y\nabla^2\psi_x - \psi_x\nabla^2\psi_y)}_{\mathrm{B}} = \underbrace{\frac{1}{\mathrm{Re}_\delta}\nabla^4\psi}_{\mathrm{C}},$$

where the values are normalized to the boundary layer thickness δ and flow velocity at infinity, U_∞ and subscript denote differentiation with respect to corresponding coordinates. The linear terms A are minimal near the wave critical layer, where $U(y_c) \approx c$ (Fig. 4.2). Therefore, it is possible to assume that the non-linear terms B become comparable with the linear ones primarily at the critical layer, where the instability wave stream function has the maximum amplitude. Thus the 'threshold' amplitude $\varepsilon_{\text{ampl}}$, when the non-linear terms become significant near the critical layer, is determined at $A/B \sim 1$. It appears that if the critical layer is well separated from the near-wall viscous region,

$$\varepsilon_{\text{ampl}} = (U_c')^{1/3}(\alpha \text{Re}_\delta)^{-2/3}$$

that usually gives $\varepsilon_{\text{ampl}} \approx 1\%$ of U_∞, as in the experiments by Kachanov (1994), while if they are close to each other,

$$\varepsilon_{\text{ampl}} = (U_c')^{4/3}(\alpha \text{Re}_\delta)^{-2/3} c^{-1}$$

and it can reach 10% of U_∞.

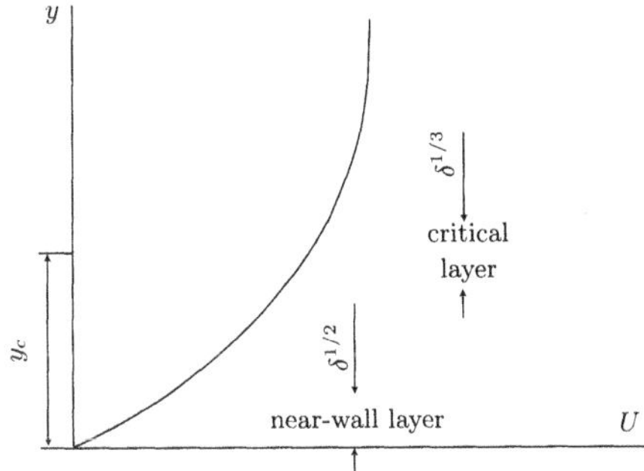

Fig. 4.2. Structure of boundary layer region at the separation of critical and near wall viscous layers (Maslowe 1985)

The further results of the asymptotic theory of the non-linear critical layer can be found, for example, in Benney and Bergeron (1968), Haberman (1972) and Zhigulev and Sidorenko (1985). The review of modern results on non-linear stability theories can be also found in Kachanov et al. (1982), Maslowe (1985), Zhigulev and Tumin (1987), Goldstein and Choi (1989), Goldstein and Lee (1992) and Moston et al. (2000).

4.1.2 Conditional stability and weakly non-linear theories

For fluid mechanics applications the significance of accounting for conditional instability (see Sect. 1.2.1) is seen from Landau's example (Landau and Lifshitz 1959). In particular, for almost neutral disturbances a multiple-scale analysis of the equations of motion is allowed. This incorporates a non-linearity assuming that the wave amplitude function A is weakly time dependent at a very slow time scale τ compared to the linear one (Stuart 1960; Watson 1960):

$$\{u, v, w\} = A(\tau)\{\hat{u}, \hat{v}, \hat{w}\}e^{i(\alpha x + \beta z - \omega t)}.$$

After substitution into the Navier–Stokes equations and certain mathematical simplifications assuming $|\text{Re} - \text{Re}_\text{L}| \ll 1$, the approach leads to the amplitude

Landau equation for temporally developing flows (Landau and Lifshitz 1959; Drazin and Reid 1981; Sen and Venkateswarly 1983; Fujimura 1991) and the Landau–Ginzburg equation for spatially developing flows (Stewartson and Stewart 1971). A detailed derivation of the Landau equation is given by, e.g., Schmid and Henningson (2000). In general form, the equation can be written as

$$\frac{\mathrm{d}A}{\mathrm{d}\tau} = \sigma A + \varkappa|A|^2 A. \tag{4.1}$$

This can be considered as the first terms in the amplitude expansion for the disturbance solution at $\mathrm{Re_L}$. Constants σ and $\varkappa$ are the Landau coefficients, which can be either positive or negative and are specific for a particular flow. Multiplication (4.1) by A^* and of its complex conjugate by A, followed by summation of the two equations yields

$$\frac{\mathrm{d}|A|^2}{\mathrm{d}\tau} = (\sigma + \sigma^*)|A|^2 + (\varkappa + \varkappa^*)|A|^4. \tag{4.2}$$

Denoting $\mu_1 = \sigma + \sigma^*$ and $\mu_2 = \varkappa + \varkappa^*$, the solution reads

$$|A|^2 = \frac{|A(0)|^2}{\mathrm{e}^{-\mu_1\tau} - \mu_2/\mu_1|A(0)|^2(1 - \mathrm{e}^{-\mu_1\tau})}.$$

Let us consider some possible variants depending on the signs of the coefficients:

$\mu_1 > 0, \mu_2 < 0$: A steady state (amplitude saturation) at $A_{\max} = \sqrt{|\mu_1/\mu_2|}$ independent of the initial amplitude $A(0)$ can be reached with time – the exchange of stability can occur.

$\mu_1 < 0, \mu_2 < 0$: The only steady state is the initial undisturbed flow.

In such a way two types of instability phenomena can be distinguished. If the non-zero steady state appears at $\mathrm{Re} > \mathrm{Re_L}$, such a system is called the *system with soft excitation*, since the state can be reached by introducing an infinitesimal disturbance (Fig. 1.4a). Otherwise, at $\mathrm{Re_G} < \mathrm{Re} < \mathrm{Re_L}$ the basic motion is *metastable*: a new steady state is reached with the help of a disturbance of finite amplitude (Fig. 1.4b). Such a system is a *system with hard excitation*.

The non-zero steady solutions which appear at $\mathrm{Re} < \mathrm{Re_L}$ and $\mathrm{Re} > \mathrm{Re_L}$ are called *subcritical* and *supercritical*, respectively. In particular, the calculations show that plane Poiseuille flow close to the critical Reynolds number $\mathrm{Re_L}$ is metastable (Stuart 1980). The Blasius boundary layer at $\mathrm{Re_L}$ represents a system with the soft excitation (Goldshtik and Shtern 1977; Rotenberry 1993). The idea of the hard disturbance excitation and metastable states is also applicable for flows linearly stable at any Re, as in Couette flow but using a slightly different formulation (Maslowe 1985; Nagata 1990).

The stability of the steady solutions can be investigated by linearizing the Landau equation about the steady states, which yields:

$$\frac{\mathrm{d}\Delta A}{\mathrm{d}\tau} = (\mu_1 + 2\mu_2 A_{\mathrm{eq}})\Delta A,$$

where ΔA is an infinitesimal deviation of the amplitude A from the equilibrium state A_{eq}. In particular, for $A_{\mathrm{eq}} = A(0) = 0$ it is seen that the stability is governed by the sign of μ_1, with the flow being stable at $\mu_1 < 0$ and unstable at $\mu_1 > 0$. The finite-amplitude state $A_{\mathrm{eq}} = \sqrt{|\mu_1/\mu_2|}$ is stable for $\mu_2 < 0$ and is unstable for $\mu_2 > 0$.

The validity of the Landau expansion only for finite, but small disturbance amplitudes is the reason to call such an analysis the weakly non-linear theory. It seems reasonable to build a higher-order theory as it was suggested originally by Landau. In such a way, to find a non-trivial solution at $\mu_1 < 0, \mu_2 < 0$, it is necessary to take into account higher order terms in the amplitude expansion (4.2). To outline this procedure, let us add formally $\mu_3|A|^6$ to the right-hand side of (4.2) and assume that $\mu_1 < 0, \mu_2 < 0, \mu_3 < 0$. In this case

$$|A|^2_{\mathrm{max}} = -\frac{\mu_2}{\mu_3} \pm \left[\frac{\mu_2}{\mu_3}\mu_1 - \frac{\mu_2^2}{\mu_3}\right]^{1/2}.$$

However, Herbert (1983a) indicated that such a procedure is of a little practical importance due to the rapidly decreasing extent of validity of the high-order solutions over the Reynolds number and amplitude ranges.

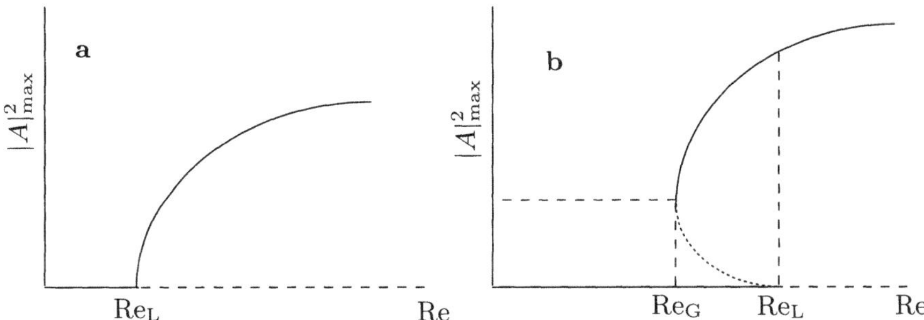

Fig. 4.3. Equilibrium amplitudes for a basic flow: **a** a system with weak excitation and **b** a system with hard excitation

One can try also to numerically find non-linear equilibrium states and study their stability based on a global bifurcation analysis of the system (Zahn et al. 1974; Nagata 1990; Ehrenstein and Koch 1991; Koch et al. 2000) or by the energy method (Joseph 1976). The latter technique is useful primarily to describe the subcritical transition at large initial disturbance amplitudes in flows that are subject to subcritical bifurcations, as in the plane Poiseuille flow.

4.2 Basic transition scenarios in quasi-two-dimensional flows

There are a variety of non-linear processes observed at the transition to turbulence. All of them are mutually dependent, and can occur more or less individually only with special adjustment of the initial conditions. This indicates the necessity to select some specific mechanisms responsible for the formation of pre-turbulent structures.

Two typical scenarios of the transition for the Blasius boundary layer and plane Poiseuille flow – with predominance of two-dimensional instability – are usually distinguished at low free-stream turbulence levels: the subharmonic regime (also called the N-regime after its discoverers, the *Novosibirsk* group) and the K-regime (after *Klebanoff*) of the transition[1].

In both cases, so-called Λ-structures are observed in the late stage of transition. They are a pair of strong shear layers of finite length directed at an angle to the flow, both in the streamwise and wall-normal directions (Kachanov et al. 1982; Herbert 1988; Sandham and Kleiser 1992; Kachanov 1994). Thus their downstream portions coalescence, forming a 'head', with their general form reminiscent of the Greek letter Λ (Fig. 4.4). However, if in the K-regime they follow each other, then in the N-regime they are staggered. The last testifies to an essential role of the subharmonic frequency in the formation of such a flow pattern. Mixed or combination types can also take place.

Studies of the transition in straight and swept-wing boundary layers have also shown that the transition in both cases is connected with the formation of the Λ-vortices (Fig. 4.5). In the straight-wing boundary layer, a symmetric Λ-vortex pattern develops that is similar to that in the flat plate boundary layer. Owing to the effect of crossflow, it becomes asymmetric at the swept wing. The fact that the structures were observed in different external and internal flows of the boundary layer and channel types, as well as in the presence of body curvature and in separated regions (see Chap. 6), indicates the universality of the mechanisms underlying their formation.

The appearance of two distinct (staggered and ordered) patterns of three-dimensional structures observed in experiments is explained by a predominance of different physical mechanisms at different intensities of the initial disturbances. The K-regime of transition is usually observed when the initial wave amplitude is strong enough, whereas the N-regime dominates at smaller disturbance amplitudes. In the latter case, the non-linear processes of gener-

[1] There exists a certain confusion related to a variety of names associated with the study of the Tollmien–Schlichting wave breakdown. In particular, Kachanov (1994) indicated that scenarios of the transition and their theoretical models are frequently mixed; e.g. the C-type (Craik 1971) and of H-type routes (Herbert 1983b, 1984, 1988) describe various resonance models, whereas both are concerned with the experimentally apparent subharmonic N-regime of the transition.

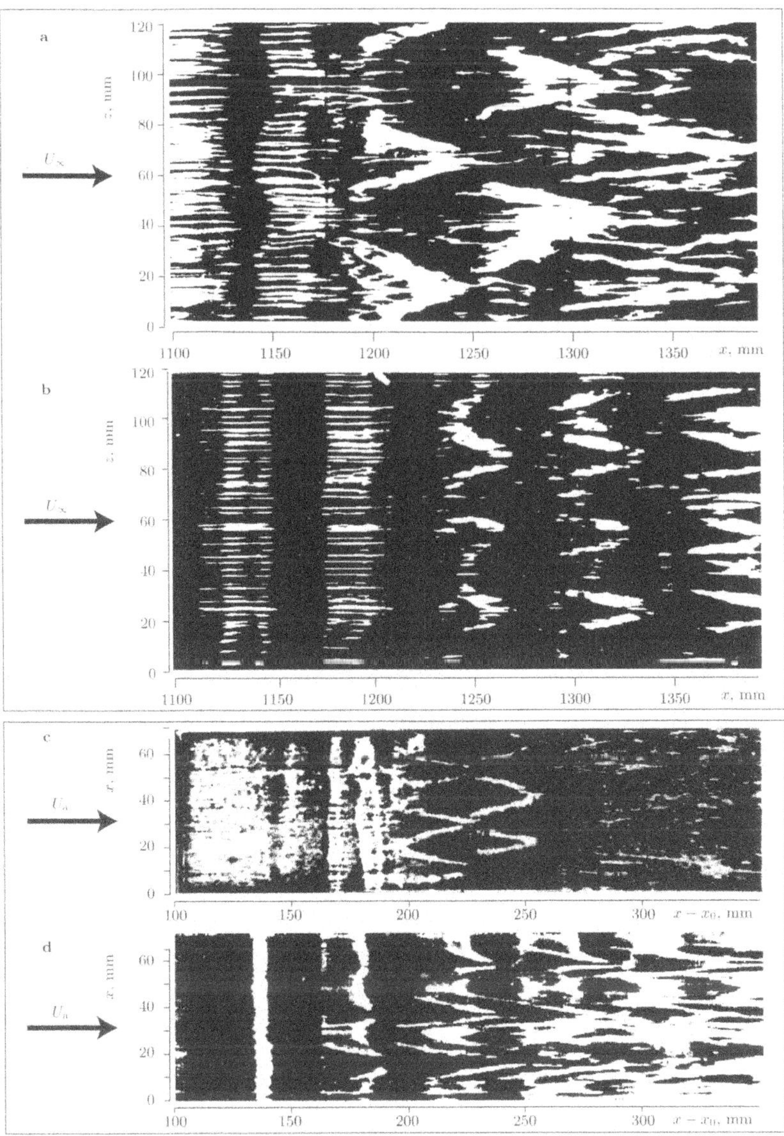

Fig. 4.4. Smoke visualization of Λ-structures in the plane parallel to the surface: **a** the N-regime in a flat plate boundary layer; **b** the K-regime in a flat plate boundary layer (Saric et al. 1984); **c** the N-regime in in plane Poiseuille flow; **d** the K-regime in plane Poiseuille flow (Ramazanov 1985)

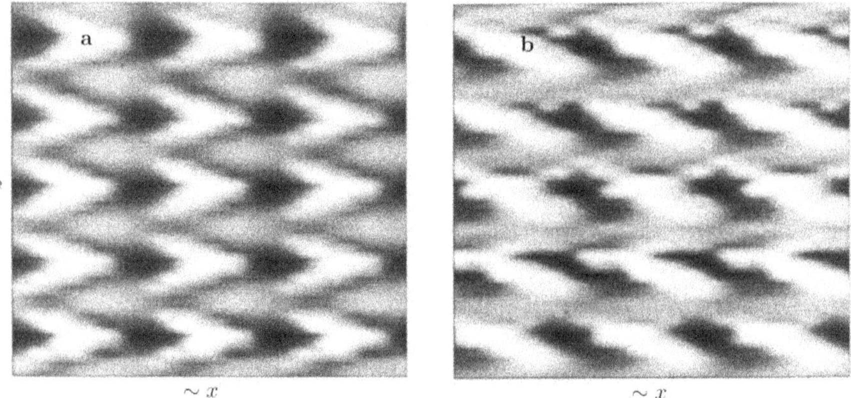

Fig. 4.5. Patterns of the ensemble-averaged streamwise velocity disturbance component in (t–z)-planes demonstrating the scenarios of transition in straight (**a**) and swept (**b**) wing flows (Chernoray et al. 2000a)

ation of multiple harmonics of the primary wave do not have time to operate, while the subharmonic comes from the intensive parametric resonance with the primary wave. The subharmonic activity is also observed at the stage of non-linear development of three-dimensional wave packets (Gaster and Grant 1975; Gaster 1990; Cohen et al. 1991). Hence, the subharmonic resonance is more probable at a low degree of background disturbances in the flow; e.g., in the conditions of glider flight.

4.2.1 K-regime of transition

The K-regime of the transition was found for the first time in the Blasius boundary layer in an early experimental study of non-linear processes by Klebanoff et al. (1962). By introducing controlled two-dimensional Tollmien–Schlichting waves, they observed the appearance of spanwise modulations of their wavefronts so that the mean and disturbance velocities developed an almost-periodic structure in the spanwise direction. Its formation was accompanied by appearance and development of strong low-velocity 'spikes' (intensive outbursts of the streamwise velocity at each period of the primary wave) on oscilloscope traces of disturbances when measured using a hot-wire. Multiple spikes (double and triple) were observed as well (Fig. 4.6). The subsequent transition seemed localized in the maxima (peaks) of the spanwise mean velocity distribution and characterized by the formation of strong shear layers. Similar results were obtained independently in the experiments of Kovasznay et al. (1962).

A typical distribution of the streamwise velocity component along spanwise coordinate z, with the characteristic peaks and valleys, is shown in Fig. 4.7. This change in the spanwise mean velocity distribution indicates

Fig. 4.6. Generation and multiplication of spikes in oscilloscope traces of velocity oscillations in the region of the outer boundary layer border. *1*, initial section; *2*, downstream section; *3*, vibrating ribbon forcing signal (Klebanoff et al. 1962)

a development of a system of streamwise vortical structures. A substantial growth of disturbance amplitude modulation in the streamwise coordinate is seen.

It appeared that the spikes were strictly deterministic and periodic formations (Kachanov 1994). They manifest themselves in frequency spectra as phase-synchronized oscillations of higher harmonics of the primary wave. The analysis of the oscillations shows that their spectrum up to the stage of developed spikes contains only higher harmonics nf, $n = 2, 3, \ldots$ (at least up to $n \approx 20$–50).

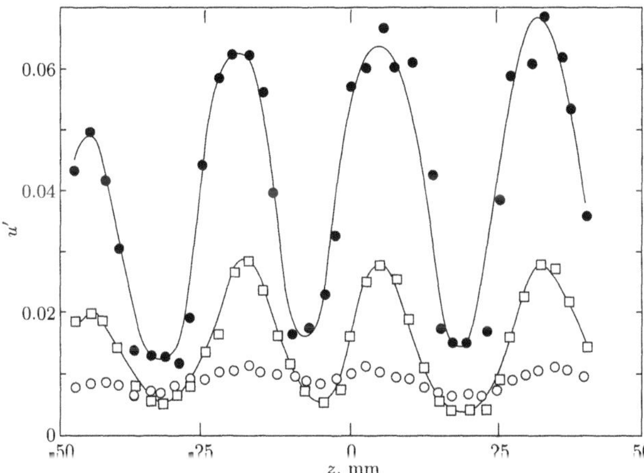

Fig. 4.7. Spanwise distribution of streamwise velocity oscillations at various distances downstream of the vibrating ribbon. $f = 145$ Hz, $y = 1$ mm. $x = 72$ (○), 142 (□) and 178 mm (•), $U_0/\nu = 0.95$ m^{-1} (Klebanoff et al. 1962)

It is possible to describe satisfactorily the initial quasi-two-dimensional stage of spike formation on the basis of a two-dimensional soliton theory of integro-differential Benjamin–Ono equation (Kachanov et al. 1993). This approach is interesting as an attempt to explain the whole process of the formation of the strong shear layer before its transformation into the Λ-structure. In fact, it is a rare example when an analytical solution of a two-dimensional non-linear theory *quantitatively* describes the behaviour of disturbances in a transitional flow. However, the theory is limited only to the initial stage of spike development, when three-dimensional character of disturbances still plays no significant role and the two-dimensional soliton approach is valid.

To describe the spike behaviour at the three-dimensional stage of their development, a resonance-wave concept of the formation and development of the spikes in the boundary layer was formulated by Kachanov (1987, 1994). Within the framework of this concept, the formation of spikes and other characteristic features of the K-regime of transition are explained as interactions of two- and three-dimensional spectral components. The concept is built on the basis of the well-established experimental fact that the amplification of three-dimensional oscillations of primary frequency f_1 in the K-regime can be caused by a four-wave resonance of the waves $(f_1, 0)$, $(2f_1, 0)$ and $(f_1, \pm 2\beta)$ (in frequency–wave-number notation). In the concept this idea is extended to the case of multiple waves (resonance cascades) such as $(nf_1, 0)$ and $(mf_1, \pm\beta)$, where $n = 2, 3, \ldots$, $m \approx n/2$.

During their formation, the spikes move quickly from the wall and propagate downstream along the outer boundary layer border with velocity close to that of the free stream, while the vortices related to them preserve a toroidal shape (Kachanov et al. 1978). The Λ-vortex development is well illustrated in the DNS 'visualization' by Rist et al. (1998) in Fig. 4.8. One can see the formation and development of several transitional structures. The right-most of them has already developed into a ring vortex and tears off from two 'legs' extending down towards the wall, while the others are at different stages of formation of pre-toroidal Ω-vortices at the tip of the Λ-structure.

Thus, comparison of the results of the hot-wire measurements, visual observations and calculations allows us to imagine the K-regime of transition up to the formation of the multiple spikes as follows. The three-dimensional distortion of a plane Tollmien–Schlichting wave leads to the formation of the system of Λ-vortices. In the process of propagating downstream, the vortex head is displaced in the direction of the outer boundary layer border and enters the region of higher velocities, tearing off the 'legs' and forming the ring-like vortex (Fig. 4.8).

4.2.2 Subharmonic transition

The subharmonic behaviour has been studied extensively in control experiments (Kachanov and Levchenko 1984; Saric et al. 1984; Corke and Mangano

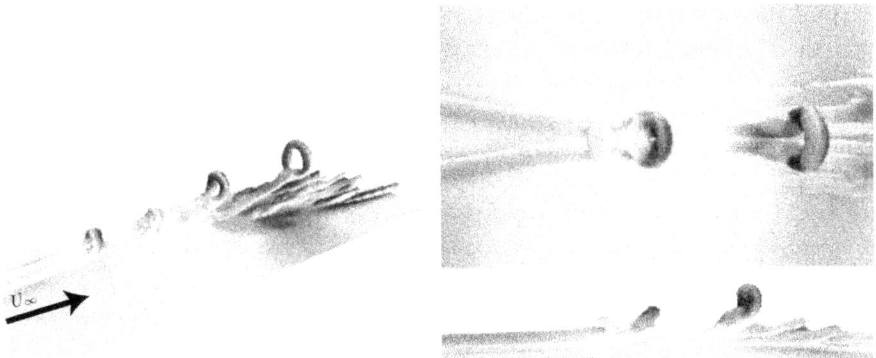

Fig. 4.8. A DNS visualization of laboratory experiments of Kachanov (1994) by Rist et al. (1998)

1989). It was found that the N-regime is characterized by a wide-band amplification of low frequencies centred about the subharmonic with frequency $f_{1/2} = f_1/2$ of the wave of dominant amplitude with frequency f_1 (the fundamental frequency). The subharmonic becomes periodic along the spanwise coordinate in the resonance regime, which corresponds to a pair of oblique waves propagating at equal angles but in opposite directions in respect to the mean flow (cf. Fig. 4.4b). Later stages of the subharmonic transition are characterized by a gradual filling of the frequency spectrum and by a growth of its overtones, produced by the wave cross-interactions. A careful study of this packet of low-frequency disturbances shows that it is dominated by quasi-subharmonic oscillations with almost constant frequency and phase, but with the amplitude randomly varying in time, so that the phase experiences 180° jumps when the amplitude passes through zero.

Such behaviour of the phase of the oscillations is characteristic for a subharmonic resonance in mechanical systems. The parametric resonance (the subharmonic growth) is observed at an appropriate phase tuning between the oscillations involved in the interaction, with the anti-resonance (the subharmonic attenuation) occurring at a large phase mismatch. The subharmonic amplitude grows super-exponentially (as a doubled exponential function) – a property which is well known for parametric resonance in mechanical systems – and it can even exceed the amplitude of the forcing wave without an essential counter-influence or feed-back. This is explained by a catalytic role played by the primary Tollmien–Schlichting wave: the subharmonic receives its energy directly from the mean flow, rather than from the primary wave (Orszag and Patera 1983; Herbert 1988).

4.3 Theoretical approaches for the onset of three-dimensionality and breakdown

The most modern physically justified numerical methods consider the three-dimensional non-linear initial-value problem (Benney and Gustavsson 1981). Its numerical solution (by DNS) (Henningson et al. 1990) or after dropping elliptic terms (by PSE) (Bertolotti et al. 1992; Li and Malik 1995) are the only feasible ways of describing the latest stages of the transition for specified initial and boundary conditions. However, the presence of characteristic transition scenarios makes it possible to get deeper into the dominating physical mechanisms. Below we consider some of them approved by practice.

4.3.1 Resonance model of the subharmonic growth

A classical weakly non-linear resonant model explaining the subharmonic growth is the so-called resonant triad model (Craik 1971). In essence, due to the quadratic-type non-linearity in the complete Navier–Stokes equations, the primary two-dimensional wave can form a resonant triad with a pair of oblique subharmonic waves. According to kinematic considerations, the wave triad can occur if the real parts of the wave vectors and frequencies of the waves satisfy the relation:

$$\omega_1 = \omega_2 + \omega_3, \qquad k_1 = k_2 + k_3.$$

For channel flow and in the boundary layer, Craik's triad is formed if one chooses $\omega_2 = \omega_3 = 1/2\omega_1$, $\alpha_2 = \alpha_3 = 1/2\alpha_1$ and $\beta_2 - \beta_3 = 0$, which can be satisfied only for a limited set of the Tollmien–Schlichting waves. The dynamical condition of realization of such triads is also quite strict, as the wave dispersion quickly destroys the synchronism unless the amplification of the subharmonic is fast enough to trigger further transition mechanisms.

In the experiments of Kachanov and Levchenko (1984) it was shown that a resonance occurs for a broad range of frequencies around $f_{1/2}$, which can be explained through a quasi-stationary treatment. If the frequency of a subharmonic is not exactly equal to half that of the primary wave, i.e. $f/2 = f_{1/2} + \Delta f$, it can be interpreted as a wave with frequency $f_{1/2}$, but with slowly varying phase and amplitude. Thus the appearance of peaks at frequencies $f_{1/2} \pm \Delta f$ in the spectrum up to quite large detuning values ($\Delta f \approx f_{1/4}$) is possible, which agrees with results of the weakly non-linear resonance model (Zelman and Maslennikova 1993).

4.3.2 Theory of linear secondary linear instability

A number of stability and transition problems were successfully approached by the theory of secondary linear instability (Orszag and Patera 1983; Bayly and Orszag 1988; Herbert 1988). The idea of this theory is a conjecture that the primary two-dimensional wave is a forcing agent for parametric excitation of some other modes. The starting point is to consider a new basic flow formed from the initial stationary velocity profile with an imposed saturated (i.e. periodic) primary Tollmien–Schlichting wave. From the local viewpoint, i.e. in the coordinate system ($\xi = x - c_r x, y, z$) moving with the wave phase velocity c_r, the flow is steady. Then the linear stability of this new flow to three-dimensional disturbances is considered.

Denoting the streamwise and normal velocity components of the initial two-dimensional wave as $\varepsilon u_{2D}(x, y)$ and $\varepsilon v_{2D}(x, y)$, respectively, where ε is a measure of a wave amplitude, and their phase velocity as c, the new basic flow reads

$$U_i = [U(y) - c]\delta_{i1} + \varepsilon[u_{2D}(\xi, y)\delta_{i1} + v_{2D}(\xi, y)\delta_{i2}],$$

where i is the coordinate index and δ_{ij} is Kronecker's delta. For three-dimensional cases, the two-dimensional analysis can be formally extended by adding spanwise mean and primary disturbance components in the equations both for the Tollmien–Schlichting and crossflow instabilities (Herbert 1988; Nayfeh 1987; Fischer and Dallmann 1991; Koch et al. 2000).

Substituting this expression for the basic flow in the linearized Navier–Stokes equations yields the following equations for the linear secondary instability of the flow:

$$\frac{\partial u}{\partial t}(U - c)\frac{\partial u}{\partial \zeta} + vU' + \frac{\partial p}{\partial \zeta} - \frac{1}{\mathrm{Re}}\nabla^2 u = -\varepsilon\left[\frac{\partial}{\partial \xi}(uu_{2D}) + \frac{\partial}{\partial y}(uu_{2D} + uv_{2D}) - u_{2D}\frac{\partial w}{\partial z}\right], \tag{4.3}$$

$$\frac{\partial v}{\partial t}(U - c)\frac{\partial v}{\partial \xi} + vU' + \frac{\partial p}{\partial y} - \frac{1}{\mathrm{Re}}\nabla^2 v = -\varepsilon\left[\frac{\partial}{\partial \xi}(uv_{2D} + vu_{2D}) + \frac{\partial}{\partial y}(vu_{2D}) - v_{2D}\frac{\partial w}{\partial z}\right], \tag{4.4}$$

$$\frac{\partial w}{\partial t}(U - c)\frac{\partial w}{\partial \xi} + \frac{\partial p}{\partial z} - \frac{1}{\mathrm{Re}}\nabla^2 w = 0, \tag{4.5}$$

$$\frac{\partial u}{\partial \xi} + \frac{\partial v}{\partial y} + \frac{\partial w}{\partial z} = 0. \tag{4.6}$$

These equations with boundary conditions $u = 0$, $v = 0$, $dv/dy = 0$ at $y = 0$ and $y \to \infty$ constitute the problem of the secondary instability theory. It can be additionally simplified as during the derivation of the Orr–Sommerfeld equation, so that the problem is reduced to the solution of two equations for u and v, decoupled from the solution for the third w-component, which is found from the continuity equation.

Since u_{2D} and v_{2D} are the solutions for the primary wave that is periodic with ξ, then in accordance with the Floquet theory it is possible to present the solution as

$$\{u, v, w\}(\xi, y, z, t) = \{\hat{u}, \hat{v}, \hat{w}\}(\xi, y)\mathrm{e}^{\gamma\xi}\mathrm{e}^{\sigma t}\mathrm{e}^{\mathrm{i}\beta z},$$

where σ and β are generally complex and $\{\hat{u}, \hat{v}, \hat{w}\}$ are periodic disturbance amplitude functions in ξ with periods $2\pi n/\beta$. They can be expanded in a periodic series, and then the expression describing the secondary disturbance amplitude function (e.g., $\hat{u}$) can be written as

$$\hat{u} = \mathrm{e}^{\sigma t}\mathrm{e}^{\mathrm{i}\beta z} \sum_m \tilde{u}_m(y)\mathrm{e}^{\alpha(\mathrm{i}m+\varepsilon)\xi}, \tag{4.7}$$

where $\varepsilon = \gamma/\alpha$ plays the role of a certain 'phase shift', so we may limit $0 \leqslant \varepsilon_i \leqslant 1/2$.

Since σ and ε are complex, one has to prescribe two of their four components to solve the eigenvalue problem. The real parts of σ and ε stand for the disturbance growth in time and space, respectively, so the choice can be based on consideration of the absolute or convective instability cases and the flow symmetry.

The limit cases give the solutions for the secondary instability of two types: harmonic ($\varepsilon = 0$), when the period of the secondary mode coincides with that of the Tollmien–Schlichting wave; and subharmonic ($\varepsilon = 1/2$), when the period of the secondary mode is doubled. Between them, mixed cases can exist. The corresponding equations are obtained at substitution of (4.7) and similar expressions for v and w in (4.3)–(4.6) and the continuity equation that reduces the problem to computing the eigenvalues of the secondary modes.

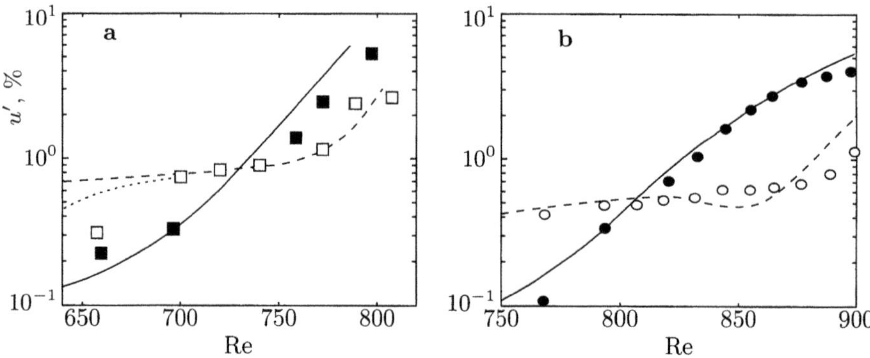

Fig. 4.9. Comparison of measured (*symbols*) and calculated (*curves*) amplification of primary (□, ∘) and subharmonic (■, •) amplitudes at the later stages of the subharmonic breakdown [after Zelman and Maslennikova (1993) (*left*) and Crouch and Herbert (1993) (*right*)]. Experiments (Saric et al. 1984, *left*) and (Corke and Mangano 1989, *right*). *Dotted line*, non-linear theory

Application of the theory to the Blasius boundary layer gives results which are in a good accordance with experimental data. In contrast to the non-linear resonances, such a secondary instability should be observed at small amplitudes of the primary wave, when the non-linear distortions are small. Figure 4.9 shows the results of comparing the computational, experimental and theoretical data for the primary Tollmien–Schlichting wave and the subharmonic disturbance. It is interesting enough that both the non-linear theory and the theory of linear secondary instability approach the experimental data quite well in this case.

4.3.3 Local high-frequency secondary instability

A major aspect of the turbulization process is the appearance of incommensurable frequencies, which due to non-linear interactions can result in a smooth turbulent spectrum. One of the mechanisms is connected to the presence of uncorrelated spectral harmonics in the initial spectrum of the Tollmien–Schlichting waves induced by external disturbances. In this case, even at a completely deterministic development of the perturbations, their initial stochasticity can produce random oscillations.

Yet another mechanism connected to a secondary instability of the flow is possible which differ essentially from that described in Sect. 4.3.2. Since in the region of the large-amplitude primary wave the instantaneous velocity distributions have inflection points along normal and spanwise directions, such profiles should be inviscidly unstable to high frequency perturbations, if the quasi-stationarity and quasi-two-dimensionality of the former is assumed. In particular, in the models of Betchov (1960), Greenspan and Benney (1963), the primary large-amplitude wave produces the inflection points in the instantaneous velocity profile that are unstable to the high-frequency disturbances.

Landahl (1972) proposed a kinematic treatment of the mechanism of the origin of outbursts of the local high-frequency secondary instability. The necessary condition of their origin is coincidence of the phase velocity of the primary wave with the group velocity of secondary disturbances. This requirement is needed for effective amplification of the secondary modes in the unstable region, which propagates with the phase velocity of the primary wave. In other words, the wave packet of the secondary disturbances must be absolutely unstable in the frame of reference moving with the crests of the primary wave. The fact that destruction of the flow occurs on scales equal to approximately one wavelength of the primary instability wave, testifies the feasibility of the local mechanism.

In the past, this was the most wide-spread theoretical idea on the mechanism of the Λ vortex breakdown in the K regime which must occur in the region of the high-shear layer formed by a Λ-structure and propagate with its velocity (Betchov 1960; Klebanoff et al. 1962; Landahl 1972; Nishioka et al. 1980). For a long time the secondary instability was associated with the spikes. However, the spikes are very deterministic and conservative formations; corresponding ring-like vortices protrude far downstream in the region, where a developed turbulent motion is observed in the boundary layer, while the high shear layer in mean velocity profiles occurs in the region of Λ-structure legs and has different velocity of propagation.

As for the high shear layer instability, Borodulin and Kachanov (1994) and Dryganets et al. (1990) showed that the criteria for the realization of the secondary instability are not satisfied at certain stages of its development. In particular, the velocities of the shear layer and that of the 'secondary' disturbances are different. Besides, the disturbance frequencies are rather low, being centred around the subharmonic frequency and the growth of irregular fluctuations is quite gradual, rather then explosive.

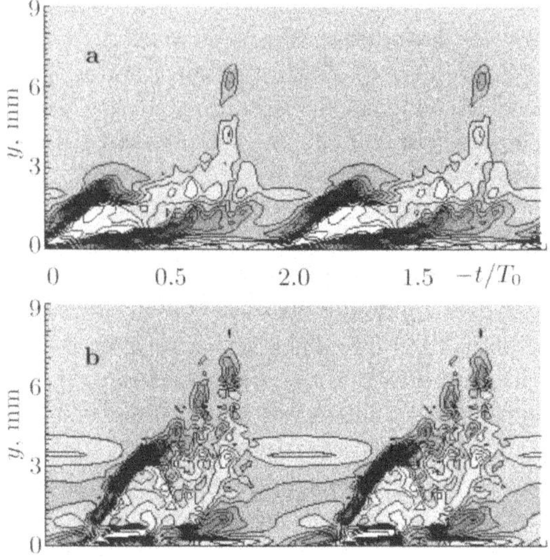

Fig. 4.10. Vorticity induced by ring-like vortices in near-wall region at late stage of transition 2 mm off the spike peak position (**a**) and at the peak position (**b**) (Meyer et al. 2000)

In contrast, it was found experimentally and numerically that the ring-like vortices associated with the spikes induce quite intensive velocity fluctuations ('stochastic anti-nodes') in the near-wall region which have the same scales as the ring-like vortices, and propagate downstream with the same (almost free-stream) speed, rather than with the speed of the high-shear layer, forming a new layer of high-shear (Fig. 4.10) (Bake et al. 1996; Meyer et al. 2000).

Nevertheless, the Λ-like vortices can appear both with and without formation of the ring-like vortices or even without previous amplification of a two-dimensional Tollmien–Schlichting wave. The latter occurs in the course of development of an isolated wave packet with strong spanwise shear (Gaster 1979) or a pair of oblique waves (Berlin et al. 1999). It is not unlikely that in these cases the transition mechanism specific for the K-regime does not operate. In Fig. 4.11, oscillograms of a wave packet and a harmonic wave

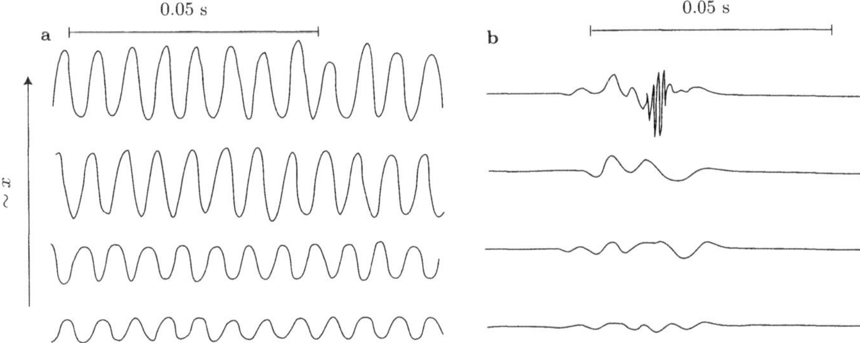

Fig. 4.11. Evidence of early appearance of outburst of secondary instability in oscillograms of Tollmien–Schlichting wave packet (**b**) in comparison with harmonic in time wave (**a**) (Gaster 1979)

for several downstream positions are shown. An outburst of high-frequency oscillations probably associated with the secondary instability is clearly seen. While the harmonic primary wave (at the left) of much higher amplitude does not lead to the secondary instability (as there is no layer of strong shear in the spanwise direction), it occurs at quite small amplitudes in the wave packet.

Experimental modelling of an isolated Λ-like vortex in the flat plate boundary layer by fluid pulses through a hole in the plate of Grek et al. (2000) showed that, depending on the amplitude of the disturbance excitation, both decaying and growing *isolated* Λ-like vortices can exist. At a combined excitation of the decaying Λ-like vortex and a small-amplitude (less than 1%) high-frequency wave, the disturbances interact with each other so that an unstable high-frequency wave packet develops at the vortex 'legs', leading to formation of a localized patch of turbulence – a turbulent spot (Fig. 4.12).

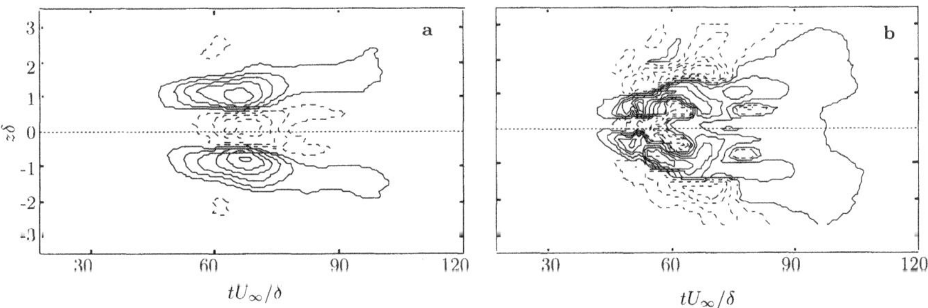

Fig. 4.12. Isolines of velocity fluctuations for decaying Λ-vortex (**a**) and its interaction with high-frequency wave (**b**) in (z–t)-plane at y close to the disturbance maximum: *solid lines*, velocity excess; *dashed line*, velocity defect, $Re_{\delta^*} = 730$ (Grek et al. 2000)

4.3.4 Direct numerical simulations

Until now, a theory which would explain the origin and *quantitatively* describe the development of three-dimensional non-linear structures up to the late stages of the transition in near-wall flows is far from completeness. It is connected first of all with the natural complexity of experimental recording of three-dimensional disturbances and their theoretical analysis even in a linear approximation, supplemented also by the effects of non-parallelity for three-dimensional waves and the importance of taking into account the modes of the continuous spectrum. Additional problems arise due to the necessity of dealing with the region close to a source of disturbances (problem of receptivity), which for low-frequency components can be rather extended in dimensional units.

The quantitative results at later stages of development of the localized disturbances can be obtained through direct numerical simulation of simplified problems for canonical flows (Henningson et al. 1990). In the past decade, DNS techniques based on numerical algorithms to solve the complete Navier–Stokes equations have become a valuable tool to gain a deeper insight into the mechanisms at work. Wray and Hussaini (1984) were the first to perform a DNS of the flat plate boundary layer transition using a temporal approach to simulate plane channel experiments by Kovasznay et al. (1962). Similar DNSs of the experiments of Nishioka et al. (1980), among others, are reviewed by Kleiser and Zang (1991). Direct quantitative comparison between spatial DNS and the experiments of Kachanov (1994) were performed by Rist and Fasel (1995), Rist et al. (1998) and Bake et al. (1996), and showed excellent agreement. The data clearly indicate that both approaches validate and complement each other.

4.4 Appearance and development of turbulence

Later stages of the transition to turbulence are characterized by the appearance of turbulent spots, turbulent bursts, and intermittency, the phenomena being universal for the near-wall flows of this book. *Turbulent spots* are spatially localized regions with turbulent fluctuations, propagating downstream in the surrounding laminar fluid. *Turbulent bursts* are intervals in oscillograms of disturbances with turbulent fluctuations superimposed on less intensive low-frequency oscillations of laminar flow. *Intermittency* in time (in space) is an alteration of laminar and turbulent regions in time (in space) at a fixed point in space (in time). Let us note that the intermittency in time (associated with the turbulent bursts) is a necessary (but not sufficient) indication of the turbulent spots. An *intermittency coefficient* determines the ratio of duration of the turbulent regime at a given point in space to the whole duration of the process. In the following section we consider these phenomena in more detail.

4.4.1 Wave combinations and intermittency

In many experimental studies of the 'natural' transition, intensive disturbances preceding the final flow turbulization were observed. The disturbances have typical frequencies much lower than that of the primary Tollmien–Schlichting wave or its subharmonic. These low-frequency fluctuations quickly amplified in the region of the formation of the spikes and turbulent spots (Kachanov et al. 1978). This sharp increase followed by a fast decay of the low-frequency fluctuations in the transition region is widely observed and is even used as a criterion of the transition at hot-wire measurements (based on the detection of the downstream maximum of the fluctuations with most of the energy concentrated at low frequencies).

The low frequencies of the very large amplitudes can be a consequence of wave combinations, for example, a result of the direct interaction of a pair of Tollmien–Schlichting waves with the generation of the difference frequency. The Tollmien–Schlichting waves observed in experiments on natural transition represent a wave packet with a central frequency f_0 detuned by frequencies less than approximately Δf_0 equal to the characteristic frequency of the low-frequency amplitude modulation (Fig. 4.13a). Kachanov et al. (1982) modelled the Tollmien–Schlichting wave packet by introducing two waves of incommensurable frequencies $f_1, f_2 \sim f_0$ of an identical amplitude detuned as $f_2 - f_1 \approx \Delta f$ (Fig. 4.13b). Their combinational interaction gradually leads to the onset of turbulence and, in essence, does not differ from the multi-wave case.

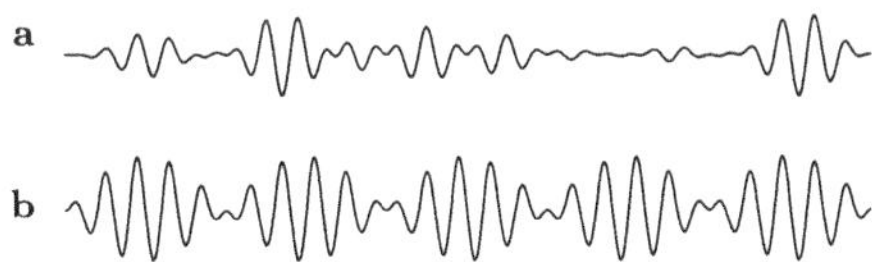

Fig. 4.13. Time traces of hot-wire signal: **a** natural transition (Arnal and Juillen 1977); **b** simulation of a Tollmien–Schlichting wave packet. (Kachanov et al. 1982)

Figure 4.14 shows a sequence of frequency spectra in the flat plate boundary layer measured in different streamwise sections at a constant distance from the wall. Initially, in the two-dimensional flow, all harmonics gradually amplify together with primary peaks f_1 and f_2 – the difference mode Δf and its harmonics $2\Delta f$, $3\Delta f$ grow most rapidly. Up to the latest stages of the transition, peaks at $f = mf_2 \pm nf_1$, where $m, n = 0, 1, 2 \ldots$, can be clearly identified, being the results of the non-linear interactions. An example of their identification is shown in Fig. 4.15.

The intermittency of bursts of the fluctuations and turbulent spots with frequency Δf is well explained within the framework of a phenomenological quasi-stationary treatment of the boundary layer unsteadiness. A superposition of two harmonic waves excited in the boundary layer with frequencies f_1 and f_2 and amplitudes u_1' and u_2' can be considered as a single quasi-harmonic wave with the frequency $f \sim f_1, f_2$, whose amplitude slowly varies in time

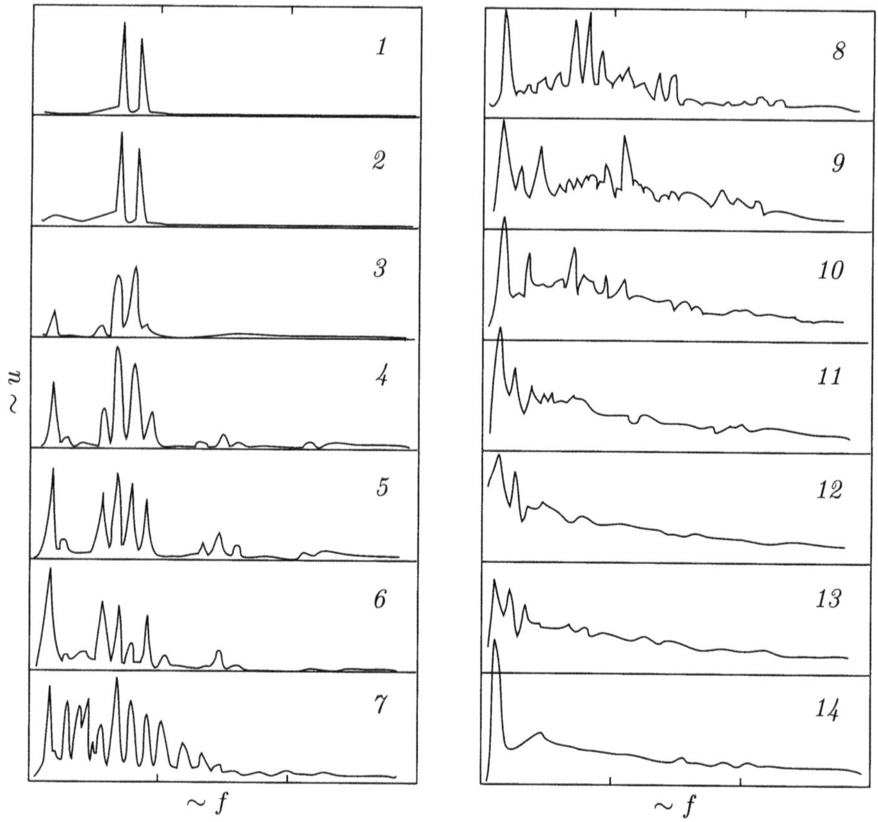

Fig. 4.14. The sequence of spectra of fluctuations for two interacting waves at different downstream positions marked by numbers *1,...,14* (Kachanov et al. 1982)

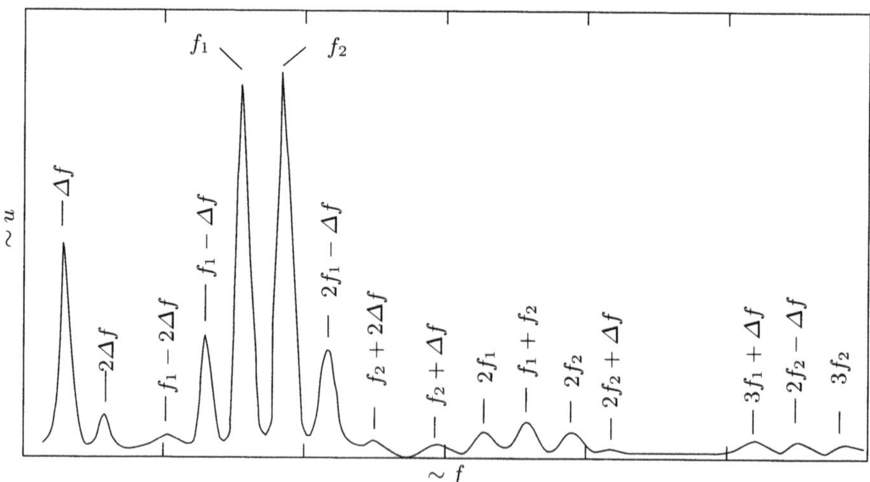

Fig. 4.15. Generation of combinational modes (Kachanov et al. 1982)

from maximum $u'_1 + u'_2$ to minimum $|u'_1 - u'_2|$, if $\Delta f = |f_1 - f_2| \ll f_1, f_2$. Thus the period of the amplitude variations $T_\Delta = 1/\Delta f \gg T_0 = 1/f$. As a result of slow, quasi-stationary alterations of the amplitude, the transition region moves periodically up and down along the body surface, producing oscillatory motions with period T_Δ. This motion is sensed by a fixed hot-wire probe located at a certain point in the boundary layer as alternating turbulent and laminar portions in time with frequency $\Delta f = 1/T_\Delta$; i.e. the intermittency is observed.

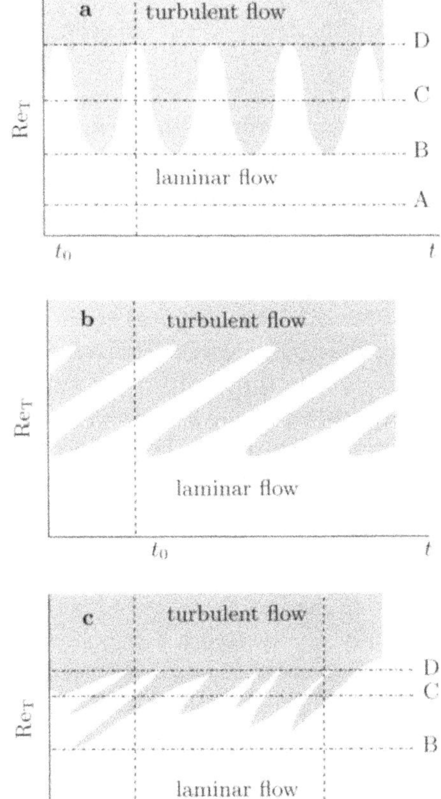

Fig. 4.16. Quasi-stationary oscillations of transitional Reynolds number $\mathrm{Re_T}$ with time due to modulations of Tollmien–Schlichting wave amplitude: **a** without appearance of turbulent spots in space; **b** with presence of turbulent spots in space; **c** in the case of natural transition with random modulation of Tollmien–Schlichting waves (Kachanov et al. 1982)

This phenomenon is illustrated in Fig. 4.16a. The region of laminar flow breakdown, associated with the transition Reynolds number $\mathrm{Re_T}$ which can be defined as, for example, the end of the region of two-dimensional development of the instability wave (Kachanov et al. 1978), periodically moves in space. The horizontal sections A, B, C, D are the positions of an observer who sees, e.g., in section C, an alternation of laminar and turbulent states in time (the intermittency). Since the difference harmonics are produced by the alternation, its amplitude reaches a maximum at a factor of intermittency of about 0.5, i.e. in a quite turbulent flow (Fig. 4.16a, section C), and decays only to the end of the transition region (section D), as was observed by Kachanov et al. (1982). At a moment t_0, Reynolds numbers $\mathrm{Re} < \mathrm{Re_T}$ correspond to the laminar region, and $\mathrm{Re} > \mathrm{Re_T}$ to the turbulent one, and there is no intermittency in space. Hence, the appearance of intermittency in oscilloscope traces as the alternation of laminar and turbulent regions in time, yet does not mean a formation of turbulent spots in space. However, the dependence of $\mathrm{Re_T}$ on t can be as illustrated in Fig. 4.16b. Then the turbulent spots are observed in space, since the regions with turbulent and laminar flow will alternate at a fixed moment t_0.

The quasi-stationary treatment of the mechanism of the formation of turbulent bursts and spots is applicable also for natural transition. Its main difference from the controlled one is aperiodicity of variations of the transitional Reynolds number $\mathrm{Re_T}(t)$, as illustrated schematically in Fig. 4.16c. These oscillations are also connected to the low-frequency modulation of the primary wave. In the crests of such composite beatings, the turbulent bursts appear similar to what occurs at the superposition of two waves. However, in oscillograms the wave amplitudes in various crests are different. In those where the amplitude reaches maximum, turbulent bursts appear earlier. Downstream they are generated also at the other crests. So the turbulent bursts appear rarely at first in section B. In section C the frequency of the turbulent portions increases, and their durations are different. In section D, the flow is almost turbulent, but laminar regions are occasionally formed.

The number and frequency of the turbulent bursts vary downstream in accordance with the modulation of the Tollmien–Schlichting wave. Results obtained by Knapp and Roache (1968), Burnel and Cougat (1972) and Arnal and Juillen (1977) confirm this conclusion. In particular, measurements of the spectrum of the low-frequency fluctuations and the spectrum of the modulations of the Tollmien–Schlichting wave amplitude carried out by Burnel and Cougat (1972) in the case of the natural transition show their conformity. Since the external disturbances determining the modulation of the Tollmien–Schlichting wave frequently have a random character, it is no wonder that the factors of intermittency are described by a normal distribution law of random quantities (Kachanov et al. 1982).

Due to the dependence of $\mathrm{Re_T}$ on time, the mean velocity profile $U(y)$ changes with frequency Δf between the turbulent and laminar states. This leads to the appearance of a characteristic profile of the low-frequency modulation along the wall-normal coordinate. This can be seen in Fig. 4.17a, where such two profiles are sketched. The difference between them, ΔU, has a maximum near the wall, decreases with y, but at the outer boundary layer border it has the other external maximum, and finally decays in the free stream. The relation of the intensity of the modulation (proportional to ΔU) is shown by

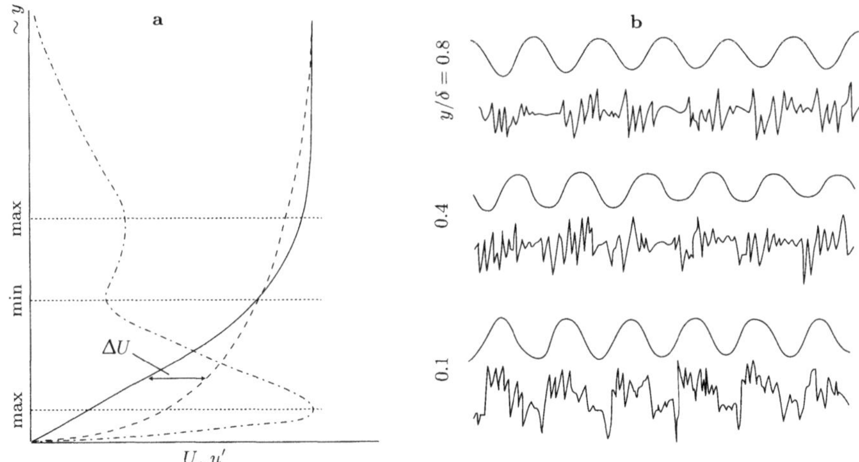

Fig. 4.17. Scheme of alternation of laminar (*solid line*) and turbulent (*dashed line*) states in a quasi-stationary treatment of transition, *dash-dotted line* corresponding profile of low-frequency fluctuations Δf (**a**) (Kachanov et al. 1982) and phase shift of difference harmonic (*upper curve*) in relation to turbulent spot on the oscilloscope traces depending on distance from the wall (**b**) (Kachanov et al. 1982)

a dash-dotted line in Fig. 4.17a. The quasi-stationary treatment predicts oscillations at the wall and at the outer boundary layer border with opposite phases, i.e. the increase of U at the wall is accompanied by a decrease of U near the outer boundary layer border, and vice versa. This phenomenon is confirmed experimentally as illustrated in Fig. 4.17b.

4.4.2 Turbulent spots

In many cases, the transition process in the Blasius boundary layer and Poiseuille channel flow passes through the formation of so-called turbulent spots (Fig. 4.18). The turbulent spot is easy to generate and visualize, for example, in a water table, by dropping on the water surface. This means that a strong-enough initial disturbance in a near-wall shear flow can produce the turbulent spot without the stage of instability wave amplification even in the subcritical region. The spot propagates downstream, grows, and eventually covers the whole near-wall region. The continuous development of several turbulent spots in space and time also leads downstream at their merging – to completely turbulent flow.

Isolated turbulent spots. The spots were observed for the first time in the boundary layer by Emmons (1951), who came to the following conclusions:

1. The spot formation is 'point-like' (actually, the turbulent spot arose in a region smaller than the local boundary layer thickness).

2. The boundary between the fluid in the turbulent spot and that in the ambient flow is sharp;
3. The growth of the spot is uniform.
4. The spots do not interact in the bulk of the boundary layer – in merging the spots produce a union rather then their superposition or a new formation.

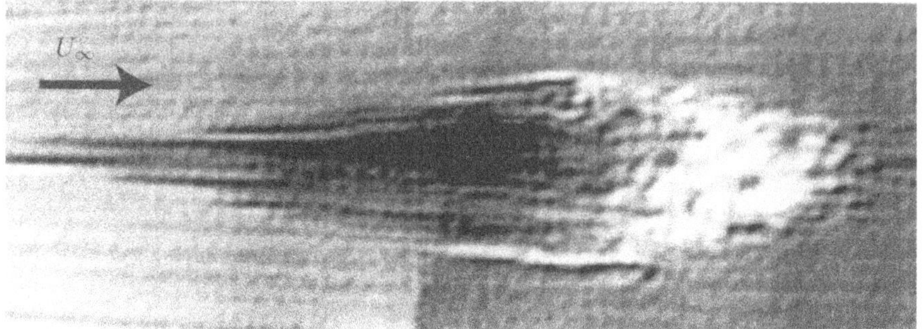

Fig. 4.18. Smoke visualization of turbulent spot in a flat plate boundary layer (Matsubara et al. 1996)

The development of turbulent spots was first studied in detail by Schubauer and Klebanoff (1956), and then by Elder (1960). Emmons' main conclusions were confirmed. It appeared that during the downstream development, the spot has conical similarity irrespective of the method and place of its excitation. Typical spanwise semi-angle of expansion of the spot is 9–12°, depending on its precise definition. The spots develop almost independently of one another – their interaction in the majority of the boundary layer is minor. The critical amplitude of the disturbance initiating the spot formation across the most part of the boundary layer appeared to be approximately equal to 20% of the free-stream velocity. Later studies by Wygnanski et al. (1975), Cantwell et al. (1978), Wygnanski et al. (1979, 1982), Narasimha et al. (1982), Barrow et al. (1984) and others were focused on the turbulent spot structure, the processes of their development and merging, as well as on separation of mean characteristics from the turbulent fluctuations.

Using an ensemble-averaging technique allowed these studies to considerably advance the knowledge of the mechanisms of turbulent spot development, their internal structure and velocities of propagation. In Fig. 4.19, an ensemble-averaged distribution of the streamwise disturbance velocity in the turbulent spot plane of symmetry measured at various distances from the wall is shown (Grek et al. 1987). The curves characterize two regions of the spot: the velocity excess $\langle U \rangle - U_{\mathrm{lam}} > 0$ and the velocity defect $\langle U \rangle - U_{\mathrm{lam}} < 0$, where $\langle U \rangle$ is the ensemble-averaged velocity in the spot and U_{lam} is the mean local velocity in the laminar boundary layer in the absence of the spot.

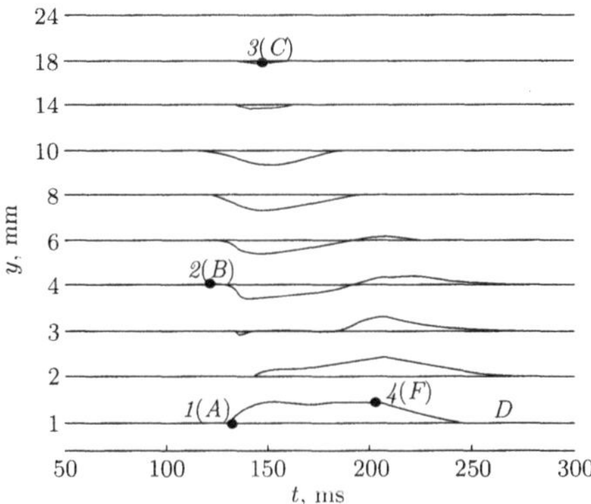

Fig. 4.19. Ensemble-averaged oscillograms across the turbulent spot in the plane of its symmetry in a flat plate boundary layer (Grek et al. 1987). Characteristic regions and fronts of the spot are marked by numbers, as in Wygnanski et al. (1982), and letters as in Cantwell et al. (1978): *1* (*A*), spot leading front near wall; *2* (*B*), leading front forward-most region; *3* (*C*), maximum spot height; *4* (*F*), spot trailing edge; *D*, region of velocity relaxation to undisturbed state

By repetition of the measurements at other streamwise positions, it is possible to determine characteristic velocities of the spot propagation (Fig. 4.20). It can be seen that the velocity of its trailing front is much lower than that of the leading front, which is the reason of growth of the spot streamwise size. The velocities of the leading and trailing fronts are 0.89 and 0.57 of U_0, respectively as was also found previously by Wygnanski et al. (1982). This coincidence of the results obtained in different experimental facilities and flow regimes testifies that the spot does not remember the 'initial' conditions and is a turbulent 'eigendisturbance' of the boundary layer. It is interesting to note that the straight lines in Fig. 4.20 intersect at the point of an 'effective' turbulent spot origin, which is not coincide with the point of the disturbance generation. This means that at the short initial stage of the spot formation (so-called 'incipient' spot stage), its self-similarity is evidently lost. It is natural that in the other conditions, this effective spot origin has other coordinates, which can be found experimentally.

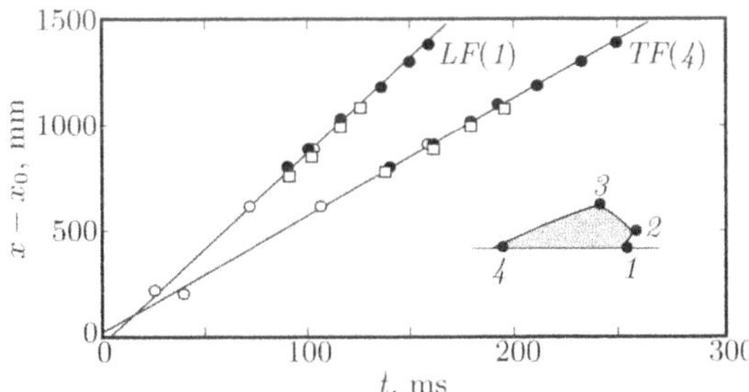

Fig. 4.20. Velocity of propagation of turbulent spot fronts. $x_0 = 300$ mm; $U_0 = 10$ m/s. LF(*1*), leading front, TF(*4*), trailing front. Experimental data: □, (Grek et al. 1987), Tu $<$ 0.04%, ○, (Grek et al. 1987), Tu = 1%; •, (Wygnanski et al. 1982). For other notations see legend to previous figure

Linear expansion of the turbulent spots in the spanwise direction in gradientless flow has been observed in many other studies (Wygnanski et al. 1975, 1982; Cantwell et al. 1978; Barrow et al. 1984; Vasudevan et al. 2001). Narasimha et al. (1982) carried out special studies concerning the influence of the streamwise pressure gradient on the development of the turbulent spots in the spanwise direction. It was found that the pressure gradient leads to a non-linear dependence of their size with respect to the streamwise coordinate.

Interaction of turbulent spots. An experimental study of an interaction of two turbulent spots was carried out for the first time by Elder (1960) in a flat plate boundary layer at $\mathrm{Re}_x = 4 \times 10^5$. The turbulent spots were simultaneously initiated by an electrical discharge at two points spaced in the spanwise coordinate. The intermittency inside the boundary layer was measured by a hot-wire probe. The result of the study confirmed the conjecture of Emmons (1951) that the spots' geometrical characteristics evolve independently of each other inside the boundary layer.

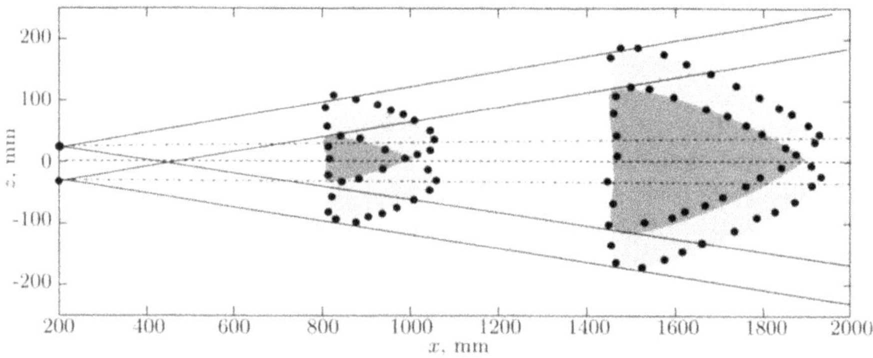

Fig. 4.21. Development and merging of two turbulent spots in (x–z)-plane spaced in the spanwise direction (Grek et al. 1987)

In experiments by Grek et al. (1987), the turbulent spots were investigated at Reynolds numbers that were twice as large as those in the experiments of Elder (1960). A scheme of the spots' merging in the boundary layer is shown in Fig. 4.21. It can be seen that as they propagate their overlap increases, while their spanwise sizes still grow linearly, as if each spot develops irrespective of the others, thereby preserving the characteristics of their individual development.

The independence of their front development at their overlap is also confirmed when the spots move one behind the other. The scheme of this interaction is shown in Fig. 4.22. The figure also presents ensemble-averaged oscillograms of developing separately and the axially aligned mutually penetrating turbulent spots measured inside the boundary layer. This demonstrates that

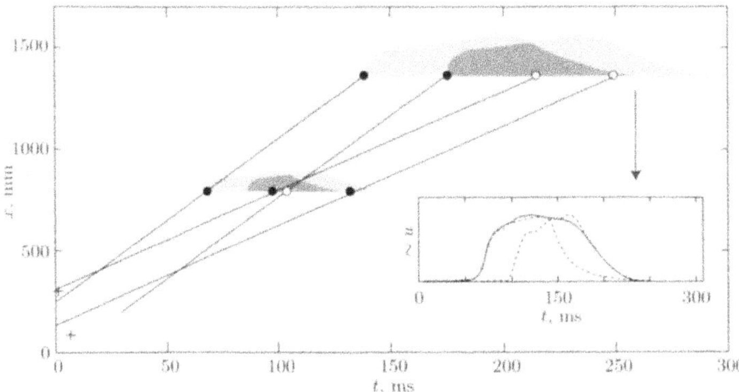

Fig. 4.22. Development and interaction of two turbulent spots that are following each other (Grek et al. 1987)

the oscillogram obtained at their interaction is virtually a simple union of the other two.

Contours of the velocity fluctuations for the interacting and isolated turbulent spots are shown in Fig. 4.23. Two characteristic regions can be indicated: the region of the velocity excess close to the wall, and the region of the velocity defect at the outer boundary layer border and above it. The maximum deflection of the velocity inside the spot is 25–30% of U_0, in accordance with previous findings. Careful examination of the experimental results shows that the regions of the velocity excess is a geometrical union of the regions which would be obtained if the spots were independent, while in the region with the velocity defect at the outer boundary layer border, some minor changes in the back part of the interacting structure can be identified.

4.5 Transition to turbulence in flows modulated by streamwise vortices

The formation of large-scale streamwise vortical structures in various types of both stationary and non-stationary flows has been observed by many investigators since ancient times. The formation of stationary vortices in water were apparently documented first by Leonardo da Vinci (Fig. 4.24a).

The streamwise vortices can arise due to a variety of reasons in shear flows. The centrifugal instability causes the appearance of Taylor–Görtler vortices (Sect. 2.3.2), the presence of a crossflow on a swept wing initiates crossflow vortices (Sect. 2.3.1), local roughness generates streaks in a boundary layer, etc. The three-dimensional flow pattern in the presence of streamwise vortical structures imposes certain difficulties in their study. In particular, the mechanism of the formation of streamwise vortical structures at a high free-stream turbulence level or behind three-dimensional roughness is still far

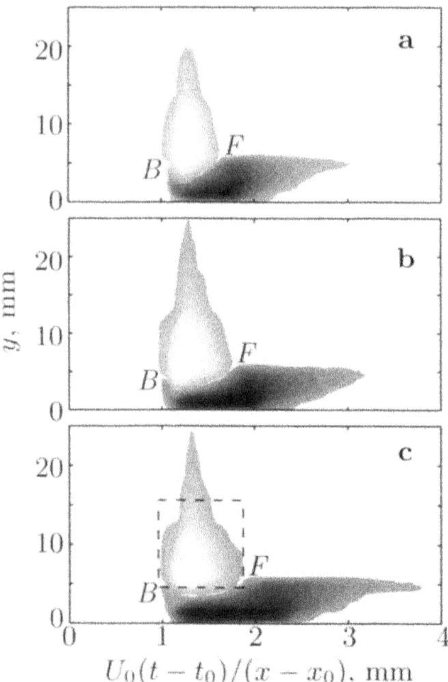

Fig. 4.23. Velocity fluctuations in turbulent spots at $x = 1400$ mm initiated separately at x_0: **a** 90 mm; **b** 300 mm; **c** simultaneous initiation. *Dashed box*, interaction region. The extreme contours of both regions are lines of velocity differing by 1% from the velocity of a laminar fluid – they represent 'boundaries' of the turbulent spot (Grek et al. 1987)

from clearness (see Chap. 5). Certain advances in their understanding can be obtained from the most recent theoretical and experimental studies within the framework of the lift-up effect (Schmid and Henningson 2000), as surveyed in Chaps. 1 and 2. In particular, the lift-up effect operates at the side edges of a three-dimensional roughness element forming a pair of the streaks, as illustrated by the scheme presented in Fig. 4.24b.

The current interest in studying the mechanisms of the origin, development, and destruction of the streamwise vortices arose from their cardinal role both in the transition process and in turbulent flows (Blackwelder 1983). It is revealed experimentally that stationary vortices of small amplitude, excited artificially in a flat plate boundary layer, can prevent growth of Tollmien–Schlichting waves and, hence, stabilize the flow (Tani and Komoda 1962; Kachanov and Tararykin 1991; Bakchinov et al. 1995). However, quite frequently they reach amplitudes of more than 10% of the free-stream velocity promoting the onset of turbulence. Nevertheless, they do not lead to turbulence directly, instead creating a spanwise periodicity in the velocity distribution, i.e. forming a new stationary flow pattern (Orszag and Patera 1983; Zhigulev and Tumin 1987). Such strong modulation of the boundary layer can serve as a source for various non-linear interactions between the stationary vortices or between them and travelling instability waves (Reed 1987; Reed and Saric 1989; Malik and Hussaini 1990; Floryan 1991; Saric 1994b).

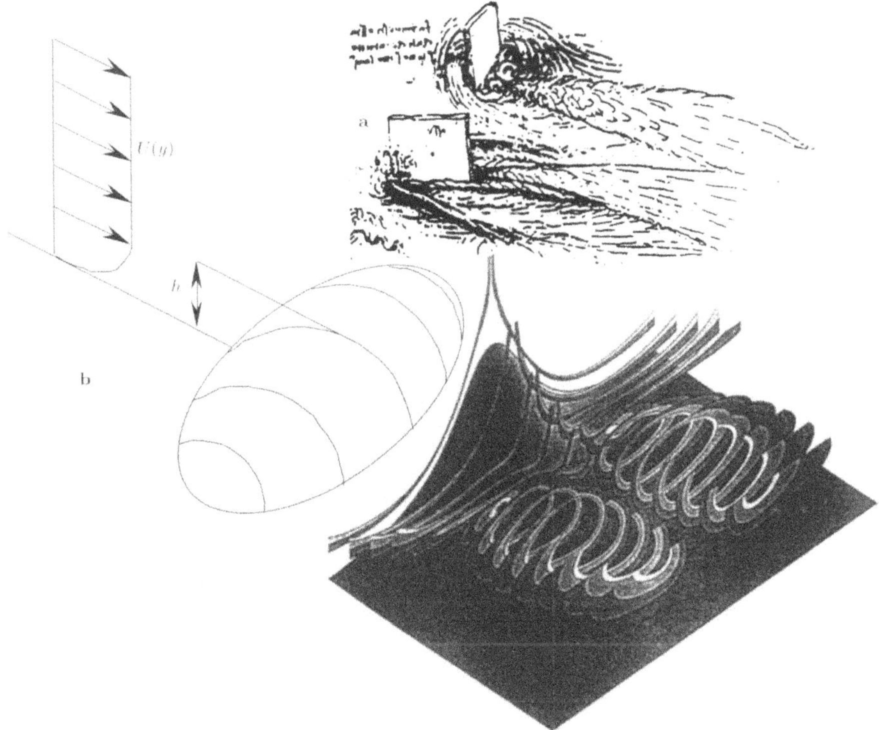

Fig. 4.24. Perhaps the first ever vortex 'visualization' by Leonardo da Vinci (**a**) and modern scheme of separated flow past a three-dimensional hump (Roget et al. 1998) (**b**)

At a large-enough amplitude of the modulation, the travelling disturbances propagating along the primary vortices, illustrated by flow visualization in Fig. 4.25, can be treated as eigenmodes of the new stationary flow formed by the initial mean flow and the vortices, and can be therefore subjected to secondary instability analysis. The secondary instability in stationary vortices is usually of inviscid type (Orszag and Patera 1983) and can lead in a number of cases to subcritical – related to the Tollmien–Schlichting wave instability – transition (Kohama 1987; Dagenhart et al. 1990; Kohama et al. 1991; Janke 2000).

4.5.1 Instability of crossflow vortices

One of the types of vortical structures are the co-rotating stationary vortices frequently observed in swept-wing boundary layers. Growing downstream according to the linear theory, the vortices reach an amplitude of 10–12% before saturation. Therefore, if the pressure gradient changes sign rapidly, the spanwise modulation of the flow caused by the vortices also determines the

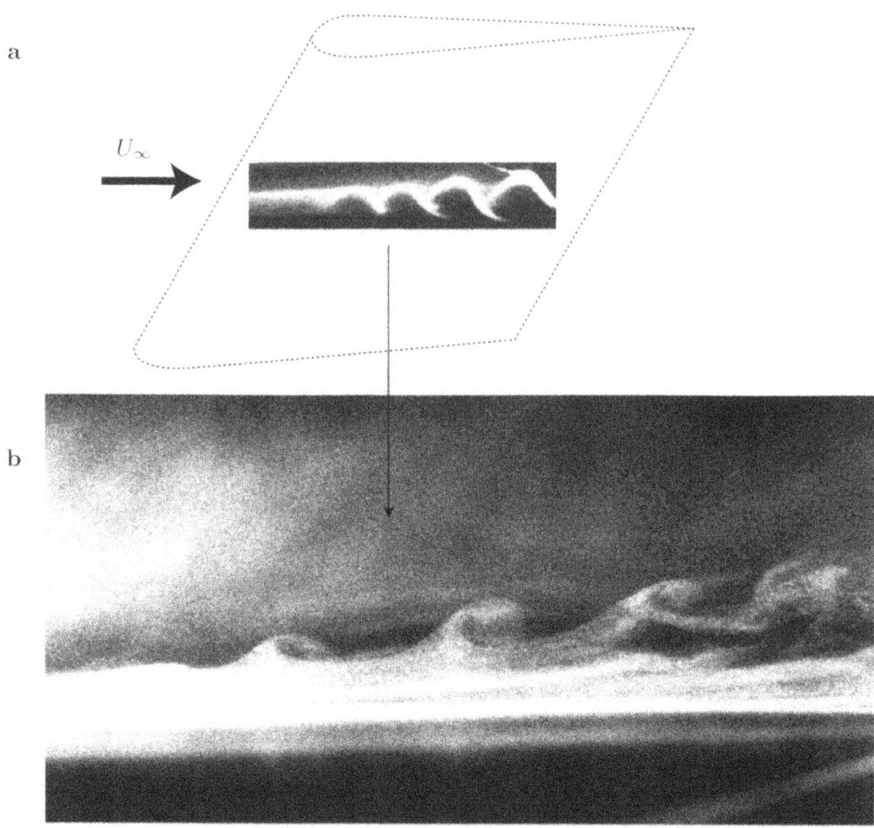

Fig. 4.25. Travelling disturbances arising on stationary vortices in a three-dimensional boundary layer (Kohama et al. 1996): **a** view parallel to the surface; **b** view normal to the surface

instability even in the region of the decelerating flow, where the mechanism of their initial linear growth is absent.

Poll (1979, 1985) was probably the first to observe the high-frequency oscillations (in comparison to the low frequencies of linearly unstable travelling crossflow modes) in the swept-wing boundary layer. Later these results were confirmed by numerous hot-wire measurements and visualizations. Based on them, Kohama (1987) assumed that the high-frequency oscillations can be caused by spiral waves arising around the stationary vortex because of inflection-point instability in its core, i.e. similar to the sinuous instability of the Görtler vortices. However, it was later revealed that the presence of Tollmien–Schlichting waves in the swept-wing boundary layer leads to a similar destabilizing effect (Kachanov and Tararykin 1990). Fischer and Dallmann (1991) and Malik et al. (1994) supposed that the travelling disturbances in the stationary crossflow vortices can be connected to the inflections in the velocity distribution. Fischer and Dallmann (1991) used experimental results on the secondary instability modes obtained by Nitschke-Kowsky and Bippes (1988) as initial data for the analysis based on the Floquet theory. They found harmonic and combinational types of instability with frequencies differ almost tenfold. At the superposition of the instabilities and by proper selection of their relative amplitudes, a good qualitative correspondence with the experiment was achieved. In the calculations of Fischer et al. (1993), several different instabilities were found, with most growing at a vortex amplitude of more than 10%.

Radeztsky et al. (1999) investigated the development of the secondary disturbances on a set of stationary vortices of small amplitude, and their influence on the transition to turbulence. This revealed that reducing the spacing between the roughness elements showed the growth of the disturbances, in accordance with results of Balachandrar et al. (1992).

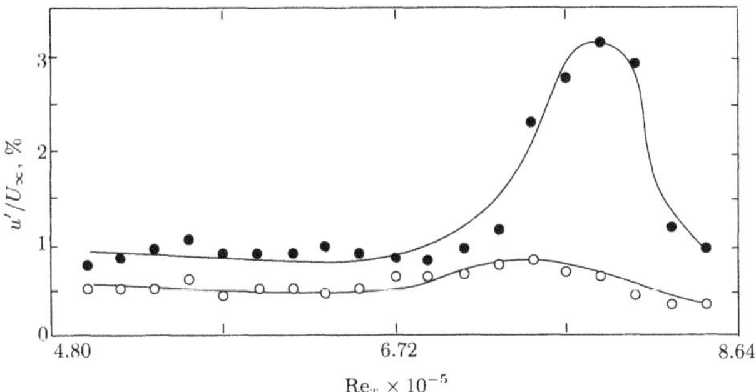

Fig. 4.26. Downstream growth rates of travelling disturbances developed in the large-amplitude stationary vortices in swept-wing flow: ○, natural perturbations; •, perturbations excited by sound (Kozlov et al. 1999)

Kozlov et al. (1999) were the first to observe a wave packet of perturbations travelling along the stationary vortices, which led to turbulence in a swept-wing boundary layer in natural conditions. It was shown that the acoustic field, with a frequency characteristic for the natural wave packet, excites travelling disturbances stimulating the laminar–turbulent transition (Fig. 4.26).

In a number of cases, non-linear interactions between the travelling and stationary crossflow instability modes is also referred as the secondary instability (Saric and Yeates 1985; Reed 1987; Nitschke-Kowsky and Bippes 1988; Meyer and Kleiser 1990; Bassom and Hall 1991; Deyhle and Bippes 1996). As a result of the interaction, small stationary vortices with doubled periodicity can appear (Saric and Yeates 1985; Reed 1987).

4.5.2 Instability of Görtler vortices

Due to the centrifugal instability, Görtler vortices with a characteristic periodicity in the transverse direction are initiated (Hall 1983; Floryan and Saric 1984). Depending on the local conditions, both a division and a merging of neighbouring pairs can occur due to their non-linear interaction (Guo and Finlay 1994). Furthermore, high-frequency oscillations (Yu and Liu 1991; Park and Huerre 1995; Bottaro and Klingmann 1996), growing downstream appear in the frequency spectrum of the disturbances, which can be analyzed theoretically using the Floquet theory for periodic base flow.

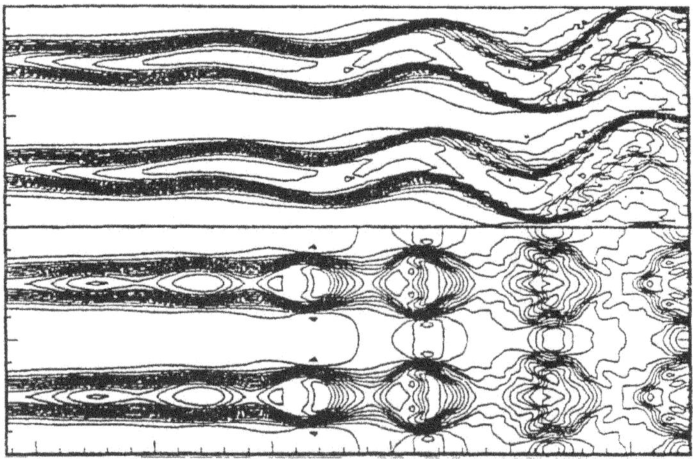

Fig. 4.27. Instantaneous velocities in (x–z)-plane: *top*, sinuous (odd) mode; *bottom*, varicose (even) mode (Li and Malik 1995)

Two characteristic types of the secondary instability were identified being in the form of either periodic 'meandering' of the vortices in the transverse direction or in the form of horseshoe vortices in the region of strong spanwise shear. Such disturbances are called sinuous and varicose modes, respectively (Fig. 4.27). Li and Malik (1995) and Bottaro and Klingmann (1996) numerically found that the varicose mode is dominant for long-wave vortices, whereas the sinuous mode prevails for short-wave ones, since the vortices with large wavelengths provide weak spanwise shear, and, on the contrary, those of small wavelength induce strong spanwise shear, which is responsible for the mode formation.

However, the flow visualizations of Swearingen and Blackwelder (1987) and Floryan (1991) show that the transition on a concave surface can be determined by instability mechanisms, which produce growing waves independently in each vortical pair, so that the neighbouring pairs can experience different types of secondary motions. In such cases, the appearance of the travelling disturbances can be attributed to a local high-frequency instability mechanism. This can be stipulated by inflections in the instantaneous mean velocity profiles in the wall-normal and spanwise directions.

The selection of the instability mode, which is excited first and grows faster, depends on initial conditions, in particular on the distance between the vortices. Blackwelder and Swearingen (1990) and Park and Huerre (1995) remarked that since the spanwise shear layers arise simultaneously with the beginning of the growth of the vortices, instability associated with them has, at other equal conditions, better chances to trigger the growth of secondary waves.

4.5.3 Experimental modelling of streamwise vortex instabilities

At free-stream turbulence levels Tu $\lesssim 0.1\%$, the stationary crossflow vortices usually dominate due to a large contribution of stationary background perturbations, whereas with growth of the free-stream turbulence level they become less in comparison with travelling crossflow disturbances (Bippes 1990). Such behaviour demonstrates a competition of various mechanisms and introduces considerable difficulty to the study of the secondary instability of the vortices in naturally developing flows.

This and similar situations in other flows make reasonable a modelling of the vortices in simplified cases to control their characteristics during experiments. Numerical analysis of Corbett and Bottaro (2000) indicates that the shapes of the 'natural' stationary crossflow vortices steamed from the linear instability of the flow are virtually indistinguishable from the streaks generated by the lift-up effect.

Streamwise vortices in controlled conditions were studied for the first time by Tani and Komoda (1962). In their experiment, the spanwise pressure gradient in the Blasius boundary layer was modulated by a row of small wings located outside of the boundary layer, that caused periodic variation of its thickness. The travelling waves were excited by a vibrating ribbon at frequencies over the range of the Tollmien–Schlichting instability. This showed that compared with the case of the undisturbed Blasius boundary layer, the instability of the flow essentially changes in the presence of stationary velocity modulations.

Kachanov and Tararykin (1991) measured different velocity components in a swept-wing boundary layer. It was modulated in the spanwise direction by alternating sucking and blowing through streamwise slits, the amplitude of the excited vortices being rather small. It was revealed that the normal and the spanwise velocity components decay downstream, whereas the modulations of the streamwise velocity are observable far from the slits. The resulting flowfield characterized by the absence of the streamwise vorticity is dominated by streaks which can lead to the appearance of the secondary instability, provided that the amplitude of the modulations is large enough.

Hamilton and Abernathy (1994) excited an isolated stationary vortex by a special obstacle at a wall. They carried out a series of measurements to define conditions at which the streamwise vortices cause the transition to turbulence, avoiding the effects of crossflow instability. A comparison of the transition in the single and multi-vortex flows did not show qualitative differences, testifying that the interaction of the vortices during the transition did not play a determining role in that case.

Bakchinov et al. (1995) modulated the boundary layer structure by roughness elements located periodically along the span of flat plate. The travelling disturbances excited by a vibrating ribbon had an inviscid nature and reminded the secondary instability of Görtler vortices of a sinuous type. The simplicity of the flow configuration allowed detailed study of the wave characteristics and the transition to turbulence. Using a similar technique, Boiko et al. (1995a, b) conducted an experimental study of travelling waves in a single vortex of large amplitude excited in a swept-wing boundary layer. It was shown that an obstacle located near the wing leading edge can effectively generate a stationary vortex, thereby providing an experimental model of a 'natural' large amplitude vortex excited by a surface non-uniformity.

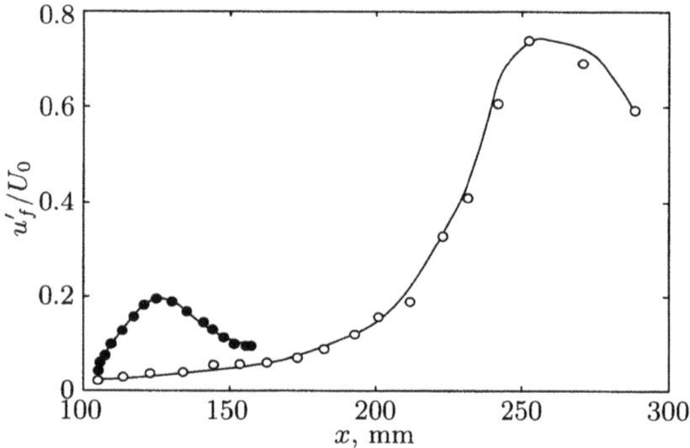

Fig. 4.28. Development of travelling high-frequency (∘) and low-frequency (•) disturbances in a swept-wing flow (Boiko et al. 1995a)

That study revealed the 'natural' development of the travelling disturbances as two packets that appear with different central frequencies and locations in the vortex. The low-frequency packet is located at the vortex core, while the high-frequency one is at the outer boundary layer border. The intensity of the high-frequency packet grows rapidly up to amplitudes about 2% and then decay, playing no direct role in the transition (Fig. 4.28). However, a strong local distortion of the mean flow leading to influences on a transition point appear upon its intensive controlled forcing. The low-frequency packet begins to grow and initiate the transition to turbulence when the high-frequency one has already decayed. The results of further studies on the travelling disturbances showed, that smoothing of the frequency spectrum occurs through amplification of higher harmonics of the waves of the low-frequency packet.

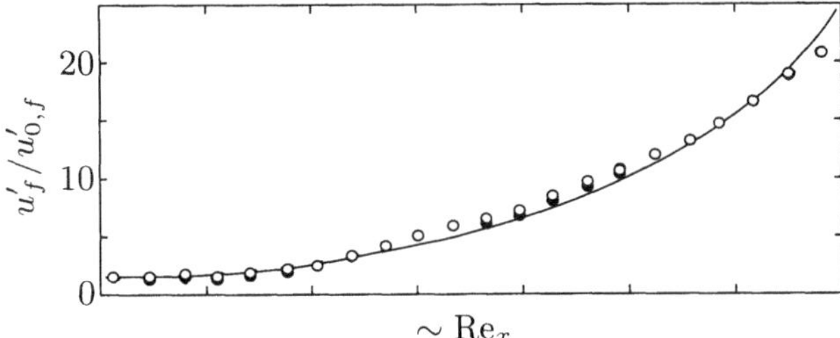

Fig. 4.29. Growth of waves excited in vortex by sound of amplitude equal to about 1% (○) and 0.5% of U_0 (•) at the end of the region shown; *solid line*, an approximation by exponential growth (Boiko et al. 1997b)

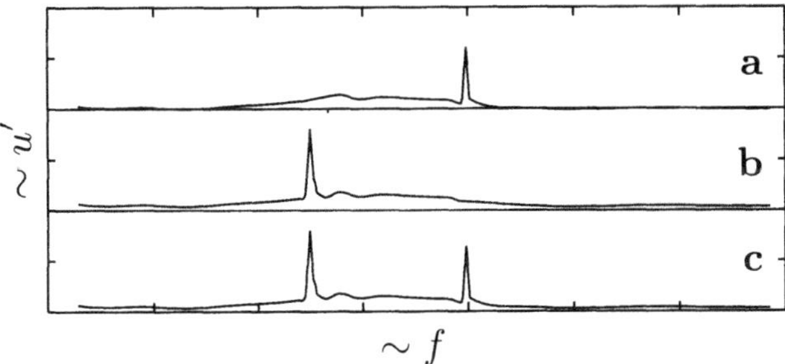

Fig. 4.30. Spectra of natural disturbances: **a** excitation by periodic sucking and blowing; **b** excitation by sound; and **c** combined excitation (Boiko et al. 1997b)

Boiko et al. (1997b) proved the linearity of the travelling disturbance growth at the initial stage of development inside the vortex. This property indicates that the investigated waves represent a secondary *linear* instability of the flow, rather than a result of non-linear interaction as, for example, in the studies of Bassom and Hall (1991) and Deyhle and Bippes (1996). In Fig. 4.29, the growth curves of the disturbances measured at two different initial amplitudes are shown, in comparison with an exponential amplification. Another indication of the wave linearity is the amplitude independence of their development along the vortex. The fulfilment of the superposition principle for the waves is demonstrated in Fig. 4.30, where the spectra are shown at separate and combined excitation of the disturbances with periodic sucking and blowing through a hole in the surface of the experimental model, and with acoustic oscillations from an external source. The spectra exhibit an absence of a noticeable interaction of the waves (combinational modes) up to a total amplitude of 1%.

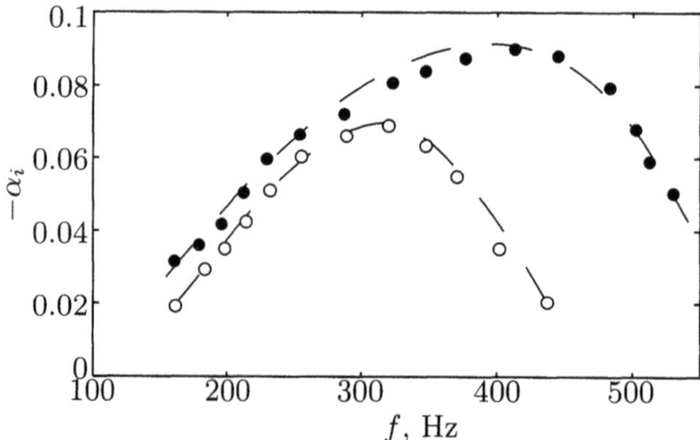

Fig. 4.31. Wave growth rates in the range $x = 200$–240 mm at two velocities: ∘, $U_0 = 6.8$ m/s; •, $U_0 = 7.8$ m/s (Boiko et al. 1997b)

It was also shown that the phase velocity of the travelling disturbances in the streamwise direction was in essence one and the same in the range of the measurements, and did not depend on the means of the excitation, amplitude and frequency of the oscillations. At different free-stream velocities, the disturbances propagated at a speed of approximately $0.6U_0$. Besides, since the phase and amplitude behaviour of the waves – excited by sound and periodic sucking and blowing – is identical, it was concluded that in both cases they are generated by a local receptivity mechanism. The wave growth rates become larger with the increase of the free-stream velocity, which shortens the downstream region of the linear disturbance growth (Fig. 4.31).

Boiko et al. (1997b, 1999) applied the methods of the experimental modelling to study the development of the travelling disturbances for a system of streamwise vortices. Mutual influence of the vortices was described, with the reduction of their spanwise spacing reducing the growth rates of the travelling waves. The last can be probably explained by redistributions of streamwise and normal velocity components in the vortices (Fig. 4.32).

The behaviour of disturbances is probably controlled by mechanisms of inflection-point (inviscid) instability, as was conjectured, for example, by Blackwelder and Swearingen (1990), Hamilton and Abernathy (1994) and Bottaro and Klingmann (1996). The validity of a local secondary instability mechanism for the case under consideration was tested by Boiko et al. (1997b). The distributions of mean velocity defect and growing disturbances as well as the position of the layer where local mean velocity is equal to the velocity of the wave propagation along the vortex – an analog of 'the critical layer' – are shown in Fig. 4.33. The core of the perturbed region is close to

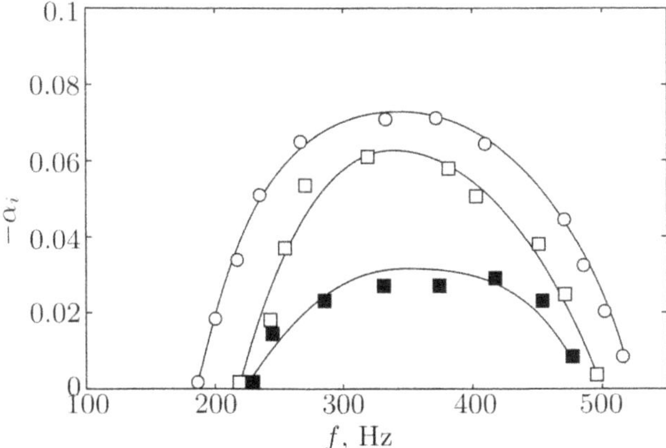

Fig. 4.32. Estimated growth rates of naturally occurring travelling waves at different spacings between the vortices: ○, isolated vortex; □, $\Delta z = 16$ mm; ■, $\Delta z = 8$ mm (Boiko et al. 1997b)

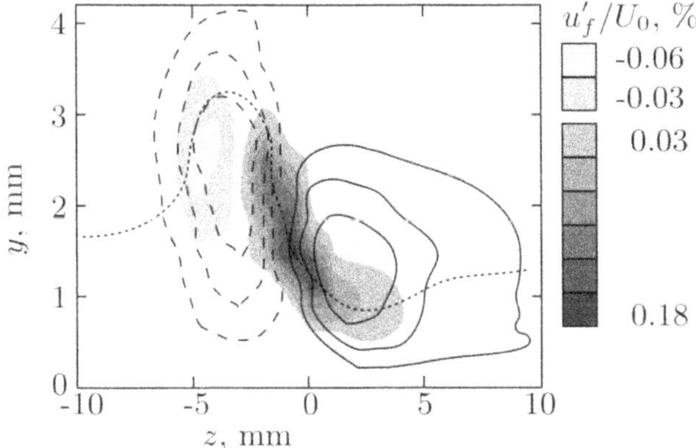

Fig. 4.33. Isolines of disturbances in the region of localization of stationary vortices: $\langle U \rangle / U_0$ and contours $u'_f/U_0\%$; *dash-dotted line*, location of the 'critical layer' $U/U_0 = 0.6$ (Boiko et al. 1997b)

the 'critical layer', and the distribution of the travelling wave amplitude correlates well with the location of local maxima of the velocity gradient along the spanwise coordinate.

To investigate the generation of the streamwise vortices by spatially-extended roughness elements, a pair of natural stationary vortices on a swept wing was modelled by Boiko et al. (2000). The characteristics of both the stationary vortices and the high-frequency travelling waves developing inside them were studied. It can be seen that the shape of the mean and disturbance velocity distributions along the spanwise coordinate are different for each vortex. This stipulates more effective growth of the travelling distur-

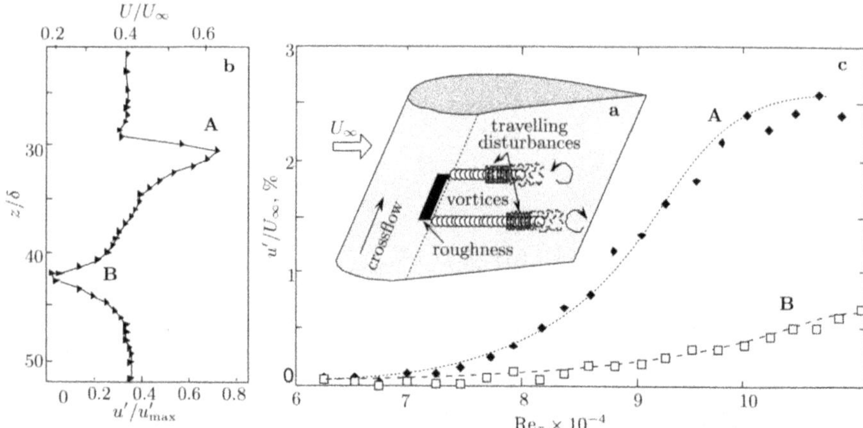

Fig. 4.34. Instability of the large-amplitude stationary vortices in a swept-wing boundary layer to travelling disturbances: **a** experimental set-up; **b** mean (*solid line*) and travelling disturbance velocity (*dashed* and *dotted lines*) spanwise distributions; **c** downstream growth of corresponding disturbances (Boiko et al. 2000)

bances in the vortex marked by 'A' in Fig. 4.34. The vortex exhibits the breakdown to turbulence much earlier than the other one ('B').

5 Laminar–turbulent transition at high free-stream turbulence level

The problem of laminar–turbulent transition is of great interest in applied engineering, for instance for the prediction of transition to turbulence on turbine blades, where the impingement of the turbulence from the wake of the stator influences the boundary layers on the rotor. Another important aspect is the effect of the free-stream turbulence on wind tunnel data[1]. Usually it is desirable to reduce the level of turbulence in order to resemble free-flight conditions. However, all wind tunnels have some background disturbances with different characteristics, so that additional knowledge about the 'dangerous' parameters of the free-stream turbulence is of interest. Nevertheless, despite being a subject of interest since the 1930s, the understanding of the effect of free-stream turbulence on the onset of transition has been significantly advanced only in the last decade.

At the free-stream turbulence level (Tu) above approximately 5%, the transition at a flat plate occurs at the minimum Reynolds number, at which the self-sustained boundary layer turbulence can exist, i.e. at $\mathrm{Re}_{\delta^*} \approx 190$ (Arnal 1992). At lower levels of Tu different experiments disagree widely concerning the position and the extent of the transition region, and there seems to be no general correlation between the free-stream turbulence level and Re_{T}.

The transition is also sensitive to a large number of other parameters, including not only the overall level of the free-stream turbulence, but also to its spatial scales, the degree of isotropy and homogeneity, conditions at the leading edge, pressure gradients, etc., each of which requires special attention. Depending on these conditions, the transition process may be dominated by Tollmien–Schlichting waves, streaks, or other as yet unknown mechanisms, which need different modelling approaches.

However, in most experiments on the transition, the turbulence level is described only in terms of a value of Tu based on streamwise disturbance velocity measurements. At the same time, in numerical studies by Yang and Voke (1991) and Voke and Yang (1995) it was found that free-stream pressure p' and normal velocity v' fluctuations are the most efficient at exciting perturbations in the boundary layer, whereas fluctuations of the streamwise velocity

[1] The notion of 'a high level' can be understood differently, for example, in aeronautics and engineering. In what follows it stands for $\mathrm{Tu} \geqslant 1\%$.

u' are rather ineffective. Unless the free-stream turbulence is isotropic, the scales and intensity of v' cannot be obtained from the frequency spectrum of u'. If a turbulence generating grid is installed in the settling chamber, as in the experiments of Suder et al. (1988) and Kendall (1985, 1990, 1991), the streamwise fluctuations are suppressed more efficiently than the spanwise fluctuations when the flow passes through the contraction. This provides an anisotropy in the free-stream turbulence, which decays along the test section. In such cases, to determine accurate characteristics of the free-stream turbulence, it is necessary to study the spanwise scales.

For engineering purposes, attempts have been made to model the transition in the presence of free-stream turbulence using transport equations. Most transport models rely on a transition criterion that is usually derived from empirical correlations for modifying the closure parameters through the transition region. Results of extensive comparison of numerical methods with experimental data by Roach and Brierley (1992) are collected in Pironneau et al. (1992). The observations indicate strong sensitivity of the transition to imposed free-stream characteristics. It seems that the transport models can barely reproduce the structure of the transitional boundary layer fluctuations needed for closure of the equations. Studies of Westin (1997) and Westin and Henkes (1997), in which the applicability of different models to the description of the boundary layer transition at high free-stream turbulence was analyzed, also confirm this conclusion. Consequently, if the transition is to be modelled without resorting to empirical correlations, it is necessary to correctly reproduce the laminar flow characteristics upstream of its final onset.

5.1 Streaks in the Blasius boundary layer

At a high free-stream turbulence level two types of phenomena are usually distinguished inside the boundary layer, both of which can be responsible for the transition to turbulence (Boiko et al. 1994): the generation of travelling waves with characteristics of local linear instability and the generation of quasi-stationary longitudinal structures or 'streaks', with characteristics probably determined by the external vortical and pressure disturbances. The streaks, which were named due to the visualization patterns produced by the flow structures, are experimentally observed quasi-stationary three-dimensional deformations of the laminar boundary layer which grow downstream in amplitude and length. The experimentally observed disturbance motion produced by the streaks is also sometimes referred to as the 'Klebanoff mode', due to the early observations of Klebanoff and Tidstrom (1959) and Klebanoff et al. (1962), who found steady and low-frequency spanwise periodic modulations of the streamwise velocity component in the Blasius boundary layer caused by a residual non-uniformity caused by wind tunnel damping screens.

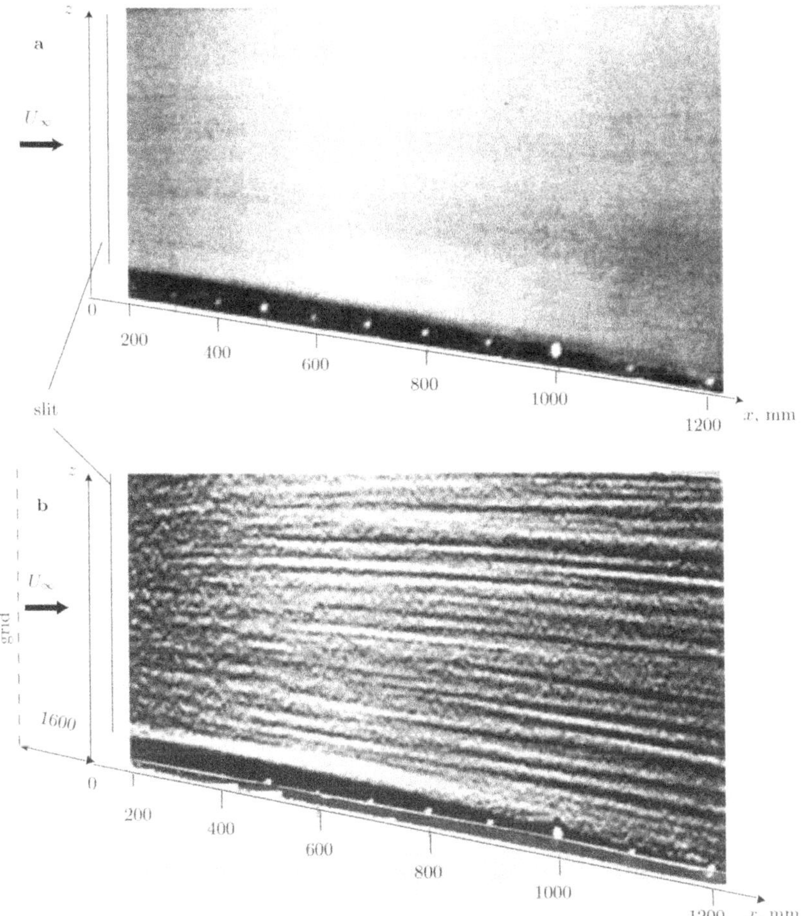

Fig. 5.1. A smoke visualization of boundary layer at flat plate surface (Matsubara et al. 1996): **a** Tu $\leqslant$ 0.02%, **b** Tu = 1.5%; $U_\infty = 6$ m/s

It is instructive to compare a smoke visualization of a flat plate boundary layer at low (Tu $\leqslant$ 0.02%) and high (Tu = 1.5%) free-stream turbulence (Fig. 5.1). In the first case the boundary layer can be characterized as undisturbed (homogeneous smoke sheet), whereas at Tu = 1.5% the flow is modulated by the streamwise structures. These structures can be also observed in visualizations of Kendall (1985) and Gulyaev et al. (1989). Real-time flow visualizations as that of Alfredsson et al. (1996) and Matsubara et al. (1996) show that the streaks are related to the low-frequency disturbances in the boundary layer, which start to develop from the leading edge of the plate.

The first detailed data on the laminar boundary layer in the presence of free-stream turbulence were presented by Arnal and Juillen (1978). They observed a downstream growth of the streamwise velocity fluctuations u' which reached amplitudes of several per cent of the free-stream velocity before the onset of transition. They also found that the energy spectrum inside the boundary layer is dominated by much lower frequencies than that in the turbulent free stream. The distributions of u' in wall-normal direction were quite different from that of Tollmien–Schlichting waves (e.g. the maximum was found near the middle of the boundary layer, rather than at the wall as for Tollmien–Schlichting waves). Perturbations with frequencies typical for unstable Tollmien–Schlichting waves were also seen on the large-scale (low-frequency) structures. However, their amplitudes were small compared with the overall fluctuation level.

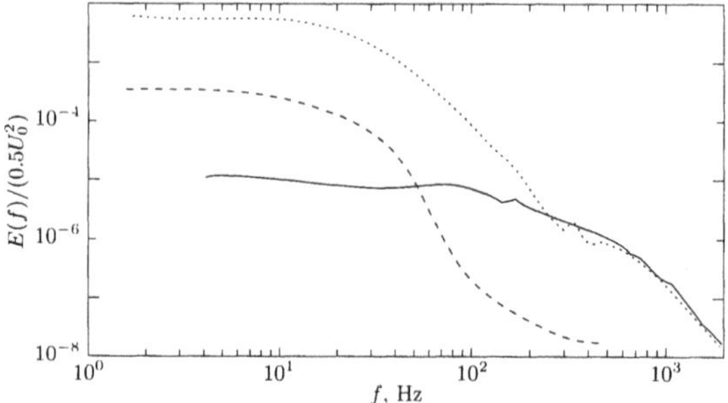

Fig. 5.2. Energy density spectra for u' at different wall-normal positions at Re = 890 in the free stream (*solid line*), in the middle of the boundary layer (*dashed line*) and in the near-wall region (*dotted line*) (Westin et al. 1994)

It should be emphasized that the boundary layer perturbations with a dominating streaky pattern, although random in time and space, are not what is usually called turbulence. The typical energy density spectra of streamwise velocity disturbances obtained in the free stream at Tu = 1.5% and in the laminar boundary layer are shown in Fig. 5.2. They demonstrate the essential difference of free-stream turbulence from large-scale boundary layer oscillations, where the high-frequency disturbances are much lower. Moreover, the spanwise and the wall-normal scales of the low-frequency structures are of the order of the boundary layer thickness, and the amplitude of normal velocity fluctuations v' inside the boundary layer is several times smaller than u' (Westin et al. 1994; Westin 1997).

5.2 Experimental and theoretical modelling of the streak development

Experimentally, the events in the boundary layer at high free-stream turbulence level can be investigated by generating controlled disturbances, and studying the subsequent development of perturbations excited inside the boundary layer. In such a way, Grek et al. (1985) excited a flat plate boundary layer by a localized periodic sucking and blowing of fluid at different intensities and frequencies. Additionally to the linear wave packets and the turbulent spots which are described in Chaps. 2 and 4, they found a localized structure with very specific characteristics, called 'puff' (Fig. 5.3). The term was borrowed from a description of disturbances observed at the transition in a round pipe (Wygnanski and Champagne 1973; Wygnanski et al. 1975). Later it become clear that the excited structure has a close relation to the streaks observed at high free-stream turbulence levels (Westin et al. 1994).

The main phenomenological distinctions between the Tollmien–Schlichting wave packets and the puffs were the following. The puffs appeared to be deterministic, localized three-dimensional formations with velocities of propagation of the leading and trailing fronts equal to 0.5 and 0.9 of U_0, respectively (as opposed to the group velocity of the Tollmien–Schlichting wave packet of about 0.3–0.4 of U_0, measured under the same flow conditions in the same experiments). They quickly elongated downstream, experienced practically no expansion in the spanwise direction and the magnitude of the maximum velocity defect continuously decayed downstream (Fig. 5.4).

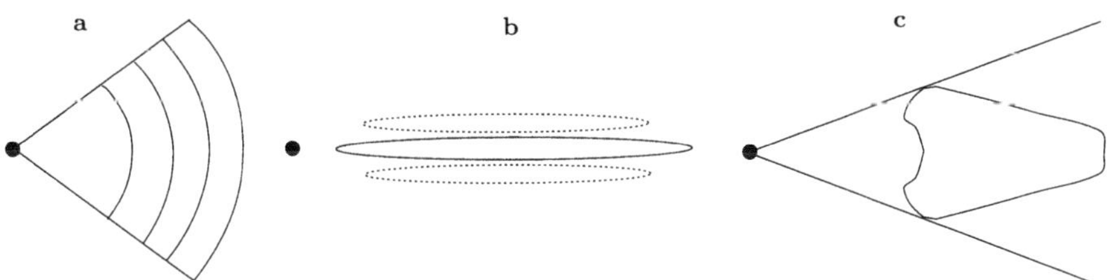

Fig. 5.3. Three types of localized disturbances observed in the boundary layer (Grek et al. 1985): **a** wave packet; **b** streak (puff); **c** turbulent spot

The results of the spectral analysis carried out by Westin et al. (1998) for the flat plate case showed that as the boundary layer puff propagates downstream, the majority of harmonics – except those at low frequencies – decay, so that the oblique waves dominate in the wave number spectra (Fig. 5.5). It is remarkable that they are observed in the region where straight Tollmien–Schlichting waves (i.e. those with $\beta = 0$) are highly damped. The maximum amplitude of the disturbances is centred at zero frequency, which is interpreted as quasi-stationary flow distortion.

Breuer and Haritonidis (1990) applied inviscid linear theory to the analysis of experimental data concerning the development of small localized disturbances in the flat plate boundary layer excited by spanwise oscillations of a membrane installed on the plate surface. The disturbances, which they excited, had characteristics similar to the puffs observed by Grek et al. (1985):

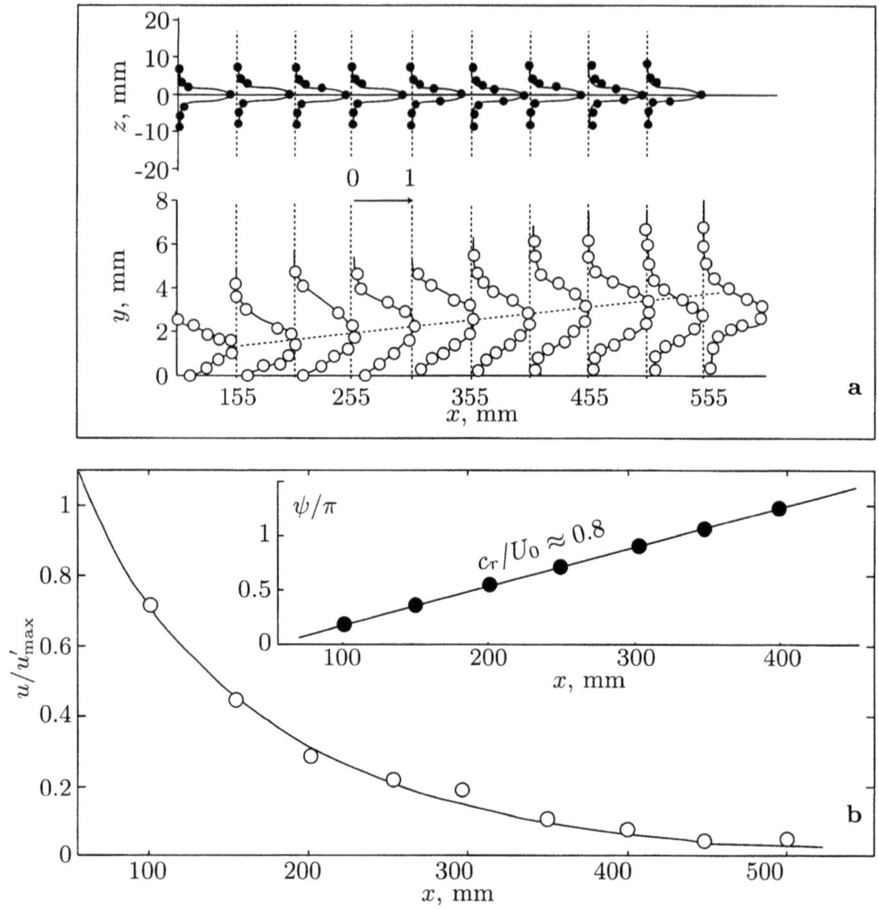

Fig. 5.4. Integral puff characteristics: **a** spanwise (along z) and wall-normal distributions of disturbance magnitude with streamwise coordinate x; **b** distributions of magnitude and phase as functions of x (Grek et al. 1985)

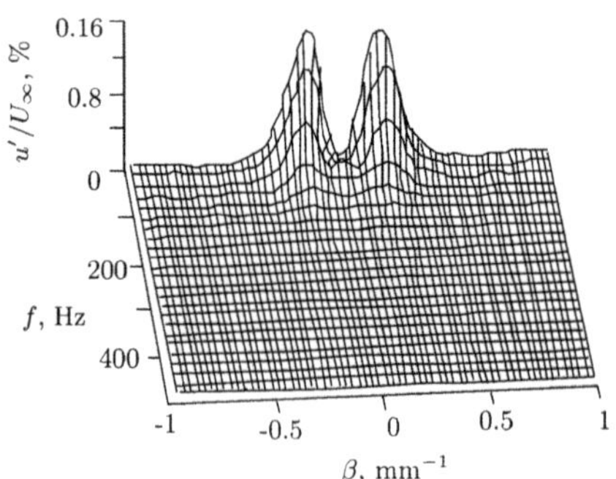

Fig. 5.5. A typical spectral content of the localized disturbance. Measurements at a subcritical Reynolds number close to the disturbance amplitude maximum (Westin et al. 1998)

the streak moved faster than the least-stable Tollmien–Schlichting waves, without strong attenuation. It was shown that the initial stage of the streak development is described satisfactorily by the inviscid lift-up mechanism. In accordance with linear theory and the experimental data, the initial development of the disturbance is divided into two parts: a broadening wave packet described by the solutions of the Rayleigh equation with waves propagating with their own phase velocities, and the convective part propagating with a local mean velocity, which due to the downstream elongation formed a streak of strong spanwise velocity shear. However, the results of the experiment show a natural dissipation of the disturbance by viscosity downstream that, obviously, cannot be obtained in the framework of the inviscid analysis (see Sect. 1.5.2).

Henningson et al. (1993) applied a DNS to plane Poiseuille flow and the Blasius boundary layer to consider primarily the transient growth and the transition associated with it. The DNS results were extended for the case of a boundary layer close to separation and to a variety of initial disturbances by Bech et al. (1998). The initial linear transient development was then extracted from the DNS results. Tumin and Reshotko (2001) used spatial, rather then temporal approach to obtain details of the transient development in the quasi-parallel case of Blasius boundary layer. Almost identical behaviour of the disturbances subject to temporal and spatial development was found. The development of optimal perturbations in adverse and favourable pressure gradient boundary layers were considered by Corbett and Bottaro (2000), in the linear stability framework. A comparative study of the transient growth in two- and three-dimensional boundary layers was performed by Breuer and Kuraishi (1994).

The predominance of very oblique waves in the streaks indicates that the non-parallel mechanisms probably play a significant role in their formation and development. To this end, the application of the approach developed initially by Libly and Fox (1964) was suggested by Luchini (1996). The basic idea is to consider the equations for three-dimensional linear disturbances derived from Prandtl boundary layer equations rather than from the Navier–Stokes equations. The solutions for the corresponding eigenvalue problem represents stationary Reynolds-number-independent two- and three-dimensional boundary layer modes, with the streamwise velocity component of the least stable three-dimensional mode experiencing algebraic growth $\sim x^{0.214}$. Their independence of the Re, provided by an appropriate scaling in the Prandtl equations, ensures that the instability occurs at any Re where the boundary layer approximation is valid. Since the linearization procedure used in this case was performed without the parallel flow assumption, the non-parallel effects are automatically included in the solution.

Later the approach was extended to the case of travelling disturbances and transient growth by Andersson et al. (1999), Luchini (2000). These results additionally proved and extended the conclusions of previous study of

Luchini (1996). It seems that the concept of 'optimum' disturbances (see Sect. 1.5.4) can describe the most energized harmonics in the spectra of localized disturbances such as puff in a flat plate laminar boundary layer. An important feature of the 'optimal' stationary modes derived with the help of the Luchini's theory is the self-similarity of the disturbances in the streamwise direction, which is also found in the experiments. It is demonstrated in Fig. 5.6: as seen, the experimental profiles follow the theoretical self-similar curve almost perfectly, indicating that these modes contribute primarily to the streaks observed in the experiment.

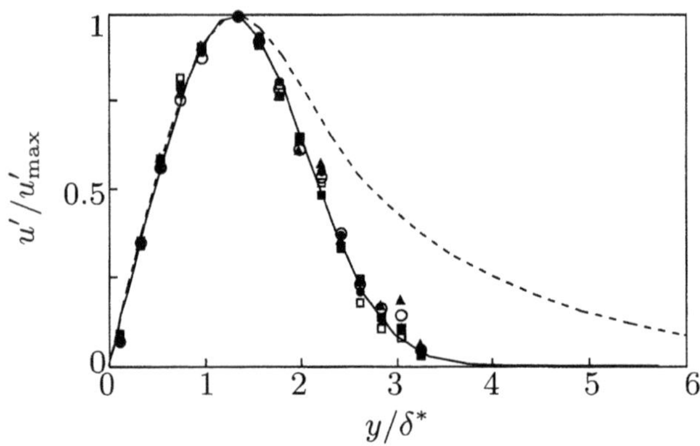

Fig. 5.6. Amplitude functions of puff wave component with $F = 27 \times 10^{-6}$, $\alpha = 0.022$ and $\beta = 0.419$ (■), 0.582 (□), 0.791 (•), 0.977 (○), 1.163 (△) (A.V. Boiko, D.S. Sboev, G.R. Grek, unpublished work, 2001); *dashed line*, Tollmien–Schlichting wave profile for $\alpha = 0.022$ and $\beta = 0.419$; *solid line*, a stationary boundary layer mode (Luchini 1996)

One problem in respect to the concept of optimal disturbances is the dependence of the spanwise scale of the excited natural streaks and the puffs on flow parameters, which exhibits a large scatter in different facilities. From dimensional considerations it may be assumed that the spanwise extent of the structures, as a rule, is close to the boundary layer thickness. However, in the wind tunnel tests of Westin et al. (1998) a localized free-stream disturbance excited the boundary layer perturbation on the same scale. The difference between the experimental data can be explained by the theoretical and numerical results of Butler and Farrell (1992), Berlin and Henningson (1994) and Herbert and Lin (1993), which show that the mechanism of scale selection is rather weak.

5.2.1 Streak generation at the leading edge

As it has been shown, the development of streaks is usually attributed to the lift-up effect, which is strongly dependent on the forcing disturbances, resulting in the receptivity rather than the stability problem.

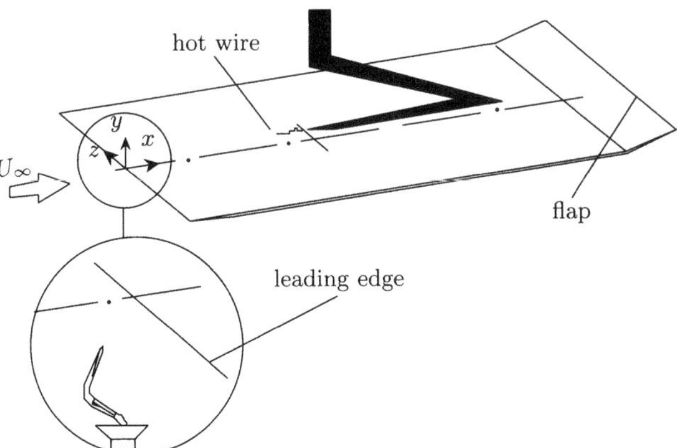

Fig. 5.7. Principle scheme of the experimental setup (Westin et al. 1998)

The first careful experiments on leading edge receptivity of the laminar boundary layer to vortical disturbances of the free stream that are localized in time and space were carried out by Grek et al. (1991a); further details were given by Westin et al. (1998). The disturbances were introduced by isolated pulses of sucking or blowing through a narrow tube located in front of the leading edge (Fig. 5.7). The contours of streamwise velocity component shown in Fig. 5.8 give an impression of the initial stage of the disturbance development. It can be seen that the disturbance is a fairly symmetric, rapidly decaying formation. The leading edge divides it into two parts. However this does not result in a qualitative reconstruction of the disturbance: the spanwise and streamwise scales remain unchanged. Meanwhile, the region of the velocity defect at the leading front of the disturbance becomes dominant in amplitude.

The produced structure of the velocity fluctuations across the boundary layer is illustrated in Fig. 5.9a. The perturbed motion reaching 15% of U_∞ is characterized by a disturbance in free-stream and a strong deformation in the shear flow. In the plane parallel to the wall, the excited structure represents two symmetric regions of a velocity defect separated by a region of a velocity excess (Fig. 5.9b). Approaching the wall, the duration of these regions grows and the intensity of perturbations inside them is increased.

At the next streamwise position (Fig. 5.9c, d), most of the free-stream part of the perturbations has already decayed and the main part of the initial disturbance develops inside the boundary layer. Thus, a decaying localized

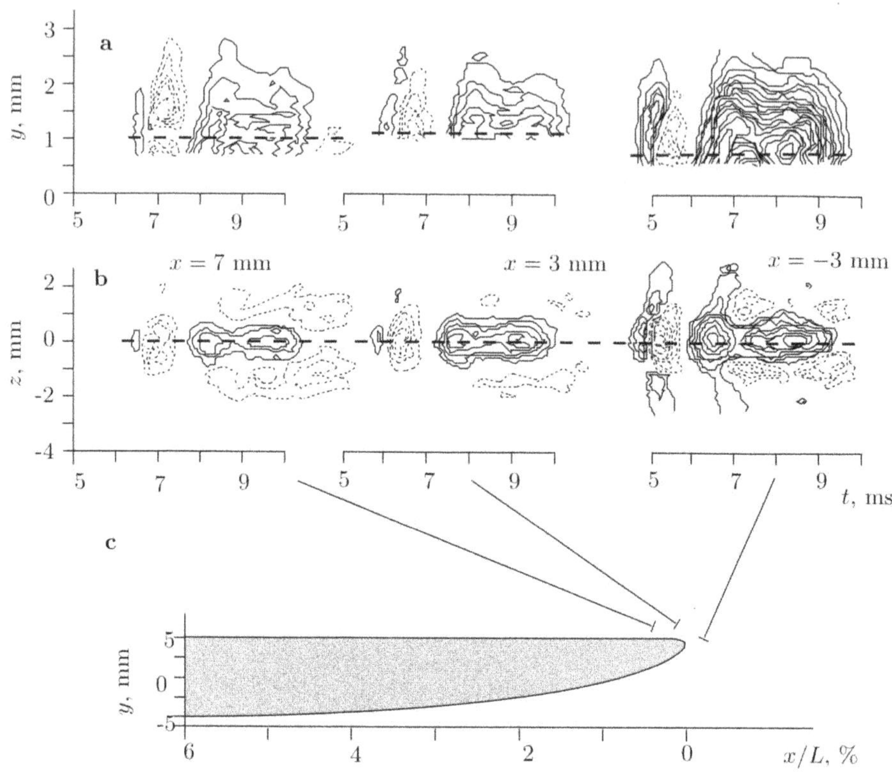

Fig. 5.8. Streamwise velocity u' distribution in wall-normal (**a**) and wall-parallel (**b**) planes at the leading-edge (**c**). Plate length $L = 2$ m. *Dashed lines*, corresponding sections through normal planes. Contour spacing: 0.01 of U_0 (Westin et al. 1998)

free-stream vortical disturbance forms a streak-like structure in the boundary layer consisting of a layer with a velocity excess and two layers with a velocity defect.

The downstream development of the streak shows no qualitative change. Figure 5.10 illustrates its different features further downstream. Most of the phenomenological characteristics of this disturbance are the same as those of the streaks observed at a high level of free-stream turbulence. The disturbance propagates with velocities equal to 0.8–0.9 and 0.5–0.55 of U_0 for leading and trailing fronts, respectively, which indicates continuous elongation of disturbance during downstream motion. The corresponding contours of the streamwise velocity fluctuations in (y–t) and (y–z)-planes are shown in Fig. 5.11. The spanwise size of the structure in this case is approximately two to three times the local boundary layer thickness δ, which is considerably smaller than its streamwise extent ($\sim 13\delta$ in last downstream measured point).

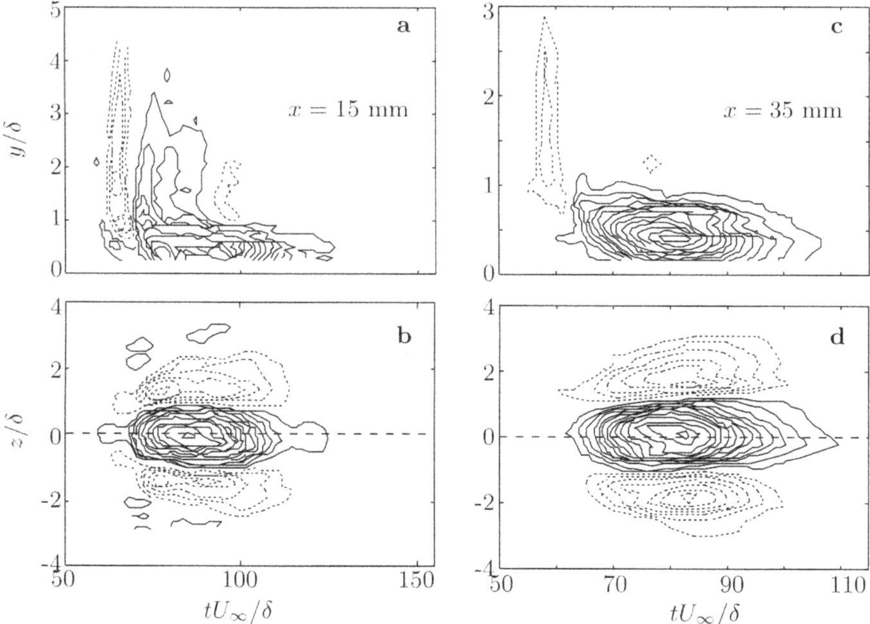

Fig. 5.9. Formation of the streak streamwise velocity component in boundary layer: (**a, c**) in (y–t)-plane at disturbance centreline; (**b, d**) in (z–t)-plane. Contour spacing: 0.01 of U_0. *solid lines* – velocity excess, *dotted lines* – velocity defects, *dashed lines*, corresponding sections through normal planes (Westin et al. 1998)

By averaging the localized disturbance in time and along the spanwise coordinate, 'typical' mean and disturbance velocity profiles are obtained, which are very similar to that widely observed in boundary layers at high free-stream turbulence (Fig. 5.12). Consequently, the majority of the streamwise velocity induced by the free-stream turbulence in the boundary layer can be associated with the puff development.

The experiments of Sboev et al. (1999a, b, 2000) showed that disturbances excited through a tube oriented at an angle to the free stream, i.e. with a predominant spanwise disturbance velocity, represent asymmetric streaks in the boundary layer. A comparison of these two cases is shown in Fig. 5.13a. In the asymmetric case, the streak can consist of only a single region of the defect and excess, with the axis of the disturbance as well as the direction of the puff propagation being inclined to the free-stream velocity vector.

To study puff generation and development in a gradient flow, controlled experiments both on straight and swept wings were carried out (Sboev et al. 1999a, 2000). The scheme of the excitation mechanism based on the studies (and closely related to the lift-up effect) for the case of flat plate and straight-wing receptivity is presented in Fig. 5.14. It can be seen that the boundary

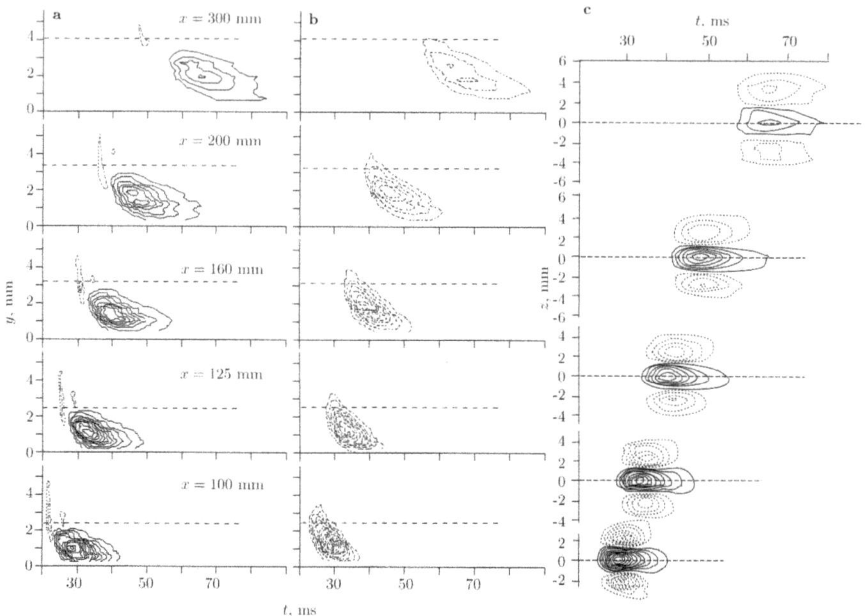

Fig. 5.10. Contours of streamwise velocity deviations in different downstream positions: in the plane of symmetry (**a**), at z with a maximum velocity defect (**b**), in wall-parallel plane at $y/\delta^* \approx$ 1–1.5 (**c**). *Dashed lines*, corresponding sections through normal planes. Contour spacing equal to 0.005 of U_∞ (Westin et al. 1998)

layer disturbance at the straight wing is much wider compared to a similarly generated disturbance on a flat plate.

In a three-dimensional boundary layer of a swept-wing model, the asymmetric puff arises under the effect of crossflow (Fig. 5.15). Comparison between the spanwise scales of the asymmetric disturbances in the boundary layers of the straight and swept wings (both models had identical radius of curvature for the leading-edge and angles of attack) shows that in the latter case the disturbance is even wider, which is evidently related in some way to the presence of crossflow.

5.2.2 Local generation of streaks from the wall

The main features of the receptivity of the flat plate boundary layer to the three-dimensional localized disturbances generated from the wall are the same, as discussed above. The response of the boundary layer to such an excitation consists of generating the streaks irrespective of the disturbance sources (by single pulses of sucking or blowing through a hole, or spanwise slot, or by surface vibration) over a wide range of excitation amplitudes (Bakchinov et al. 1997; Chernorai et al. 2000b; Spiridonov and Chernorai 2000). The detailed hot-wire measurements showed that the inner structure

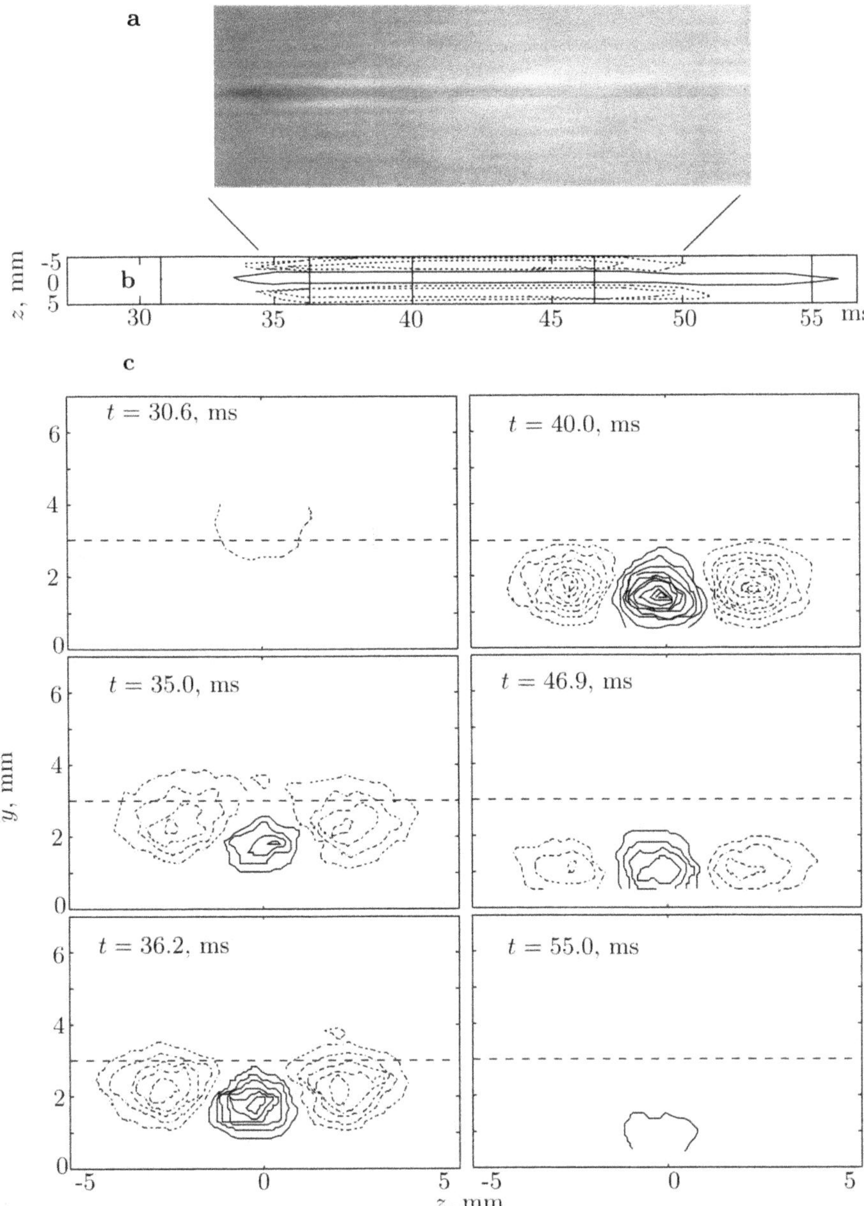

Fig. 5.11. Localized disturbance excited from the free stream: **a** smoke visualization of the disturbance in a boundary layer (Alfredsson et al. 1996); **b** contours of the disturbance streamwise velocity u' (Westin et al. 1998) in $(t$–$z)$-plane; **c** spanwise cross-sections in $(y$–$z)$-plane of streamwise velocity disturbances at some time instances (Westin et al. 1998). *Dashed line*, boundary layer outer border; contour spacing equal to 0.005 of U_0; x=160 mm, U_0=6 m/s

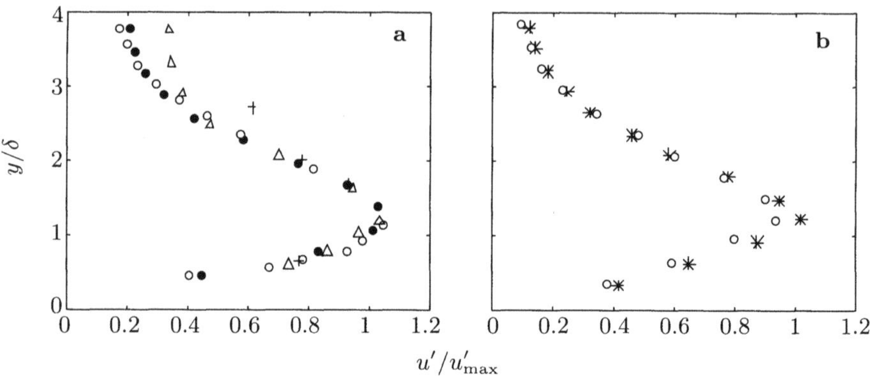

Fig. 5.12. Influence of free-stream turbulence Tu = 1.5%. Normalized profiles of u' measured at $x = 250$ (∘), $x = 500$ (•), $x = 800$ (△), $x = 1000$ mm (+) (**a**), and extracted from the localized disturbance at $\mathrm{Re}_{\delta^*} = 460$ by averaging over t and z (**b**). Averaging over spanwise distance corresponding to two (∗) and three (∘) streaks. The curves are normalized to the maximum values obtained when averaging over two streaks (Westin et al. 1994, 1998)

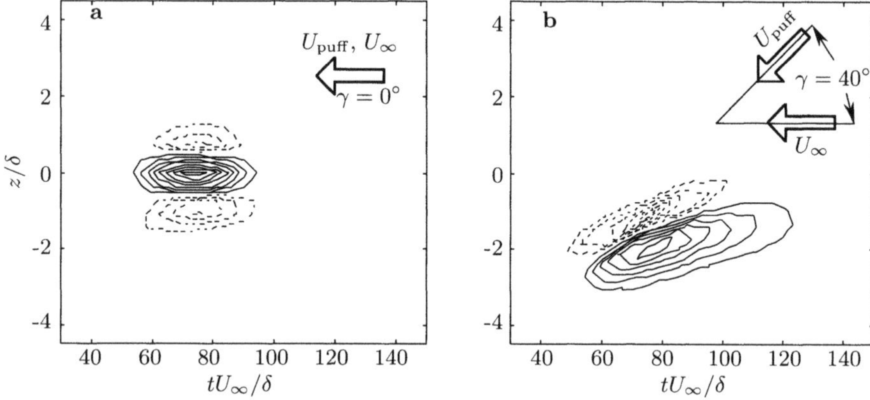

Fig. 5.13. Contours of u' for symmetric (**a**) and asymmetric (**b**) puffs in the flat plate boundary layer in the wall-parallel plane of disturbance maximum (Sboev et al. 1999b). $x = 114$ mm, $U_\infty = 6.6$ m/s. *Solid* and *dashed lines* show the excess and defect of velocity, respectively. Contour spacing equal to 0.01 of U_∞

of the boundary layer perturbations is qualitatively unaffected by the excitation magnitude, free-stream velocity and parameters of the disturbance source.

Similar results were obtained in the computational study by Breuer and Landahl (1990) and visualizations by Grek et al. (1990). The theoretical and experimental studies of, for example, Konzelmann (1990), Fasel (1990), Breuer and Haritonidis (1990) and Cohen et al. (1991) also revealed that the detailed structure of the localized disturbances in the boundary layer gener-

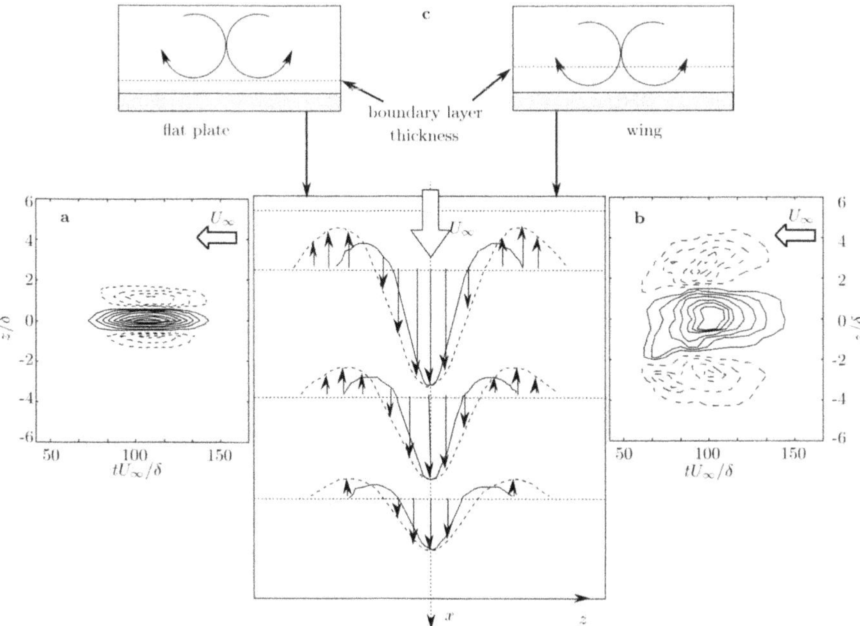

Fig. 5.14. Comparison of the boundary layer receptivity for flat plate (**a**) and straight wing model (**b**); scheme of spanwise velocity profiles over a flat plate (*solid lines*) and straight wing (*dashed lines*) (**c**); *dotted line*, boundary layer thickness (Sboev et al. 1999a). Contour spacing equal to 0.005% of U_∞ (Sboev et al. 1999a, 1999b)

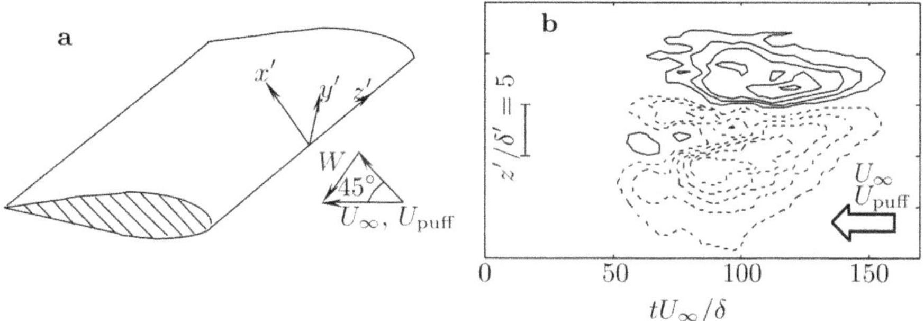

Fig. 5.15. Overview of the experimental set-up on a swept wing (**a**), and the asymmetric puffs structure (**b**). *Solid lines*, velocity excess; *dashed lines*, velocity defect. Contour spacing equal to 0.0025% of U_∞ (Sboev et al. 2000)

ally consists of narrow and intensive streaks of the velocity defects as well as wave packets of Tollmien–Schlichting waves. As the streaks spread faster than the waves, they separate from the wave packet and develop independently.

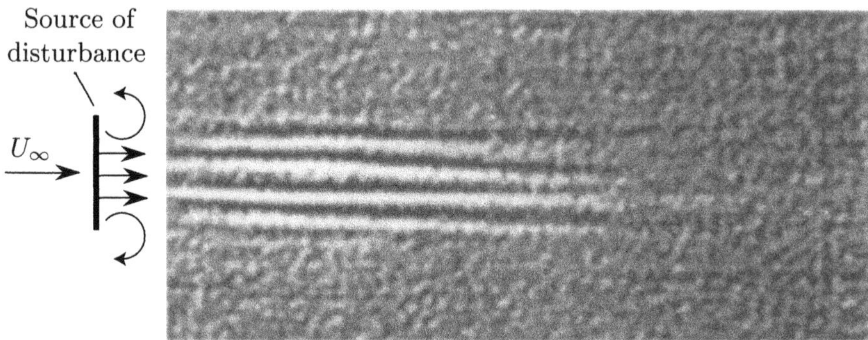

Fig. 5.16. Smoke visualization of the streamwise localized structures generated in a boundary layer of a flat plate by blowing air through a spanwise slot; Tu = 0.02% (Alfredsson et al. 1996)

Smoke visualization of the disturbance generated by blowing through a spanwise slot shows that narrow streamwise streaks appear in the boundary layer Fig. 5.16. Their generation occurs at the slot edges as it is outlined in that figure. Spectral analysis shows that the response of the boundary layer to blowing or sucking through a short spanwise slot results in three kinds of disturbances with variable periodicity in the spanwise direction. They include the streaks generated at the slot edges with their most energy concentrated at $\beta \approx \pm 0.5/\delta$, oblique Tollmien–Schlichting waves ($\beta \approx \pm 0.2/\delta$) and a two-dimensional Tollmien–Schlichting wave ($\beta = 0$) which rapidly decays in the region of subcritical Reynolds numbers. The oblique waves are generated at the streak periphery. The identification of these characteristic spectral components is shown in Fig. 5.17.

Qualitatively similar results were obtained for the disturbances generated by an impulse-like boundary layer suction (Bakchinov et al. 1997). However, the amplitude of the excited streaks during sucking is more than three times smaller than during blowing, and the regions of velocity excess and defect are exchanged (Fig. 5.18). Moreover, the streamwise structures during blowing are much narrower, because of differences in the streamlines at the sides of the slot. A corresponding scheme of the process of generation and downstream development of the streaks initiated by the suction and blowing through a spanwise slot is shown in the same figure.

The streaks initiated by Chernorai et al. (2000b) with vibration of a rectangular membrane on the flat plate surface are shown in Fig. 5.19, and for a straight wing by Spiridonov and Chernorai (2000) in Fig. 5.20. In contrast to the flat plate case, the oblique waves at the edges of the streaks can not be identified in the region of the favourable pressure gradient (Fig. 5.20b). However, at the adverse pressure gradient, Fig. 5.20c, the waves at the leading and trailing edges of the streak grow, whereas the intensity of the streak itself decreases.

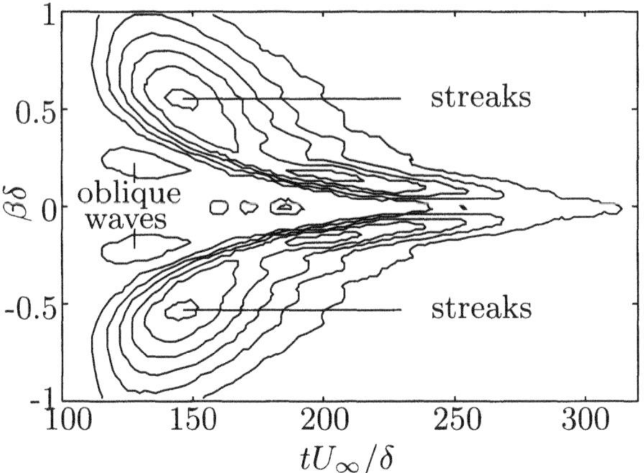

Fig. 5.17. Wave number spectra of disturbances $u'(\beta, t)$ (Bakchinov et al. 1997)

The effect of the free-stream velocity on the disturbance behaviour was considered by Bakchinov et al. (1997). As the velocity increases, the spanwise distance between the regions of the velocity defect and excess becomes smaller. A reduction of the transverse spacing between the naturally occurring neighbouring streaks with increasing flow velocity is also a characteristic feature that is observed at high free-stream turbulence (Fig. 5.21). The effect is obviously related to a decrease of the boundary layer thickness with increasing velocity (Alfredsson et al. 1996). However, the general characteristics of the development of the disturbances during sucking exhibit no major quantitative changes both in their structure and downstream evolution at different flow velocities. The studies of the integral characteristics of the streaks carried out later by Westin et al. (1998) demonstrate their reproducibility and qualitative identity in different wind tunnels at different free-stream velocities.

5.2.3 Distributed receptivity

The identity of the observed response of the boundary layer to the three-dimensional free-stream disturbances over a broad range of their amplitudes, experimental models, and facilities indicates a generality of the mechanism of the boundary layer distortion. However, the observed continuous decay of the amplitude of the locally introduced disturbance in all these cases is in contrast to the continuous downstream growth of the streamwise disturbance

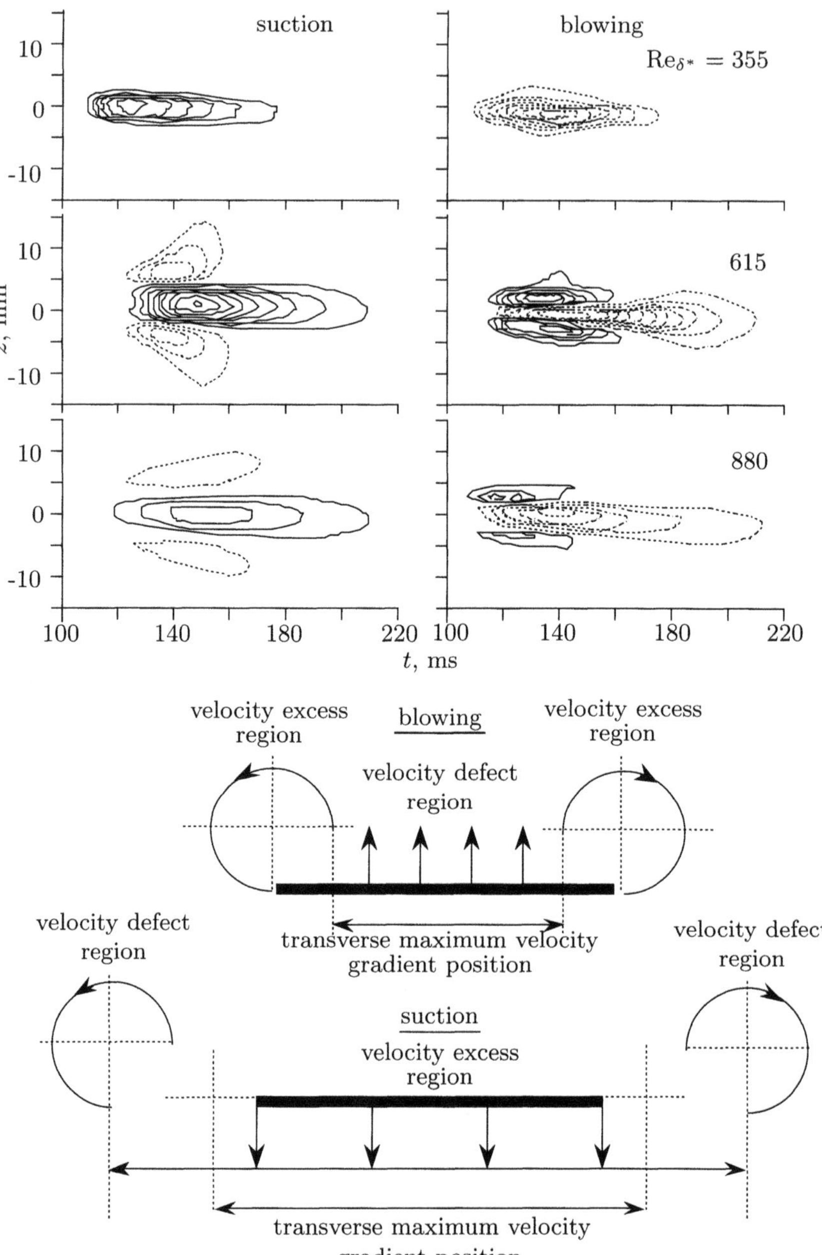

Fig. 5.18. Spanwise distribution of velocity u' measured at the disturbance maximum in a boundary layer: (*top*) (Bakchinov et al. 1998b); contour spacing at the *left* and the *right* plots are the same. Scheme of streak generation at the spanwise slot edges by pulses of fluid suction and blowing (*bottom*) (Bakchinov et al. 1997)

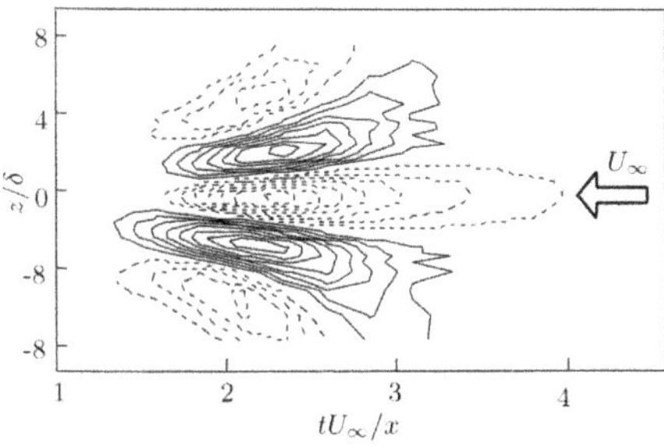

Fig. 5.19. Localized disturbances generated by membrane vibration on a flat plate. Localized disturbances structure (velocity excess and defect is 2% and 1.75% of U_∞, respectively). *Solid lines*, velocity excess; *dashed lines*, velocity defect (Chernorai et al. 2000b)

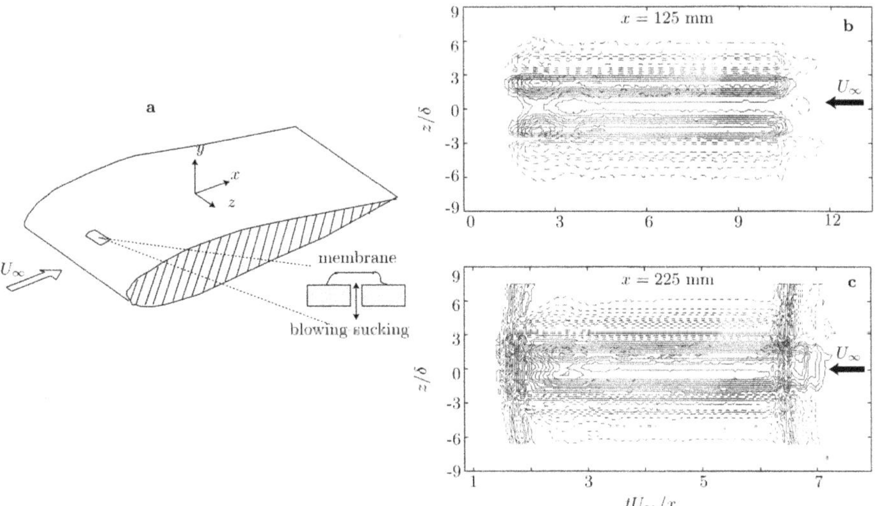

Fig. 5.20. Localized disturbances generation by membrane vibration on a straight wing: **a** overview of experimental set-up; **b** the structure of localized disturbances at a favourable pressure gradient (velocity excess and defect is 3.6% and 2.1% of U_∞, respectively); **c** the structure of localized disturbances at an unfavourable pressure gradient (velocity excess and defect is 4.2% and 3.6% of U_∞, respectively). *Solid lines*, velocity excess; *dashed lines*, velocity defect. $U_\infty = 6.6$ m/s, $y = y(u'_{\max})$ (Spiridonov and Chernorai 2000)

velocity observed in the presence of free-stream turbulence that supports the idea of distributed receptivity of the boundary layer in respect to free-stream vortices at high level of turbulence.

Similar conclusions regarding the inability of the localized receptivity mechanisms to explain the observed streak growth in the Blasius boundary layer were given by Bertolotti (1997), based on a consideration of both the experiments of Kendall (1991) and the asymptotic theory of Goldstein et al. (1992). Kozlov et al. (1990) and Kendall (1991) measured no difference in the disturbance amplitude for two different values of the leading-edge bluntness; i.e. it was independent of the flow structure in that region. It was also

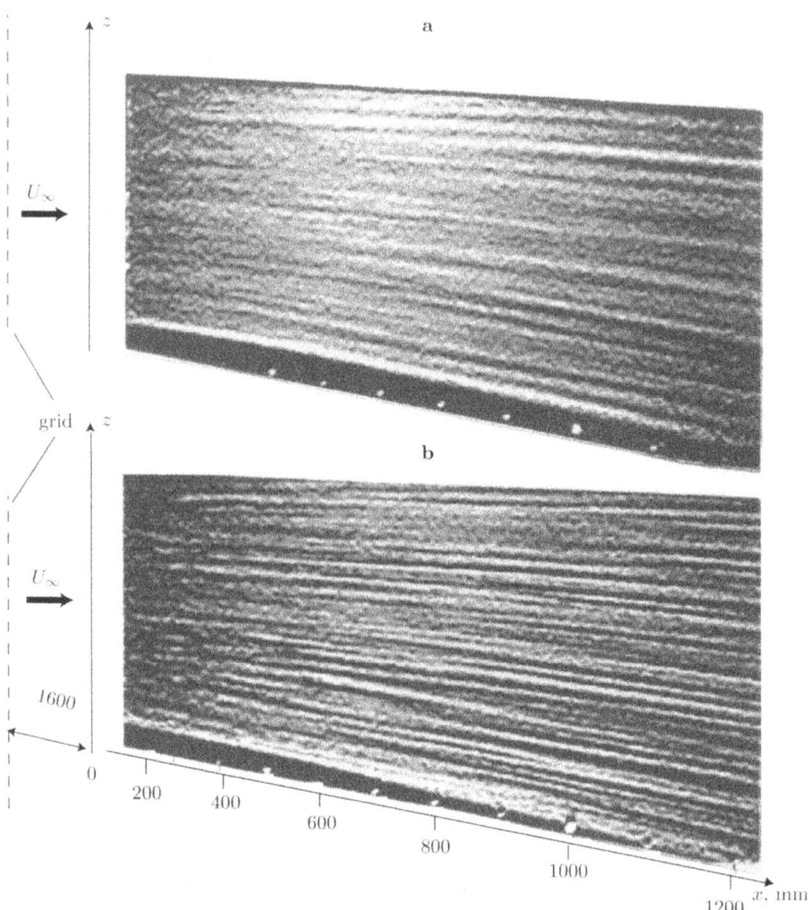

Fig. 5.21. Structure of a perturbed boundary layer at Tu = 1.5% near the a surface of a plate at various velocities of the free stream: **a**, 3 m/s; **b**, 9 m/s (Alfredsson et al. 1996)

argued that disturbance growth based on the local receptivity mechanism of the asymptotic theoretical model of Goldstein et al. (1992) occurs at the flow velocities and wind tunnel conditions of the reference experiments of Kendall (1985, 1990, 1991) and Westin et al. (1994) only at extremely large distances from the leading edge.

It is noteworthy that the free-stream three-dimensional inviscid disturbances are assumed to have scales comparable with the streaks and that they penetrate deep into the boundary layer, i.e. the mechanism of the wave-scale reduction is not necessary or is of less significance than for Tollmien–Schlichting wave excitation (Hultgren and Gustavsson 1981). It means that a quite effective distributive forcing of the streaks along the whole region of boundary layer development is not unlikely.

As a consequence, Bertolotti (1997) excluded the leading-edge effect from his receptivity theory. He linked the streak generation to the free-stream turbulence by incorporating non-parallel effects, and studied the distributed response of the Blasius boundary layer to two- and three-dimensional vortical modes satisfying the linearized Navier–Stokes equations in the free stream, which were used to represent some key features of low-level turbulence at low frequencies.

An experimental way to introduce certain controlled 'representative' disturbances to investigate the distributed receptivity from the free stream was found by Bertolotti and Kendall (1997). The idea is to produce a free stream vortex at the tip of a micro-wing. By varying the wing's angle of attack and the free-stream velocity, the vortex strength can be controlled. In contrast to wake-induced disturbances as in the studies of Kyriakides et al. (1999) and Wu et al. (1999), such an approach provides the fundamental possibility of investigating in detail and in a quiet environment the development of the stationary disturbances independently of the travelling-wave modes. The successful comparison between the Bertolotti's theory and the Kendall's measurements of the response of the boundary layer indicates that the receptivity of the flow to quasi-stationary disturbances could be modelled in such a way.

In the experiment of Boiko (2000), the micro-wing was located above the plate, rather then in front of it, so that the leading edge region was escaped. It was found that the excited stationary boundary layer disturbance has the same phenomenological characteristics as those which appeared in the boundary layer under the effect of high free-stream turbulence levels. A comparison with the study of Bertolotti and Kendall (1997) shows that the local processes at the leading edge can play no dominant role in the streak growth.

For this kind of non-modal distributive receptivity it is not easy to construct a local receptivity function, so that a description of basic experimental data seems more appropriate. The free-stream velocity distributions excited behind a micro-wing as obtained by Boiko (2000) are illustrated in Fig. 5.22. The distributions measured at different downstream locations indicate a slow downstream decay of the vortex accompanied by little expansion of its core, which is in accordance with previous observations (see e.g. Bippes 1977) and the theoretical model of Batchelor (1964).

The boundary layer response after a spatial Fourier transformation along the spanwise coordinate at the same downstream positions is shown in Fig. 5.23. It can be seen that the disturbance maximum is concentrated at fixed $\beta = 0.082$ and $y/\delta^* \approx 1.3$. This indicates that in the framework of the current experimental conditions, the physical scale of the excited structure does not depend on the boundary layer thickness in the region of the streak generation and propagation ($\delta^* = 1.09$–1.99). This is in accordance with the

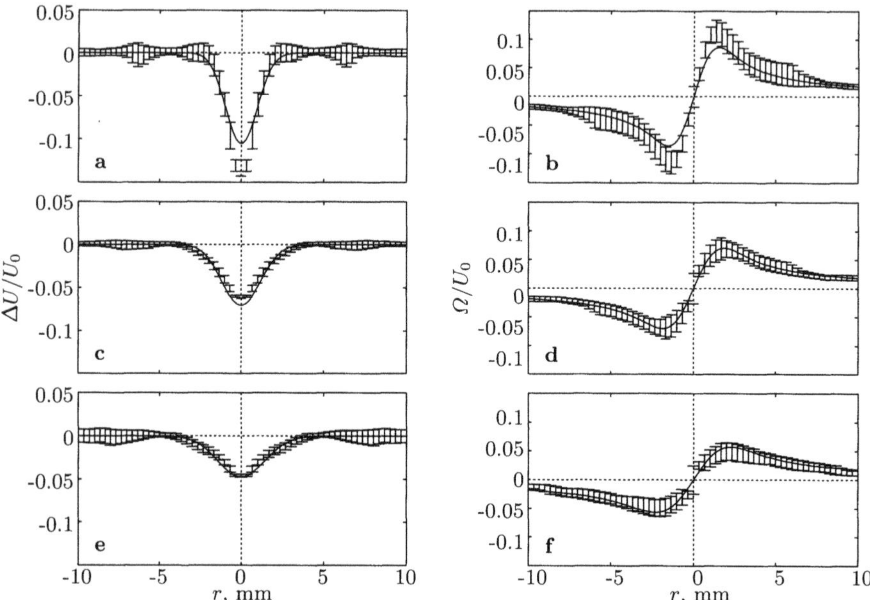

Fig. 5.22. Streamwise (**a, c, e**) and spanwise (**b, d, f**) velocity defects of the tip vortex at downstream distances from the micro-wing $\Delta x = 40$, 130 and 265 mm, respectively. $U_0 = 5.9$ m/s. *Points*, measured mean values; *bars*, error estimations (Boiko 2000). *Solid lines*, data approximation by the model of Batchelor (1964); $r = 0$ corresponds to the distance 15 mm to the wall; micro-wing is located at $x_0 = 215$ mm

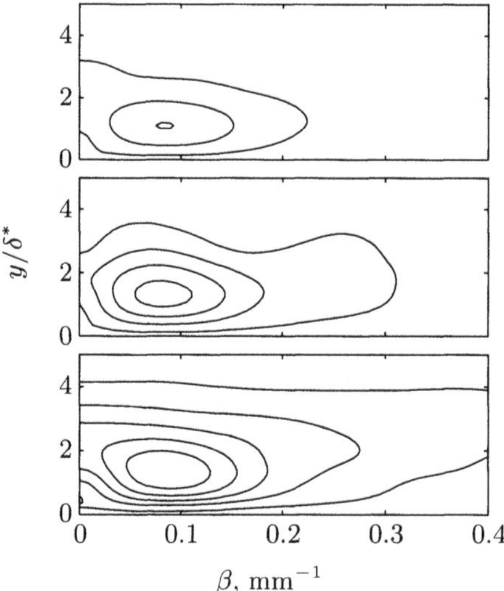

Fig. 5.23. Spectral distribution of ΔU at $U_0 = 5.9$ m/s: $\Delta x = 40$, 130 and 265 mm (from *top* to *down*). Isolines with equidistant spacing from 0.1 to 1.1 (Boiko 2000)

measurements of Matsubara and Alfredsson (2001) at Tu $\approx$ 1.5%, where stationary velocity modulations maintained their spanwise scale during the propagation downstream. Similarly, in a model experiment of Westin et al. (1998), excited boundary layer streaks did not scale with δ^*.

Amplitude profiles for different values of β together with data from some other experimental and theoretical studies are shown in Fig. 5.24. They demonstrate a self-similarity of the profiles and their identity to that of the optimal disturbances and the streaks excited experimentally by different other means (cf. Figs. 1.19 and 5.6). This remarkable similarity of the disturbance shapes found in different experimental facilities and flow conditions, as well as in different theoretical models, has been mentioned repeatedly in the literature (Breuer and Haritonidis 1990; Westin et al. 1994; Bertolotti and Kendall 1997) and discussed extensively by Luchini (1996).

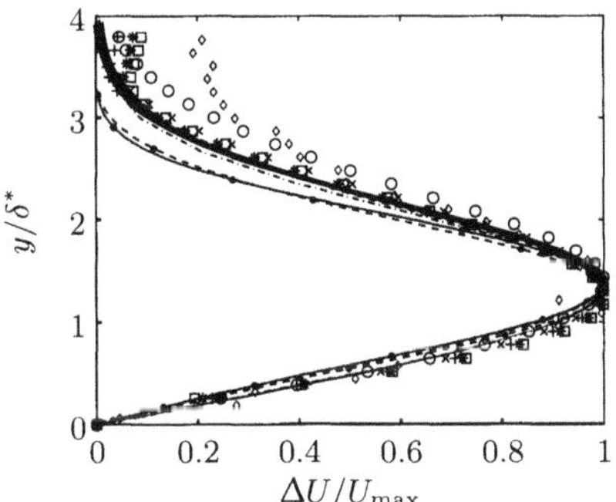

Fig. 5.24. Disturbance velocity at U_0 = 5.9 m/s, Δx = 130 mm for individual β: $\circ$, 0.050; $\times$, 0.075; $+$, 0.100; $*$, 0.125; $\square$, 0.150 mm^{-1} (Boiko 2000). *Bold solid line*, amplitude profile of quasi-stationary variations of boundary layer thickness; *solid line*, Breuer and Haritonidis (1990); *dash-dotted line*, data of Andersson et al. (1999); $\diamond$, Westin et al. (1994)

Klebanoff's observation that the streamwise velocity component of the natural streaks at a high level of free-stream turbulence resembles quasi-stationary modulations of the Blasius mean velocity profile, when the local boundary layer thickness experiences small changes (Kendall 1985), referred to as the 'breathing'. Such breathing may be caused by, for example, small-amplitude temporal oscillations of the stagnation line at the leading edge or of the free-stream velocity (Taylor 1936). The shape of this small-amplitude two-dimensional breathing has an analytical expression, $y\partial U/\partial y$ (Libly and Fox 1964; Stewartson 1969), being a solution of linearized boundary layer equations (see Sect. 2.1.3). It is shown in Fig. 5.24 normalized to its maximum. It can be seen that it repeats the amplitude shapes of individual three-dimensional stationary 'modes' almost perfectly.

It is quite impressive that the shape of the stationary modes can be obtained from such a simple two-dimensional quasi-stationary treatment. However, the two-dimensional approach can explain neither the observed disturbance amplitude nor their growth, since the Blasius boundary layer thickness depends on the square root of the free-stream velocity and the distance from the leading edge, and small changes in these values are felt less and less by the boundary layer further downstream. This means that the two-dimensional breathing is always decaying, with the possible exception of the region very close to the leading edge, where the Blasius boundary layer approximation is invalid. This once again indicates that the observed growth is purely three-dimensional effect.

5.3 Transition mechanisms at high free-stream turbulence

Whereas the streak characteristics are quite well documented, their role in the transition is less clear. For a long time it was considered that the transition at high free-stream turbulence is closely related to local boundary layer separations induced by free-stream fluctuations at leading edges, so that the turbulent spots are generated and their spatial evolution results in a completely turbulent boundary layer. Such a hypothesis was expressed by Taylor (1936) and supported by the experiments of Dryden (1934).

U_∞

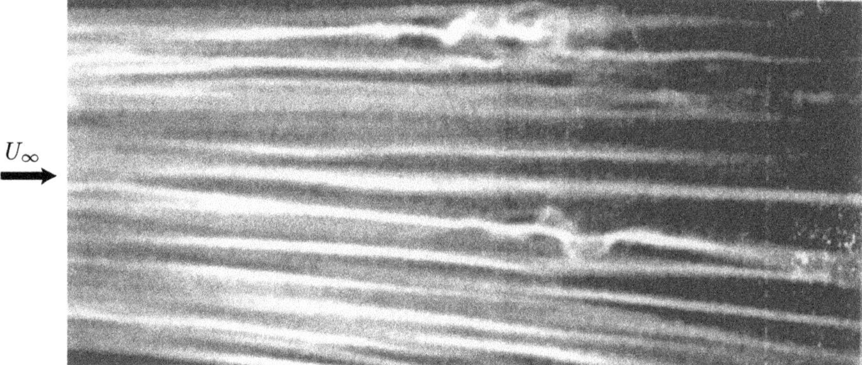

Fig. 5.25. Smoke visualization of generation of the high-frequency wave packets during natural transition on a flat plate at Tu = 1.5% (Alfredsson et al. 1996)

However, provided that the initial amplitude of a streak is large enough, it is also capable of transforming into a turbulent spot. In the smoke visualization (Fig. 5.25) the streaks begin to oscillate in the spanwise direction before the final flow becomes turbulent. The oscillations can be attributed to a secondary instability of the streaks to certain travelling disturbances. The similar spanwise oscillations are also observed in the last stages of subcritical transition in Couette and Poiseuille flows (Henningson 1995). One can conjecture that the streak breakdown mechanism caused by the disturbances could be similar to the secondary instability of Görtler or crossflow vortices (see Chap. 4).

The flow visualization of Kendall (1985) demonstrates the evolution of naturally occurring streaks into pairs of oblique shear layers and their rolling-up in Λ-like vortices. The wedge-shaped core and the strong shear layers of the puffs excited in controlled conditions are also reminiscent of the Λ-structures in K- and N-regimes of the transition (see Chap. 4). Analogously, the numerical simulation of boundary layer under the effect of a high free-stream turbulence of Rai and Moin (1991) provides examples of the development of strong shear layer, which seems very similar to that found in K- and N-regimes of the transition, and in the transition process directly provoked by the lift-up effect. As at the transition stipulated by Tollmien–Schlichting waves, one can expect in this case the existence of competing non-linear mechanisms responsible for the transition to turbulence.

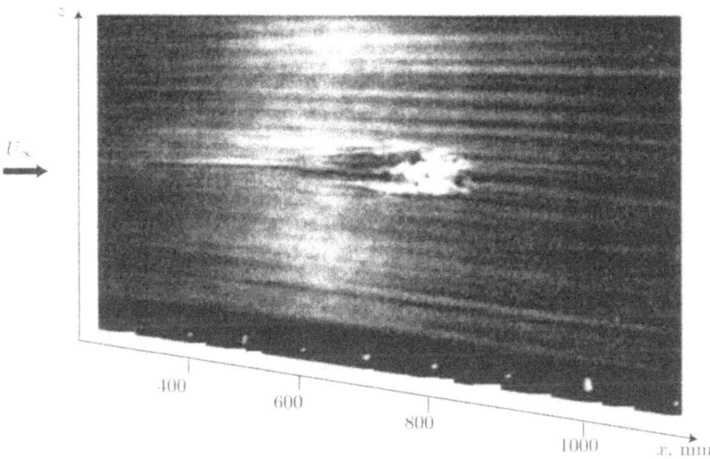

Fig. 5.26. Incipient spot smoke visualization at the natural transition on a flat plate Tu = 1.5% (Alfredsson et al. 1996)

A smoke visualization of pre-turbulent 'incipient' spot resulted from non-linear activity of the streaks at the 'natural' transition at high level of Tu is shown in Fig. 5.26. such incipient spots have been extensively studied theoretically, experimentally and numerically (Amini and Lespinard 1982; Grek et al. 1985, 1990; Breuer and Landahl 1990; DeBruin 1990; Klingmann 1992; Henningson et al. 1993; Bakchinov et al. 1994b). Below we consider the mechanisms of streak interaction with other transitional structures, which are can lead to formation of such structures in more detail.

5.3.1 Tollmien–Schlichting waves in the presence of streaks

'Natural' Tollmien–Schlichting waves were detected at moderate levels of free-stream turbulence, by Arnal and Juillen (1977, 1978), Kosorygin and

Polyakov (1990), Kendall (1985, 1990, 1991), Suder et al. (1988), Arnal (1992) and others. These studies showed that the transition in a boundary layer on a flat plate at Tu $\leqslant 0.7\%$ can occur through the development of Tollmien–Schlichting wave packets, which were observed together with the streaks discussed earlier. However, the instability waves packets were not found for a long time in experiments at Tu $\geqslant 1\%$. It was suggested that at free-stream turbulence levels above 1%, the waves play no role in transition and another, 'bypass' mechanism is at work which was associated with the presence of the streaks (Morkovin 1984; Suder et al. 1988).

This led to a discussion concerning the role of Tollmien–Schlichting waves in transition induced by high free-stream turbulence. The growth rate of the low-frequency fluctuations, in general, increases with Tu, but the their intensity at the onset of transition differs widely between different experiments, even between experiments carried out at similar values of Tu. This means that the transition point cannot be directly inferred from the intensity of the streaks, and probably not only the scale of the free-stream disturbances, but also some other perturbations like Tollmien–Schlichting waves affects their breakdown.

Generally, the extent to which the Tollmien–Schlichting waves play a role in a particular case may depend on a number circumstances, such as the efficiency with which they are triggered from the free stream, the relevant Reynolds number range, pressure gradients and leading-edge conditions. As it has been discussed in Chap. 2, pressure gradients along the flow strongly affect the amplification of Tollmien–Schlichting waves. The low-frequency fluctuations induced by high free-stream turbulence, however, seem to be insensitive to pressure gradients and leading edge conditions (Kozlov et al. 1990; Kendall 1991). It is therefore noteworthy that in an experiment at Tu $= 2\%$, Blair (1992) observed a significant delay in the onset of transition with increase of negative angle of attack of the plate. If the insensitivity of streaks to the pressure gradients is of general validity, the effect observed by Blair (1992) seems to be a result of an efficient suppression of Tollmien–Schlichting waves by the negative pressure gradient. This gives an indication of the important role that Tollmien–Schlichting waves may also play in the transition of a flat plate boundary layer at high free-stream turbulence levels.

The possibility of generating and studying the Tollmien–Schlichting wave under controlled conditions in both gradientless and gradient flows at free-stream turbulence level up to Tu $\approx 4\%$ of U_∞ was shown for the first time by Grek et al. (1987, 1989, 1991b, c). It was revealed that the streaks and turbulent spots appearing between the Tollmien–Schlichting wave in fact affect neither the main averaged characteristics of its wall-normal amplitude and phase profiles, nor the downstream propagation velocity. It was found that the boundary layer receptivity to the disturbances over a wide range of free-stream turbulence levels remains practically unchanged both with favourable and unfavourable pressure gradient flows.

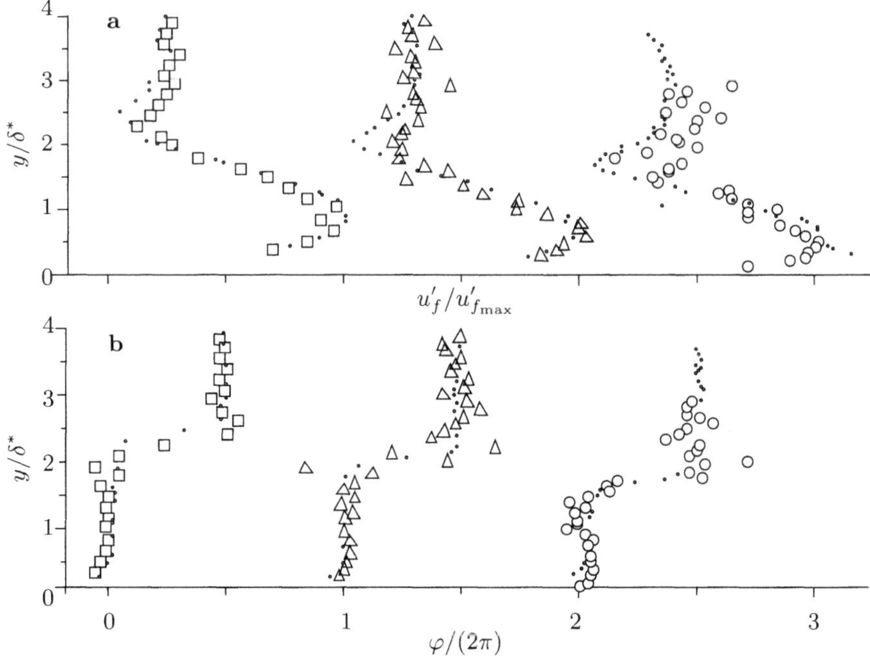

Fig. 5.27. Amplitude (**a**) and phase (**b**) profiles of a Tollmien–Schlichting waves at high free-stream turbulence level: $F = 200 \times 10^{-6}$, $\mathrm{Re}_{\delta*} = 345$ (□), 525 (△), 715 (○); *points*, profiles measured without a grid (Boiko et al. 1994)

These results were advanced in the experiments by Boiko et al. (1994). In this study, development of a small-amplitude two-dimensional Tollmien–Schlichting wave excited by a vibrating ribbon in the presence of high free-stream turbulence was investigated. The amplitude and phase distributions of the waves, extracted using a phase filtering technique, are shown in Fig. 5.27. At Tu = 1.5% the overall value of u'/U_∞ in the middle of the boundary layer was 5% of U_∞ at the most downstream position. In spite of this, the extracted wave exhibits the same characteristic features as the Tollmien–Schlichting wave in an undisturbed boundary layer. The contribution of the random fluctuations increases downstream, resulting in a certain scatter in the data points. This is particularly obvious close to the amplitude minimum and the corresponding phase shift. It was found that in these and similar measurements, the dependence of the phase on the streamwise coordinate could be approximated by straight lines, and the wave phase velocity was the same as at low Tu.

Amplification curves for the wave at low and high free-stream turbulence are shown in Fig. 5.28. The study indicated that the decrease of the amplification rate is associated with a delicate mechanism caused by the presence of the streaks. Moreover, the amplification rates of the Tollmien–Schlichting

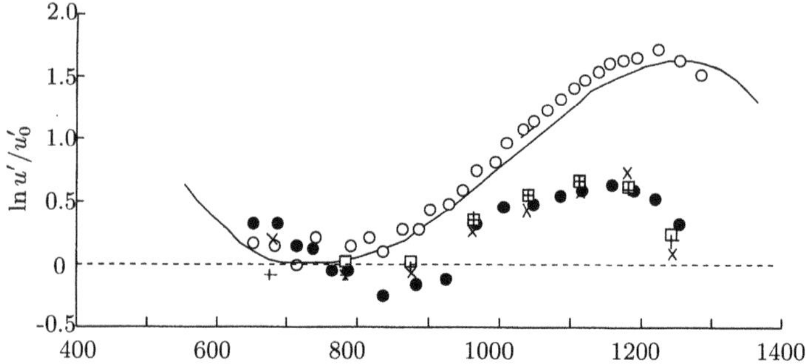

Fig. 5.28. Amplitude evolution of the Tollmien–Schlichting waves normalized to wave amplitude at branch I of the neutral curve: ∘, measurement without a grid; •, □, ×, +, measurements with the grid at different amplitudes of excitation; *solid line*, calculation by PSE; $F = 100 \times 10^{-6}$ (Boiko et al. 1994)

wave appeared to be independent of the experimental conditions of the excitation amplitude (Fig. 5.28b). Hence, the waves seem to amplify linearly. Although the vibrating ribbon initially sets up a two-dimensional wave front, randomly moving large-scale structures may lead to three-dimensional deformations of it, which cause local changes in both the frequency and phase of oscillations.

The above observations demonstrate that different kinds of interactions may be present which could transfer energy between the waves, or between large-scale flow structures and the waves. Some indication of the nature of such interactions may be obtained by analyzing power spectra of disturbances obtained in the boundary layer.

Figure 5.29a shows the spectra at two downstream positions without forcing, where the oscillations decay in the frequency band of the Tollmien–Schlichting waves. Figure 5.29b shows the corresponding spectra, where the amplitude of the forcing wave was relatively small. It is seen that the energy increases in the frequency band where the Tollmien–Schlichting waves are unstable. However, with growth of the forcing amplitude as in Fig. 5.29c, the frequency bandwidth of the increasing forcing disturbances is also increased. For the highest excitation amplitude, as in Fig. 5.29d, a change of the disturbance energy is seen at the upstream position. It is noteworthy that at both positions, the amplitude of the phase-coherent part of the wave is only about 0.35% of U_0, and the amplification rate does not seem to be affected by non-linear effects (Fig. 5.28b). Thus, as the forcing wave amplitude grows, the energy is increased over a broad frequency range; that is the energy increase can be attributed to the generation of Tollmien–Schlichting waves at frequencies other than the forcing one.

During the measurements it was observed that the number of turbulent spots and transitional flow structures at $\mathrm{Re}_{\delta^*} = 1260$ increased drastically with growth of excitation amplitude. For the highest excitation amplitude,

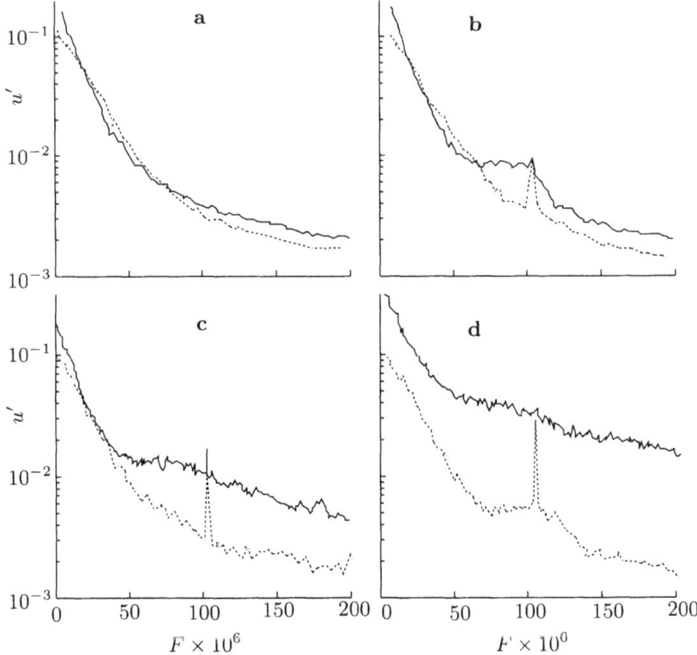

Fig. 5.29. Comparison between upstream and downstream spectra at $y/\delta^* = 0.5$ (Boiko et al. 1994): **a** without forcing; **b**–**d** different forcing amplitudes (in order of increasing amplitude) for $F = 100 \times 10^{-6}$. *Dotted line*, $\mathrm{Re}_{\delta^*} = 890$; *solid line*, $\mathrm{Re}_{\delta^*} = 1260$

90% of the collected records were considered to be affected by intermittent turbulence. The observed energy growth therefore represents an increase of random fluctuations in the laminar portions between the turbulent spots. Similar interaction processes may be expected to also occur when Tollmien–Schlichting waves are induced naturally by high free-stream turbulence. If these occur randomly in time and space, turbulence will not appear gradually, but rather in the form of localized turbulent spots.

This indicates that unsteady and three-dimensional effects have to be taken into account. If the large-scale perturbations are considered as randomly occurring streaks, the boundary layer may be considered locally as a spanwise-modulated flow. It could therefore be instructive to compare with experiments made in boundary layers possessing periodic spanwise non uniformity (see Sect. 4.5). It is known, that in such cases the mode shapes, phase velocities and amplification rates depend not only on the frequency and Reynolds number but also on the intensity and spanwise scale of the modulation. This means that in a boundary layer subjected to high free-stream turbulence, the Tollmien–Schlichting waves amplification rates may be expected to depend on the amplitude and typical spanwise scale of the disturbance in a boundary layer, which in turn depend on the free-stream characteristics.

5.3.2 Interaction of the streaks with Tollmien–Schlichting waves

The interaction of a natural streak with a two-dimensional forced Tollmien–Schlichting wave is illustrated by the smoke visualization in Fig. 5.30. As a result of the interaction, a non-linear wave packet occurs on the streak. The resulting disturbance transforms downstream into the turbulent spot.

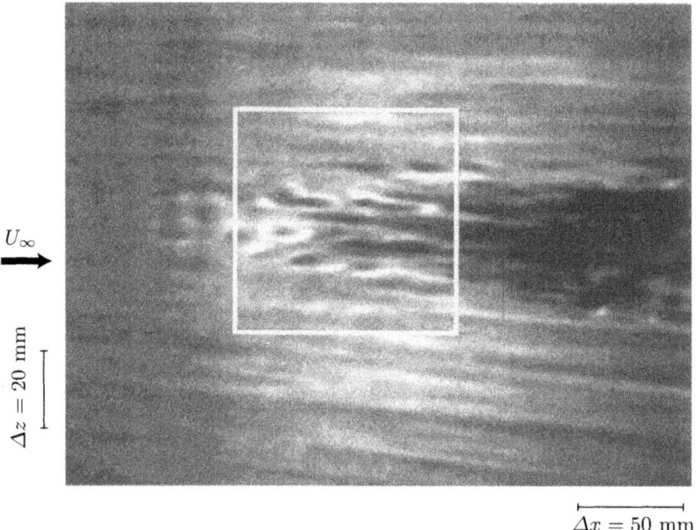

Fig. 5.30. Smoke visualization of the interaction between a forced Tollmien–Schlichting wave and a natural streak in a boundary layer: **a** general plan and **b** enlarged image of the generated disturbance at Tu = 1.5% (Matsubara et al. 1996)

An interactions between model streaks and the Tollmien–Schlichting waves was studied by Grek et al. (1991c), Grek and Kozlov (1992) and later by Westin and Henkes (1997). The interaction also produced a non-linear wave packet which gradually developed into the turbulent spot. This process was also observed when both the model streak and the waves decayed when they were generated separately.

In the experiments of Grek and Kozlov (1992), when the Tollmien–Schlichting wave amplitude was relatively small, its interaction with the oblique waves of the streak, indicated in Fig. 5.17, generated a non-linear wave packet with central frequency close to the subharmonic of the Tollmien–Schlichting wave. The wave packet was generated in the region of the streak trailing

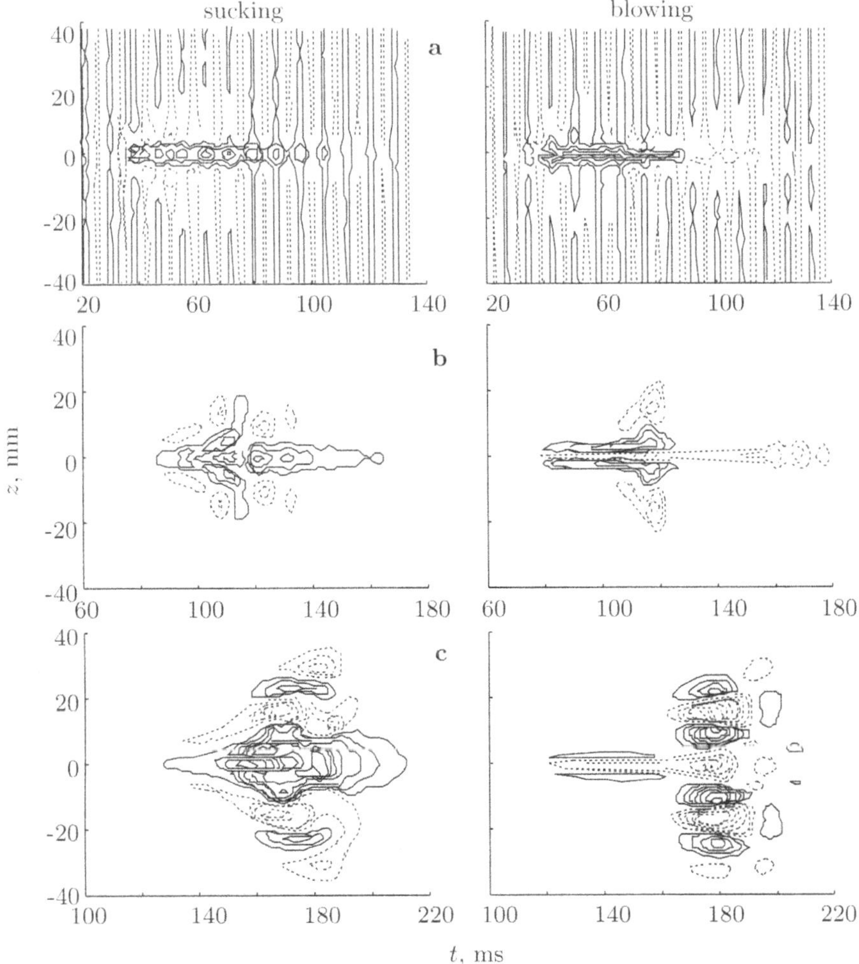

Fig. 5.31. Contours of the velocity fluctuations at the interaction of the localized disturbances and the Tollmien–Schlichting wave: **a**, $x = 200$; **b**, 400; **c**, 600 mm. Disturbance source was positioned at $x = 95$ mm, $U_0 = 6.6$ m/s. Levels of isolines equal to 0.01 from U_0; $y = y(u'_{\max})$ (Bakchinov et al. 1998b)

edge, and since it possesses a smaller propagation velocity, lags behind it. This illustrates one possible mechanism of the interaction.

However, this mechanism seems specific for single streak development, and it can hardly be dominant in a natural case. Due to this, another mechanism is preferred for when – provided the Tollmien–Schlichting wave or streak amplitude relatively large – the packet was generated inside and travels together with the streak, rather than behind it. Such interaction between the other streak spectral component (see Fig. 5.17) and the Tollmien–Schlichting waves was studied by Grek et al. (1991a) and Bakchinov et al. (1998a). Interestingly enough, due to the different intensity of the streaks generated by blowing or sucking, there are such regimes, when during blowing, the Tollmien–Schlichting wave interacts with oblique waves, and during suction, the interaction occurs directly on the streak (Fig. 5.31).

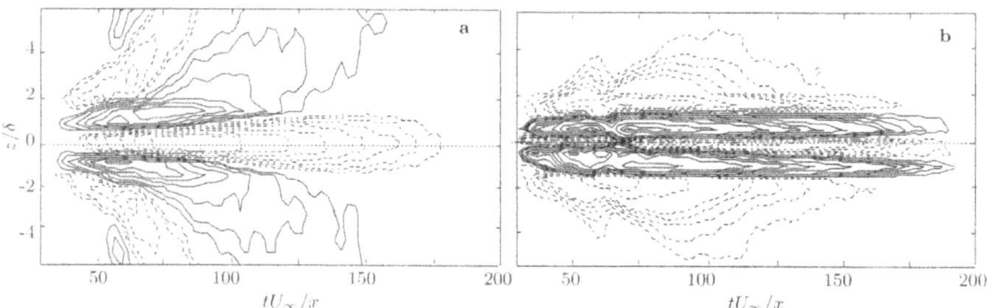

Fig. 5.32. Contours of the velocity fluctuations for the localized disturbance (**a**) and interaction between the localized disturbance and the high-frequency disturbance (**b**) in the (z–t)-plane: *solid lines*, velocity excess; *dashed lines*, velocity defect (Bakchinov et al. 1998a)

Figure 5.32 shows the structure of disturbance that appears downstream during separate and simultaneous (before it final transformation into turbulent spot) development of the wave and the streak. It can be seen that the intensity of the streak is increased more than 10 times due to the interaction. It is characteristic that the maximum of the disturbances excited by the interaction is located in the region of the maximum spanwise velocity shear, that is in the region subject to secondary instability of stationary vortices considered in Sect. 4.5. Hence it can be supposed that the breakdown mechanisms both of the streamwise stationary vortices and propagating streaks have much in common, and are connected to the origin and development of the secondary high-frequency disturbances.

5.3.3 Turbulent spots

Grek et al. (1987) investigated the development of turbulent spots in a boundary layer disturbed by Tollmien–Schlichting waves and streaks. They showed that the presence of the disturbances does not affect the averaged turbulent spot characteristics such as its shape and size, indicating that the spots are boundary layer structures that have similar characteristics in the presence of different external disturbances (see Fig. 5.33). Later Bakchinov et al. (1994a) also detect no influence of the turbulent spots upon such Tollmien–Schlichting waves characteristics as the growth rates and phase velocities. The same conclusion was made by Alfredsson et al. (1996) and Matsubara et al. (1996) in their measurements at $\mathrm{Tu} < 0.02\%$ and $\mathrm{Tu} = 1.5\%$.

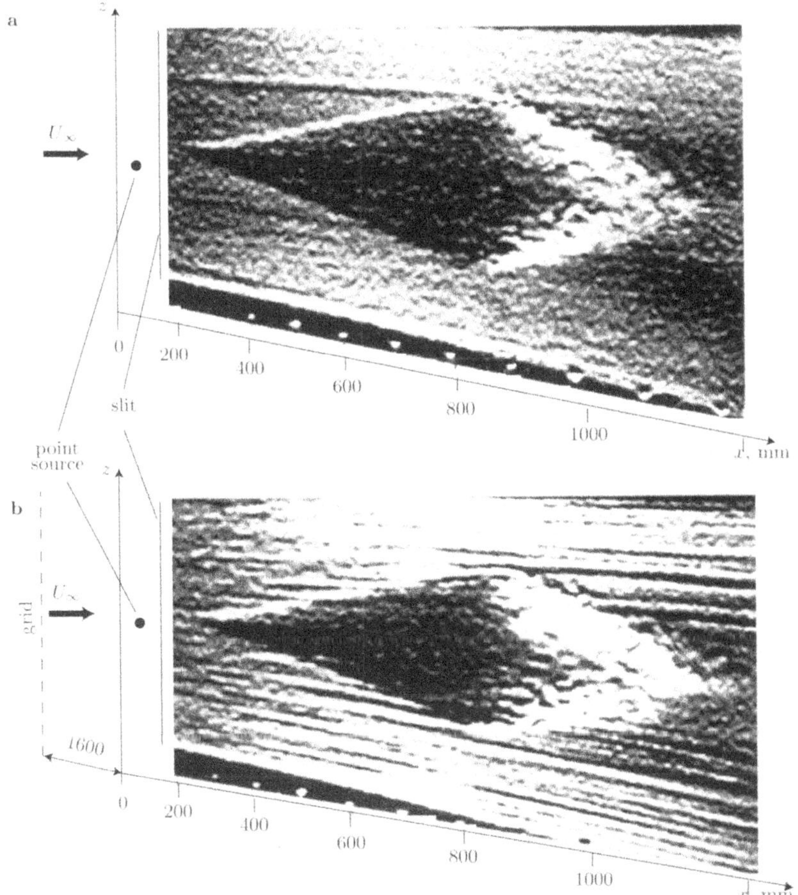

Fig. 5.33. Smoke visualization of the turbulent spot in a flat plate boundary layer at low (**a**); and high (**b**) free-stream turbulence (Matsubara et al. 1996)

Figure 5.34 shows the structure and propagation velocity of the ensemble-averaged turbulent spot fronts at low and high free-stream turbulence measured in its plane of symmetry for different streamwise positions (Grek et al. 1987). The turbulent spot structure in the plane of symmetry for both situations includes the velocity excess located near the wall and the velocity defect spreading above the boundary layer. The velocity defect region protrudes far above the laminar boundary layer bound δ. It can be seen that the fronts of the turbulent spot in the both cases practically coincide. The presence of the other spots does not affect the propagation velocities of its leading and trailing edges.

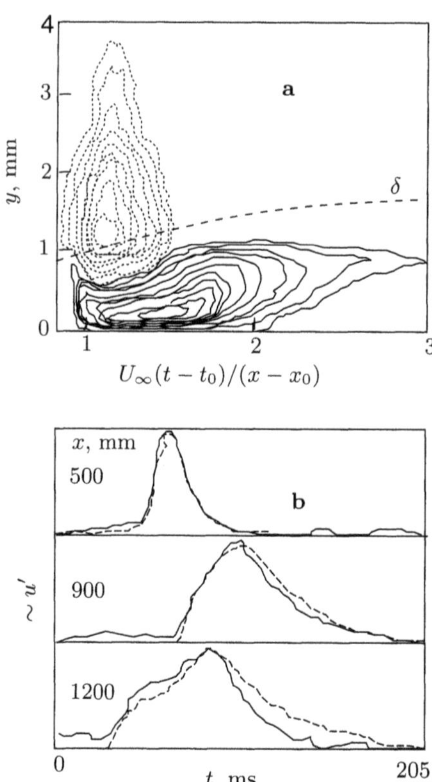

Fig. 5.34. Development characteristics of the ensemble-averaged turbulent spot at low and high free-stream turbulence: **a** the turbulent spot structure along its centreline; **b** oscillograms demonstrating propagation characteristics of the turbulent spot fronts. *Solid line*, Tu $\leqslant$ 0.04%, *dashed line*, Tu = 1%, x_0 = 300 mm, U_0 = 10 m/s (Grek et al. 1987, 1988)

5.3.4 Transition scenario at a high free stream turbulence

Based on the described investigations, a scenario of the transition at high free-stream turbulence may be proposed as a sequence of stages, as is done for transition at low free-stream turbulence (Fig. 5.35). The first of them is the formation of streaks by free-stream-localized vortical disturbances. The streaks modulate the boundary layer in the spanwise direction. The second stage includes the following streak development accompanied by the generation of high-frequency wave packets and incipient spots due to different non-linear mechanisms including the interaction with Tollmien–Schlichting waves and secondary instability. The third stage of the transition includes development and interaction of the turbulent spots which merging finishes the laminar–turbulent transition.

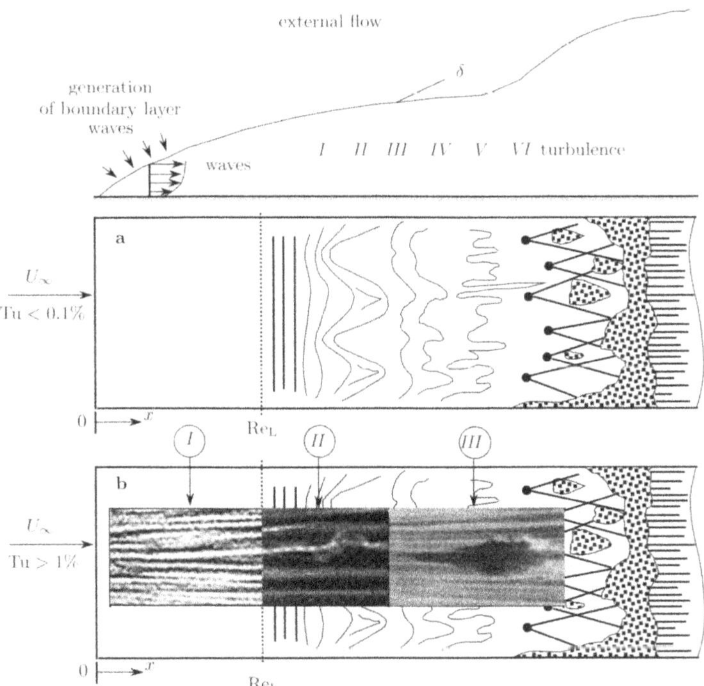

Fig. 5.35. The transition scenarios at low (**a**) and high (**b**) free-stream turbulence (Alfredsson et al. 1996). **a** *I*, linear instability stage of the Tollmien–Schlichting waves; *II*, development of three-dimensional Λ-structures; *III*, formations of the streamwise vortical structures; *IV*, appearance of strong shear layers; *V*, region of the turbulent spots origination; *VI* interaction and merging of turbulent spots. **b** *I*, development of the streaks; *II*, origination of incipient spots; *III*, formation, interaction and merging of turbulent spots

6 Transition to turbulence in separation bubbles

6.1 Problem formulation, substantiation and approaches

This chapter focuses on instability and laminar–turbulent transition in local regions of boundary layer separation or 'separation bubbles' in the steady flow of an incompressible fluid. The present topic applies to aerodynamics of aerofoils and wings at low Reynolds numbers, boundary layers affected by steps, humps and other surface imperfections, flow separation at sharp edges, etc.

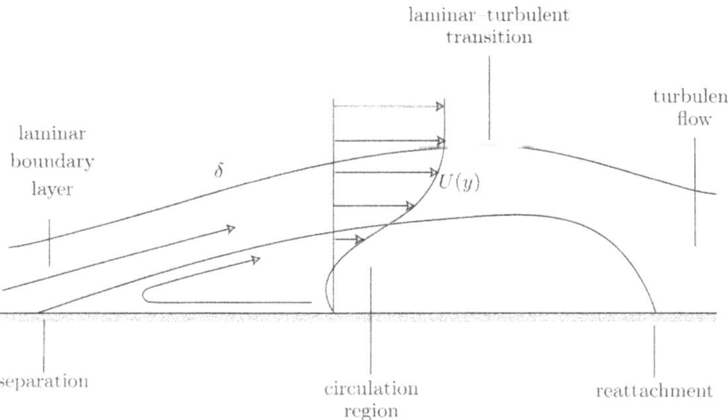

Fig. 6.1. Sketch of a transitional separation bubble

The first reason for considering the flow instability in separation bubbles is their destabilizing influence upon near-wall layers. Already at rather small Reynolds numbers, separation provokes an increase of velocity perturbations and laminar flow breakdown, taking place in the separation region or close to it. As a result, the transitional separation region occurs. Its main features are sketched in Fig. 6.1; for other details on the flow structure see, for example, the papers by Ward (1963), Horton (1967) and Brendel and Mueller (1988). Basically, rather fast turbulization behind the point of separation is caused by instability which is much greater than that in the attached boundary layer.

In terms of a local stability analysis, this is explained by the appearance of mean velocity profiles with an inflection point (see Chap. 1). In an asymptotic theory, interrelation between separation and onset of instability was shown by Smith (1979b) and Zhuk and Ryzhov (1980). Numerically, one can not obtain stationary solutions of the Navier–Stokes equations when the Reynolds number and local pressure gradient inducing flow separation increase (Briley 1971; Bestek et al. 1989, 1993; Pauley et al. 1990; Tafti and Vanka 1991; Dallmann et al. 1995).

Another motivation for transition research is the need for reliable prediction of the behaviour of separation bubbles. Transitional separation regions have a long history of being studied, first of all through correlations of their characteristics with base flow parameters and conditions at separation; see the monograph by Chang (1970), reviewing papers by Ward (1963), Tani (1964) and Eaton and Johnston (1981), and original results of Mueller and Batill (1982), O'Meara and Mueller (1987), Weibust et al. (1987), McGhee et al. (1988) and Azad and Doell (1990). The experimental data, including those on the location of the transition in separation bubbles, are used in semi-empirical models (Briley and McDonald 1975; Crimi and Reeves 1976; Kwon and Pletcher 1979; Roberts 1980; Vatsa and Carter 1984; Davis et al. 1987; Dini and Maughmer 1990; Choi and Kang 1991). Generally, such an approach to the investigation of separated flows is quite reasonable. However, in the transitional case the empirical results of different authors are not consistent. The fundamental reason for this is that the formation of a separation region depends on the transition process which is sensitive to small external flow variations. Thus one expects new possibilities in the modelling of separated flows through detailed studies of their instability and laminar–turbulent transition.

There is a good deal of experimental data indicating streamwise amplification of natural laminar flow disturbances behind the point of separation. Gaster (1967) performed an extensive investigation of transitional separation regions, observing spatially growing waves in the separated shear layer. Arena and Mueller (1980) examined the flow over an aerofoil with the leading-edge separation bubble, and explained their visualization results by amplification of instability waves. Similarly, wavy disturbances in separation regions were found by means of visualization in the experiments of Rannacher (1969) and Gates (1980). These observations correlate with hot-wire data obtained for boundary layer separation on aerofoils, testifying to the streamwise growth of wave packets in frequency spectra of perturbations in the transition process (Cousteix and Pailhas 1979; Dovgal and Zanin 1982; Brendel and Mueller 1988; Leblanc et al. 1991).

The experimental results strongly support the idea of laminar–turbulent transition in separation bubbles, starting from spatial amplification of small-amplitude vortical disturbances, and they substantiate the physical model of transition process used by theoreticians. In calculations it is supposed

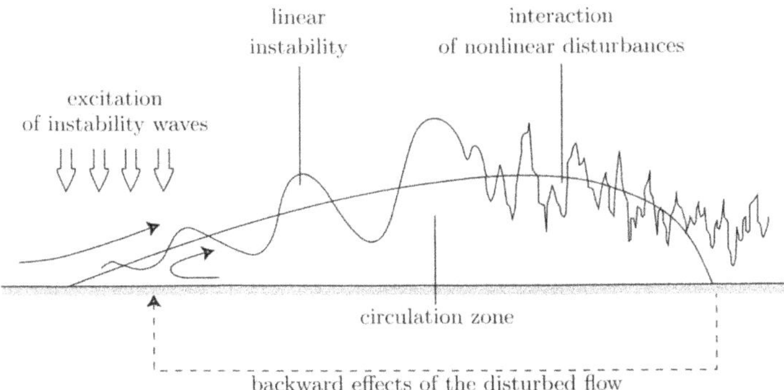

Fig. 6.2. Aspects of laminar–turbulent transition in the separation region

that separation bubbles are convectively unstable with perturbations that are dominated by local stability characteristics. Thus one expects that the initial stage of transition can be described by a linear stability theory similarly to that in near-wall and free boundary layers. If so, the transition problem for separation regions can be approached in the same way as other convectively unstable systems by focusing on several items (Fig. 6.2). Backward interactions in the figure that are additional to the usual aspects of the transition process may have much influence upon the latter. The point is that the flow patterns over the whole bubble depend on perturbations of the reattaching shear layer.

We notice here that the above approach to the problem comes from a simplified physical model of transitional separation bubbles which is applicable, first of all, to 'small' circulation regions at low Reynolds numbers and does not take into account other instability phenomena initiated by separation. Among these are low-frequency oscillations or 'flapping' observed in different flow configurations. In some cases – including separation induced by a large external pressure gradient behind high steps, humps on the surface and sharp edges – the transition to small-scale chaotic motion is substituted by coherent vortices shedding into the reattached boundary layer. One more comment is that linear stability considerations seem reasonable for separation bubbles at a low level of environmental perturbations. While free-stream turbulence Tu $\lesssim 0.6\%$ does not qualitatively affect the initial stage of transition (Dovgal and Zanin 1982), when it increases to 1.5%, linear but not exponential growth of disturbances is observed (Häggmark et al. 1997) which makes the feasibility of the classical stability theory doubtful.

6.2 Instability of separated flows to small-amplitude disturbances

To date, the major results on linear instability have been obtained in plane configurations for nominally two-dimensional separation bubbles which are quite simple for experimental examination, analysis and numerical simulations. The characteristics of small-amplitude perturbations in such flows are discussed in Sects. 6.2.1–6.2.3. Further, in Sects. 6.2.4 and 6.2.5 the problem complicated by the axial symmetry and crossflow component at the separation of a three-dimensional boundary layer is considered.

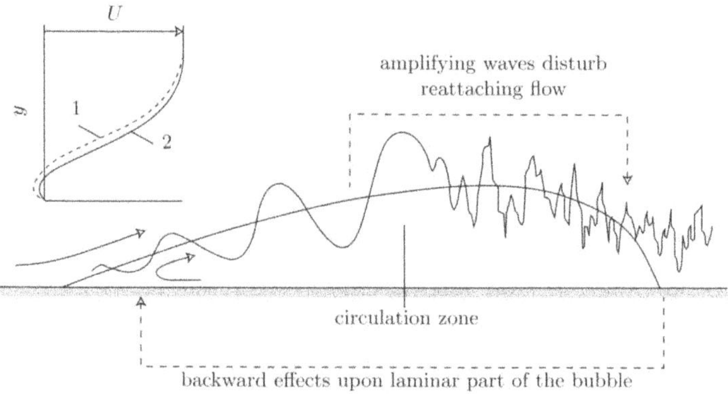

Fig. 6.3. Backward effect of disturbances in a separation bubble. Mean velocity profile under natural conditions (*1*) and at excitation of perturbations (*2*)

Before considering instability properties we should emphasize that the validity of the linear theory for transitional separated flows is not immediately obvious. The point is that the whole separation region, including its laminar portion where instability waves grow, is affected by the disturbed flow in the regions of transition and reattachment. As a result, the excitation of perturbations causes a mean flow variation, depending on their initial amplitudes (Fig. 6.3). This makes the stability analysis which deals with oscillations superimposed on a prescribed base flow and linearity of perturbations a non-trivial feature of locally separated flows. Nevertheless, when applied to separation bubbles, the linear stability theory appears really meaningful, which is supported by comparisons of experimental and numerical data with theoretical results.

An explanation of the above paradox is given by experimental data on separation bubbles instability. The influence of the laminar–turbulent transition upon separation region consists, first of all, in variations of its length and depth, while local mean flow properties in the separated shear layer that are responsible for amplification of instability waves are not affected (Dovgal

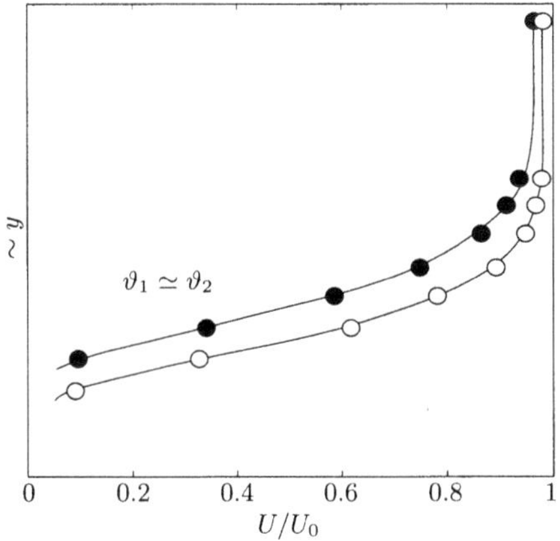

Fig. 6.4. Mean flow profiles in the laminar part of separation bubble (Dovgal and Kozlov 1983b). Natural conditions (•); under excitation of instability waves (○)

and Kozlov 1983b). Figure 6.4 shows separated mean velocity profiles behind a two-dimensional surface inflection measured under natural conditions of a wind tunnel and artificial generation of an instability wave. The excitation modifies the transition process and diminishes the separation region so that the distance between the separated boundary layer and the wall becomes smaller. However, the velocity profiles close to the inflection point near the edge of the circulation region (actually, the momentum thickness ϑ) are almost the same. Similar observations were made by Zaman and McKinzie (1991) for separation in the mid-chord region of an aerofoil under external acoustic excitation. Thus, at small-enough variations of the velocity profile in the separated shear layer and of the distance between the layer and the wall, the experimentally determined characteristics of instability waves in the upstream part of the separation region do not depend on their amplitude.

6.2.1 Waveform

The amplitude and phase distributions of an instability wave across a separated flow are shown in Fig. 6.5. The hot-wire results are taken from the experiments of Boiko et al. (1988), see also Dovgal and Kozlov (1990) for the boundary layer separation at a two-dimensional hump on a flat plate surface. In the example in Fig. 6.5 the disturbances were generated in the upstream boundary layer by a vibrating ribbon. The profiles show the streamwise evolution of the laminar flow oscillations in the separation bubble, excepting the backflow region close to the wall where the experimental method could not be used for accurate measurements.

When neglecting the streamline curvature around the hump, the amplitude distributions are those of the streamwise component of disturbance. Penetrating into the separation bubble, the perturbation transforms to the amplitude profile with three maxima near the outer edge of separated boundary layer, at the position of largest mean velocity gradient and close to the

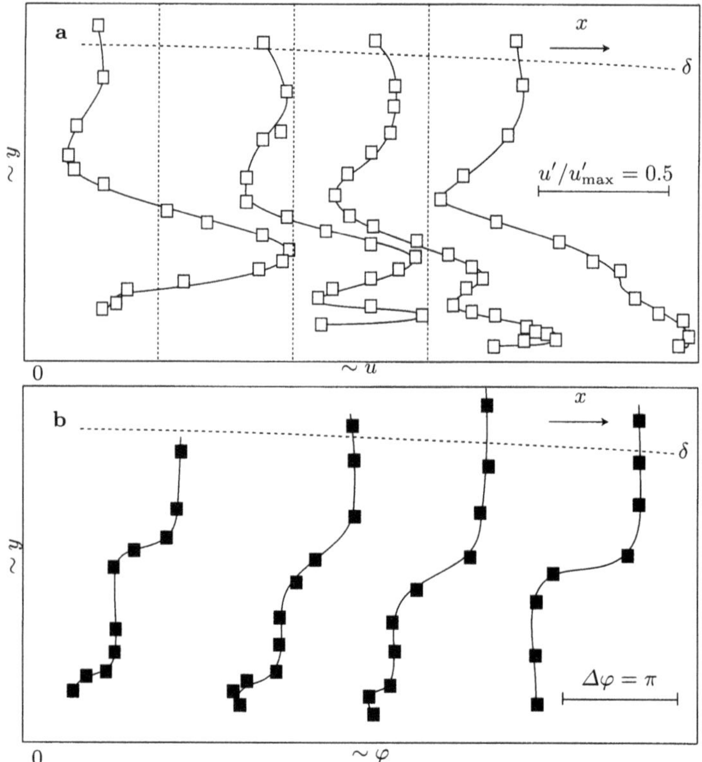

Fig. 6.5. Amplitude (**a**) and phase (**b**) profiles of the instability wave, propagating in the separation region behind a two-dimensional bump on a flat plate surface (Boiko et al. 1988). The Reynolds number based on the bumps height is $\mathrm{Re}_h = U_\infty h/\nu = 730$ (h is the hump height). From *left* to *right* at $x/h = 6.1$, 24.2, 42.2 and 78.8 from separation; *dashed lines* show the outer boundary layer border

wall in the circulation region. Further downstream at laminar reattachment (under the experimental conditions the transition to turbulence occurred behind the separation bubble), the disturbance turns back into the instability wave of the attached boundary layer; see the right-hand section in Fig. 6.5.

In controlled conditions, similar amplitude profiles of two-dimensional instability waves were also observed in other configurations, including boundary-layer separation behind the inflection of a plate surface (Dovgal and Kozlov 1983b), at a small backstep (Boiko et al. 1990) and in a corner (Dovgal and Kozlov 1984; Dovgal 1985). Haggmark (2000) and Haggmark et al. (2000) found three maxima in the amplitude profiles of a separation region generated on a plate surface by an external pressure gradient under controlled and natural conditions of a wind tunnel. The same data for natural perturbations of separation bubbles on an aerofoil and behind a backstep are reported by Cousteix and Pailhas (1979), Sinha et al. (1981).

The amplitude distributions of two-dimensional waves with normal coordinate were calculated using linear stability theory by Nayfeh et al. (1988) and Michalke (1991). Michalke (1991) examined instability of the mean velocity profiles close to those of a pre-separating boundary layer, and obtained inviscid solutions in a parallel flow approximation. Nayfeh et al. (1988) dealt with separation at a hump on a plate surface, performing stability analysis for a quasi-parallel flow at a finite Reynolds number. Comparing qualitatively the theoretical results of these studies with experimental data one can find a correlation. Particularly, the streamwise evolution of amplitude profiles in the separation region as was calculated by Nayfeh et al. (1988) is in a good agreement with the wind tunnel results shown in Fig. 6.5.

Similar amplitude distributions were found using direct numerical simulations. Navier–Stokes solutions of Gruber et al. (1987) for two-dimensional instability waves in a separation bubble caused by a local external flow pressure gradient are in accordance with experimental and theoretical data. The same flow configuration was investigated by Maucher et al. (1994), who compared numerical results with solutions of the Orr–Sommerfeld equation for the mean flow extracted from those of the Navier–Stokes equations. A good agreement was found for perturbations of small amplitudes.

Phase profiles of the two-dimensional instability waves are distinguished by two 'jumps' across the separation region (instead of one in a Tollmien–Schlichting wave of the Blasius boundary layer) with a phase difference between oscillations near the wall and in the separated layer close to 180 degrees; see Fig. 6.5. This feature was also found in the wind tunnel study of Dovgal and Kozlov (1984) and predicted by the linear stability calculations of Masad and Nayfeh (1992b) and Masad and Nayfeh (1993), who reported approximately the same character of phase distributions as that observed by Boiko et al. (1988).

6.2.2 Growth rates

The destabilizing influence of separation upon a near-wall laminar flow appears as an increase of amplification rates of disturbances in a boundary layer, both approaching the point of separation and behind it. This shows a general trend that is supported by a series of studies. Among those were systematic stability calculations for gradient boundary layers by Levchenko et al. (1975). One more indication is given by Gertsenstein (1966), who performed a stability analysis for the flow past a roughness element immersed in the shear layer with a linear variation of mean velocity normally to the wall. The frequencies and – especially – growth rates of perturbations in separation bubbles behind two-dimensional humps on a surface were higher than those in the attached flow in the theoretical analysis of Nayfeh et al. (1988). The same conclusion follows from results of numerical simulation of boundary layer separation (Maksimov 1979; Gruber et al. 1987; Rist 1994). The first experimental evidence that flow destabilization at separation is caused by rapid amplification of instability waves was probably obtained by Klebanoff and Tidstrom (1972), who explored stimulation of the boundary layer transition by a two-dimensional roughness element on a flat plate.

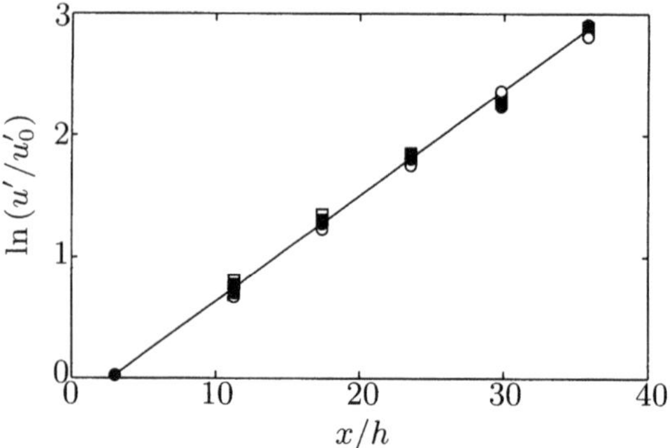

Fig. 6.6. Streamwise growth of disturbances in the separation bubble behind a two-dimensional bump on a flat plate (Boiko et al. 1988). $\mathrm{Re}_h = 730$; initial amplitudes u'_0/U_0=0.014 (□), 0.026 (○), 0.047 (△), 0.080% (•)

Variations of the maximum amplitude of two-dimensional harmonic disturbances with the streamwise coordinate in a separation region are shown in Fig. 6.6. The data were obtained in the experiments of Boiko et al. (1988) for controlled oscillations behind a hump on a flat plate surface. The amplification curves are plotted for perturbations of their different amplitudes at the same frequency. When the perturbation is small (less than about 1% of the external flow velocity U_∞), the growth of oscillations does not change within the experimental accuracy and they therefore behave as linear perturbations.

At some distance behind the separation point the disturbances amplify almost exponentially, where – approximating experimental results – one can derive a constant spatial growth rate $-\alpha_i$. Determined for different frequencies they make a linear instability range in the spectrum of perturbations. Obtained for two-dimensional separation bubbles in different configurations, the growth rates are shown in Fig. 6.7. In the figure the data are normalized by U_∞ and a mean value of the momentum thickness ϑ at the stage of exponential growth. For each experimental curve the shape factor $H = \delta^*/\vartheta$ is given, averaged over the region of amplification rates determination. Frequencies and increments of the disturbances are compared with theoretical predictions for the free-shear layer with a tanh-like profile (Monkewitz and Huerre 1982) and the Blasius boundary layer (Levchenko et al. 1975).

The experimental data show that the flow in separation bubbles becomes more unstable with their enlargement (increase of the shape factor). Qualitatively this observation agrees with the theoretical results which indicate that the larger the distance between the inflection point in the mean velocity

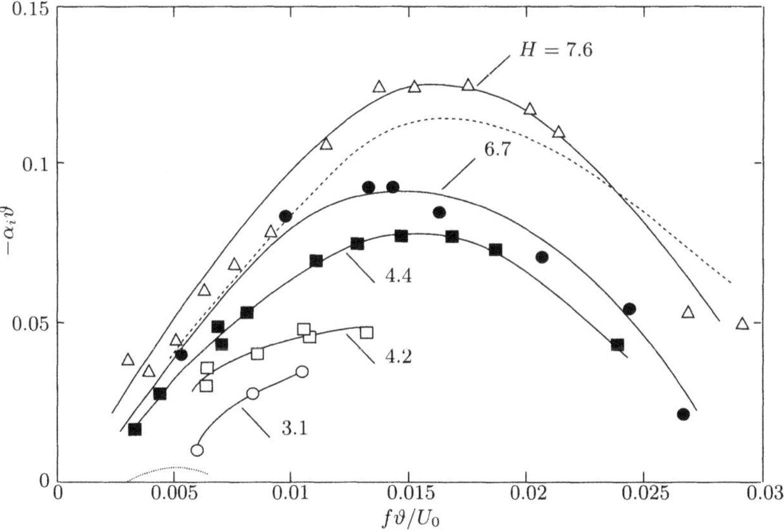

Fig. 6.7. Growth rates of two-dimensional small-amplitude disturbances. Experimental data for separation bubbles in different configurations: behind an inflection of the wall (Dovgal and Kozlov 1983b) (△), in a corner (Dovgal and Kozlov 1984) (•), past a two-dimensional bump (Boiko et al. 1988) (■) and backward-facing steps (Boiko et al. 1990) (□, ○). Results of calculations for the mixing layer $U \sim \tanh y$ (Monkewitz and Huerre 1982) (*dashed line*), and for the Blasius boundary layer at $Re_{\delta^*} = 1320$ (Levchenko et al. 1975) (*dotted line*)

profile and the wall, the higher the growth rates of small-amplitude perturbations (Taghavi and Wazzan 1974; Michalke 1990; Nayfeh et al. 1990). This conclusion was also made by Michalke (1990) for boundary layer profiles close to separation, while Taghavi and Wazzan (1974) obtained similar result for a family of solutions of Falkner–Skan equations with backflow (Stewartson profiles). The same was inferred by Nayfeh et al. (1990) in a stability analysis of the separated flow behind a hump on a plate surface. The above tendency revealed both experimentally and theoretically is further confirmed by the direct numerical simulations of Rist et al. (1996).

The maximum values of experimentally determined amplification rates in Fig. 6.7 are comparable with those calculated in the inviscid limit for a free-shear layer. Viscosity effects become more important when a separation bubble gets smaller. The flow behind the small backstep in the depth of the laminar boundary layer ($H = 3.1$) is stable to low-frequency oscillations, but unstable to them in the larger separation regions ($H = 4.2$ to 7.6) as well as in inviscid solutions. The wind tunnel data correlate with a theoretical result of Smith and Bodonyi (1985), who found that the inviscid instability in a flow past a low obstacle on the surface may not occur in spite of the inflection point in the mean velocity profile. They are also in line with the calculations of Michalke (1991) which showed that as the inflection point approaches the wall, the effect of a finite Reynolds number increases, first of all for long-wave perturbations.

Empirical data on growth rates of two-dimensional instability waves are well predicted by linear stability calculations. Michalke (1991), using a parallel-flow approximation, obtained stability solutions for the base flow behind a backstep which was examined in the experiments of Boiko et al. (1990). Solving the inviscid problem, he obtained a reasonable agreement between experimental and theoretical results which was further improved by taking into account flow viscosity. Masad and Nayfeh (1992b, 1993) calculated spatial amplification rates of the disturbances in separation regions at forward- and backward-facing steps on a flat plate for the experimental conditions of Boiko et al. (1990). The analysis was performed using a quasi-parallel stability theory for mean flow solutions obtained by an interacting boundary layer formulation. An overall good agreement between theory and experiment was found. Hildings et al. (Haggmark 2000) came a similar conclusion, when comparing the experimental and theoretical data for a separation bubble induced on a flat plate by an external pressure gradient.

On the other hand, the validity of a local stability theory to the description of the initial stage of transition in separation regions is confirmed by comparison of the theoretical data with those of direct numerical simulations and solutions of the parabolized stability equations. The theoretical and numerical results on spatial increments of two-dimensional instability waves were compared by Gruber et al. (1987) and Bestek et al. (1989, 1993) and Hildings et al. (Haggmark 2000). Rist (1994) and Rist and Maucher (1994) did the same for two and three-dimensional oscillations. In these studies a good agreement was found between the solutions of Navier–Stokes equations and the data from local linear stability analyses. A correspondence between the results of local theory, application of the parabolized stability equations and direct numerical simulation was pointed out by Hein et al. (1998) and Hein (2000). Similar solutions of the Navier–Stokes and the parabolized stability equations were obtained by Perraud (1998). Numerical investigations, which are somewhat different from the above series of studies, were carried out by Elli and Van Dam (1991) and Van Dam and Elli (1992), with conclusions about importance of non-linearity and non-parallel effects.

Considering the growth rates of the small-amplitude disturbances, we notice that most of research data have been obtained for two-dimensional waves which have rather high increments in the wave number spectrum of oscillations and, thus, are the main contribution to perturbations in a separation region. Furthermore, the two-dimensional waves are not necessarily the most amplified in the region of linear instability. The results of an asymptotic theory (Stewart and Smith 1987) and experimental data (Gilev et al. 1988) show that larger growth rates at laminar boundary layer separation may have three-dimensional spectral components.

6.2.3 Propagation velocities

A feature of locally separated flows that is found in experimental and theoretical work on instability is the relatively small dispersion of propagation velocities of small-amplitude perturbations. This applies to the two-dimensional waves travelling in the external flow direction with phase velocities that are almost independent of their frequency, and to the three-dimensional plane waves of a fixed frequency whose streamwise wave number varies slightly with the angle of propagation over a broad bandwidth.

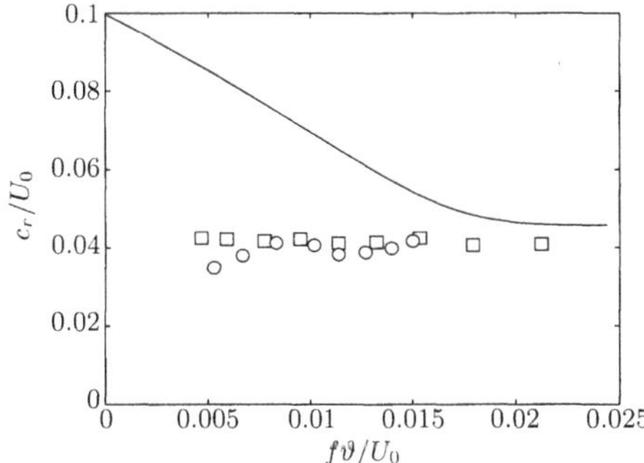

Fig. 6.8. Phase velocities of the two-dimensional disturbances. Experimental data for separation bubbles behind a bump on a flat plate surface (Boiko and Dovgal 1992) (□) and on an aerofoil (Boiko et al. 1989) (○). Results of calculations for the mixing layer $U \sim \tanh y$ (Monkewitz and Huerre 1982) (*solid line*)

Experimental data on the phase velocities of the two-dimensional instability waves obtained in separation bubbles behind a rounded hump on a flat plate surface (Boiko and Dovgal 1992) and on an aerofoil (Boiko et al. 1989) are shown in Fig. 6.8. The results are compared with the dispersion curve in a free-shear layer calculated by Monkewitz and Huerre (1982). One can observe that in the separation regions the phase velocity is almost constant over a wide frequency range, which is considerably different from the free-shear layer when the wall is removed to infinity. The wind tunnel data correlate with stability solutions of Michalke (1990, 1991), indicating diminution of phase velocities at low frequencies of the oscillations when the distance of the shear layer to the wall is reduced. A good agreement between the wave numbers of two-dimensional disturbances was found by Bestek et al. (1993) when comparing results of linear stability analysis and numerical solutions of the Navier–Stokes equations.

Wind-tunnel data on dispersion of plane three-dimensional waves are shown in Fig. 6.9 for separation at a hump on a flat plate (Boiko et al. 1991a) and on an aerofoil (Gilev et al. 1988), where spatial evolution of wave packets of small-amplitude harmonic perturbations were investigated. For a fixed frequency of oscillations, each curve in the figure shows a variation with the wave angle of the streamwise wave number normalized by that of the two-dimensional disturbance. Similar results of Kachanov (1985) for the Blasius boundary layer are plotted for comparison. It turns out that the oblique waves – amplifying in separation bubbles – have a fairly small dependence of

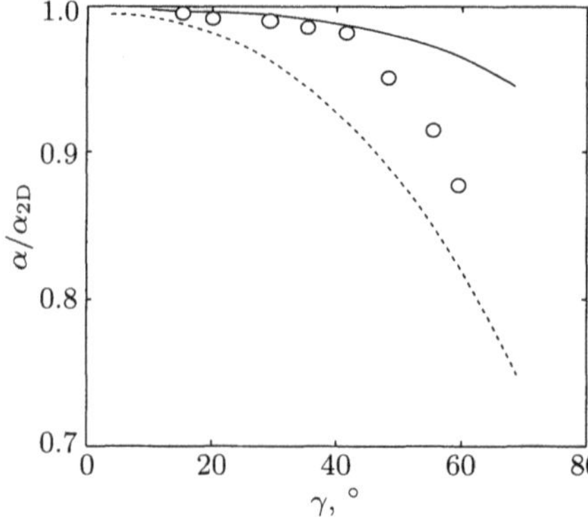

Fig. 6.9. Variation of the streamwise wave number of plane three-dimensional waves with the wave angle. Separation bubbles behind a bump on a flat plate surface (Boiko et al. 1991a) (○), on an aerofoil (Gilev et al. 1988) (*solid line*) and in a Blasius boundary layer (Kachanov 1985) (*dashed line*)

the streamwise wave number on the angle between the wave vector and the mean flow direction.

The above peculiarities of the small-amplitude disturbances are noteworthy in respect of the non-linear stage of transition in separation regions, coming after the initial exponential growth. The weak dispersion of the two- and three-dimensional waves means constancy of phase coherence between different components of the frequency and wave number spectra at their streamwise evolution, stimulating their subsequent resonant interactions.

6.2.4 Instability of an axisymmetric flow

When turning to separation of an axisymmetric flow in the stability problem, an additional curvature parameter appears as the ratio of the viscous-layer thickness to the body radius. Effects of the axial symmetry on small-amplitude perturbations at separation were investigated theoretically and experimentally by Michalke et al. (1995) and Dovgal et al. (1995).

Michalke et al. (1995) performed a stability analysis of the mean velocity profiles, modelling the base flow in the boundary layer before and behind the separation point, using a parallel flow approximation. It was found that the axisymmetric flow, similarly to that in plane configurations, becomes more unstable with increasing separation bubble thickness; that is, the maximum growth rate and the range of unstable frequencies get higher when the inflection point moves from the wall. Generally, the axial symmetry makes the flow more stable and the number of amplifying helical modes is reduced. However, the quantitative effect on an instability mode depends on mean flow parameters and the frequency of oscillations. In particular, with increases of the curvature parameter, a higher separated-flow instability in respect to the first helical mode may occur.

A wind tunnel study of the unstable flow in an axisymmetric separation bubble, and comparison with theoretical results, was carried out by Dovgal et al. (1995). That work examined the disturbances generated by external acoustic excitation at a backstep on a body of revolution. The experimental data on separated-flow instability to the axisymmetric mode of perturbations confirmed the theoretical conclusion about the stabilizing effect of the axial symmetry. The quantitative comparison of experimental observations with stability solutions confirmed their good agreement, as in other two-dimensional cases.

6.2.5 Flow instability at separation of a three-dimensional boundary layer

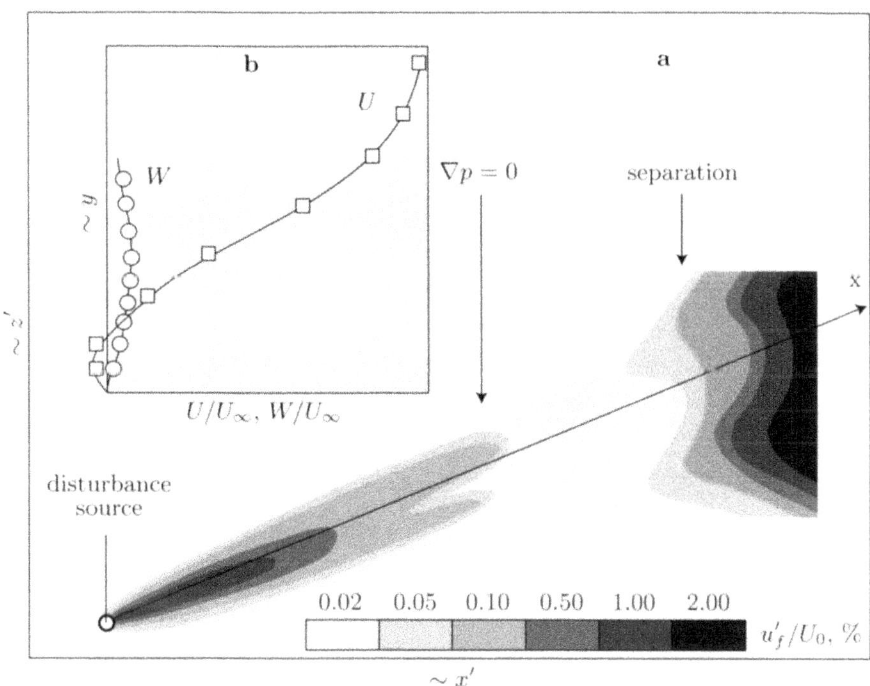

Fig. 6.10. Wave packet in a swept-wing flow (Dovgal et al. 1988a). Sweep angle is 30°, chord Reynolds number $\mathrm{Re}_l = 2.6 \times 10^5$. Contours of disturbances amplitudes as percentage of the oncoming flow velocity (**a**) and mean flow profiles in the separation region (**b**)

The research results discussed in previous sections support the quite simple notion that the initial stage of laminar–turbulent transition in two dimensional separation bubbles can be considered as amplification of small-amplitude perturbations of the separated shear layer prescribed by local mean flow characteristics. The transition problem is more complex to apply to three-dimensional boundary layers, where a wider range of instability mechanisms compete with each other. A related topic is flow instabilities and transition on a swept wing of infinite span. In this case an approach to the problem consists of decomposing the mean flow pattern into the main (in the external flow direction) and the crossflow velocity components. Normally, the crossflow instability predominates in the upstream section of a wing in the region of the negative pressure gradient. Another area where the crossflow component is large enough to switch on growth of velocity perturbations is the region of adverse pressure gradient where boundary layer separation may occur. In the aft part of the wing the mean flow is also considerably unstable because of the inflection point in its velocity profile. Increasing instability of both the main and the crossflow components before the separation point on a swept wing was observed in the calculations by Mack (1982).

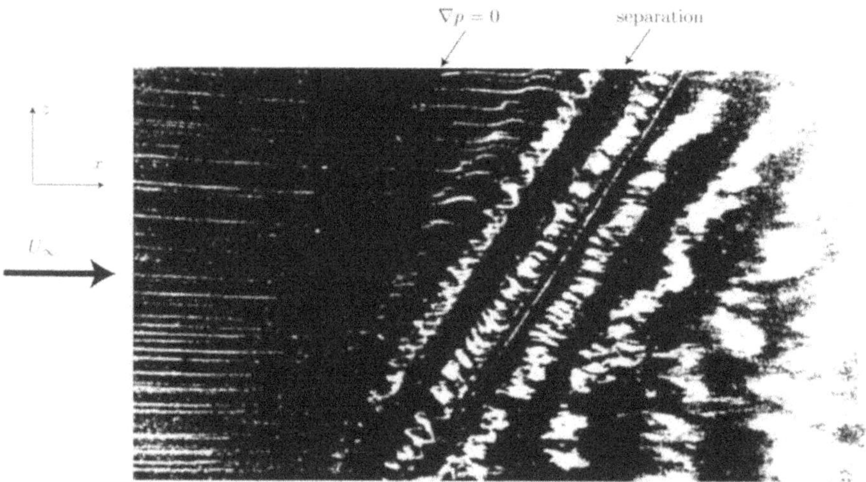

Fig. 6.11. Smoke-wire visualization of perturbations, amplifying in a swept-wing separation bubble (Dovgal et al. 1987). Sweep angle is 30°, $\mathrm{Re}_l = 2.6 \times 10^5$

Experimental studies on small-amplitude disturbances growing in the laminar separation of three-dimensional boundary layers were performed by Dovgal et al. (1988a, b). Figure 6.10 shows a wave packet of controlled harmonic oscillations, spreading in a swept-wing flow, which were excited by periodic blowing through a point-like hole in the surface of a test model (Dovgal et al. 1988a). The excitation frequency was close to that of the perturbations with maximum amplification in frequency spectra during natural transition to turbulence. The disturbances generated in the accelerated flow initially decay, and then grow in the adverse pressure gradient boundary layer and further downstream in the separation region. The selection of waves caused by a variation of amplification rates of the three-dimensional spectral components with the wave-angle, results in spanwise widening of the packet, especially when separation is approached. In the amplitude contours of Fig. 6.10 one can see that behind the separation point the perturbations are dominated by the oscillations aligned with the wing span. Their smoke-wire visualization is shown in Fig. 6.11, where the disturbances appear as a row of transverse 'eddies'.

A quantitative comparison between the instability of two- and three-dimensional separation regions is made available by the experimental data of Dovgal and Kozlov (1983b) for separation at a two-dimensional inflection of a plate surface, and by Dovgal et al. (1988b), who examined the same configuration at a 30° sweep angle. Spatial increments of the most amplified wave-spectrum components along their wave vectors, which were determined in these studies, are shown in Fig. 6.12. The data were obtained when the local streamwise mean velocity profiles in two- and three-dimensional flows, and Reynolds numbers were close to each other. In both cases the instability range is the same, which means that amplification of disturbances in the region of three-dimensional boundary layer separation is principally initiated by instability of the main flow component. An explanation of this result follows from the data of Fig. 6.12, which shows that the spatial growth rate of perturbations is an order of magnitude over one wavelength, which is typical for inviscid instability of a separated boundary layer. As well as in the two-dimensional problem, the disturbances of the three-dimensional flow evolve in the shear layer at the edge of separation bubble where the instability is dominated by an inflectional mean velocity profile, while the maximum of the crossflow component is closer to the wall inside the region of circulation. Then the observed difference between the two- and three-dimensional cases, which consists of a prevalence of the oblique waves at separation of the three-dimensional boundary layer, is due to the particular flow configuration rather than the effect of crossflow instability.

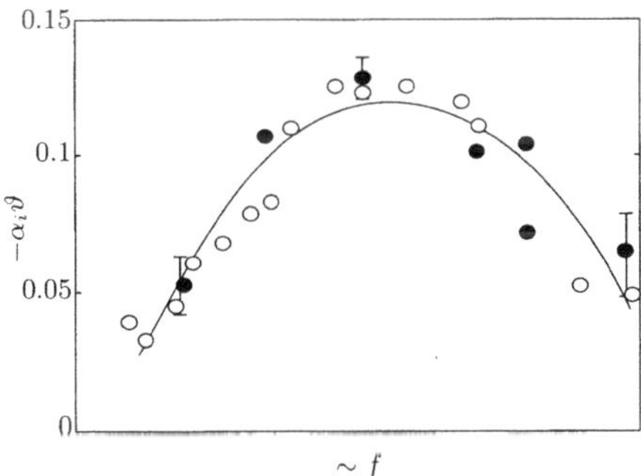

Fig. 6.12. Spatial growth rates of disturbances. Two-dimensional separation bubble behind an inflection of the wall (○) (Dovgal and Kozlov 1983b) and separation region in the same configuration at the 30° sweep angle (•) (Dovgal et al. 1988b)

6.3 Excitation of instability waves in separation bubbles

When dealing with boundary layer receptivity as a special topic in the transition problem, the flow is assumed to be convectively unstable (see Chap. 3). Thus, applying the same approach to separation bubbles, one also expects their convective instability following from the results on linear disturbances of separation regions which were discussed in the previous section.

Theoretical studies have investigated the receptivity of boundary layers at separation to various types of external disturbances. Models of instability-waves generation by long-wave oscillations were developed by Goldstein (1984) and Goldstein et al. (1987) for separation from smooth surfaces, and by Ruban (1985) and Bodonyi et al. (1989) for boundary layers at roughness elements. Excitation of a separated flow behind a two-dimensional hump on a plate by a local source of harmonic oscillations was focused on by Michalke and Al-Maaitah (1992). Considering simplified models of the mean flow in an adverse-pressure-gradient boundary layer and in separation regions, Michalke (1993, 1997) calculated the receptivity to perturbations generated on the wall and in the external flow. In the first of these two studies the effects of compressibility and cooling were also included. A theory of the instability-waves excitation by a point source of harmonic perturbations in an axisymmetric flow was constructed by Michalke (1995), with further comparison between theoretical and experimental data appearing in Dovgal et al. (1996). Numerical solutions of the Navier–Stokes equations in the receptivity problem were obtained by Maksimov (1979), who calculated a two-dimensional backstep flow in an oscillating stream and determined the ratio between amplitude of the instability waves and the strength of imposed disturbances. The general conclusion that one can draw from the theoretical results is that the boundary layer approaching separation becomes more receptive to external disturbances. Normally, efficiency of generation gets higher with an increase of the stationary perturbation of the near-wall fluid which is produced by the separation region (Bodonyi et al. 1989; Michalke and Al-Maaitah 1992; Michalke 1995, 1997).

Experimental data on receptivity of separation bubbles were obtained by Dovgal and Kozlov (1983a), Boiko et al. (1988, 1990), Dovgal et al. (1989) and Asai and Kaneko (1998) for excitation of the instability waves by external acoustic oscillations. Summarizing results of these studies, one can distinguish between two main routes of generation depending on the conditions under which a separation occurs. One of them is the transformation of disturbances evolving in the pre-separated boundary layer into the instability waves of the separated flow, and the other is excitation just at separation.

The first way is typical for separation on aerofoils and wings operated at low Reynolds numbers where, at a moderate angle of attack, a separation bubble occurs in the mid-chord region due to the adverse pressure gradient. Such a case was focused on by Dovgal and Kozlov (1983a) and Dovgal et al. (1989), when investigating the receptivity of separation regions on a wing

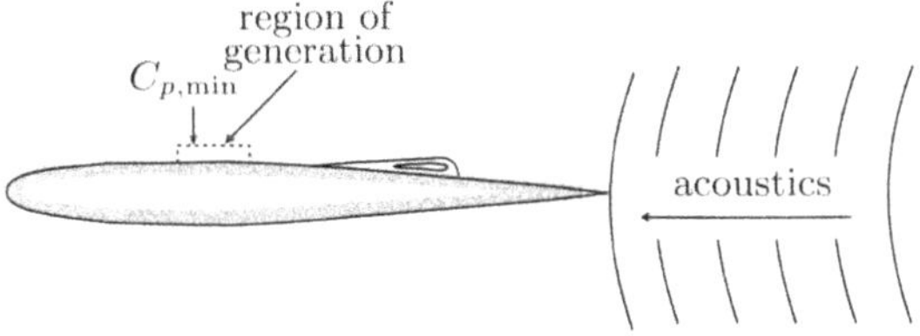

Fig. 6.13. Excitation of instability waves at the separation from a smooth surface (C_p is the pressure gradient parameter)

for variations of the sweep angle. In these studies, when switching on sinusoidal acoustic waves of appropriate frequencies, the generation of amplifying boundary layer disturbances was observed. By comparison of their characteristics with those of the instability waves injected by a vibrating ribbon well upstream of the separation point, the region of acoustic receptivity was determined; see Fig. 6.13. These disturbances, which caused the transition in separated flow, were excited at the beginning of the adverse pressure gradient, while at separation generation was not observed or, at most, was negligible. At first this was found at the transformation of the acoustic oscillations into the boundary layer disturbances at the natural roughness distributed along the wing surface. Subsequent examination of scattering of the acoustic waves on controlled roughness elements indicated that just the boundary layer perturbations excited at the minimum static pressure along the wing chord grow to the largest amplitude before entering the separation bubble with the major contribution to laminar–turbulent transition.

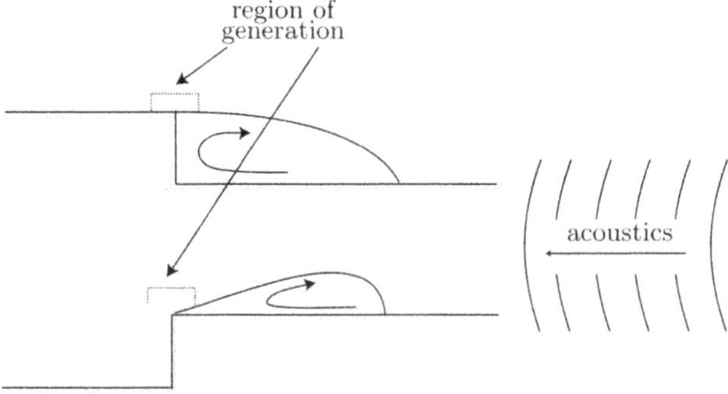

Fig. 6.14. Excitation of instability waves at the separation induced by wall imperfections

The second route applies to laminar boundary layer separation caused by local variations of a surface which, at the same time, convert the external flow acoustic into vorticity perturbations, binding the separation with the generation of instability waves (Dovgal and Kozlov 1990). Wind tunnel data on this matter were obtained by Boiko et al. (1988, 1990) for the receptivity at two-dimensional surface variations including steps and humps of a height comparable with the boundary layer thickness. Similarly, Asai and Kaneko (1998) dealt with acoustic excitation of the flow separating behind a two-dimensional edge on a plate. In their flow configurations, Boiko et al. (1988, 1990) observed the generation of instability waves in the vicinity of surface irregularities, while downstream the excited waves evolved from the instability of separation bubbles only; see Fig. 6.14. They also found that the initial amplitudes of the generated disturbances were strongly dependent on the shape of surface variations. This was also concluded by Asai and Kaneko (1998), who observed different receptivity at separations on sharp and rounded edges. One expects that this should have a pronounced influence upon the transition process in a separation region, or downstream in an attached boundary layer. To a certain degree, the above experimental data correlate with the results of flight measurements (Holmes et al. 1986) which indicate that the transition Reynolds number $\mathrm{Re_T}$ is affected by the shape of surface imperfections. Finally, we notice that at separation on geometrical surface variations, contributions to the transition process may have the disturbances excited both close to the separation point and in the upstream boundary layer, as it was in the previous case. This depends on the interplay between generation at separation and the pre-history of the boundary layer, including its receptivity and stability properties.

6.4 Wave interactions

The linear instability approach, justified for small-amplitude disturbances evolving in a separation bubble at a low level of external perturbations, fails with streamwise growth of the oscillations. The exponential stage of amplification is replaced by non-linear region with wave interactions. In a spectral representation of the transition process, the non-linearity starts from breakdown of a linear wave packet; see Fig. 6.15. At a distance comparable with the wavelength of the disturbance most amplified in the linear region, the laminar flow transforms into the turbulent one with a broad spectrum of perturbations. Naturally, the non-linear stage of the transition in boundary layers is a complex phenomenon which includes a number of non-linear mechanisms which are the focus of theoretical and experimental studies. Some of these mechanisms found in locally separated flows through experimental modelling and calculations are discussed in this section. The non-linear features at flow separation revealed in an asymptotic theory were reviewed by Smith (1987).

6.4.1 Subharmonic excitation

A key mechanism of the 'randomization' in free and wall boundary layers is the generation of subharmonic disturbances of a fundamental wave with frequency f_0 amplified in the linear region (Sato 1959; Browand 1966; Miksad 1972; Kachanov and Levchenko 1984). For a separation bubble this path

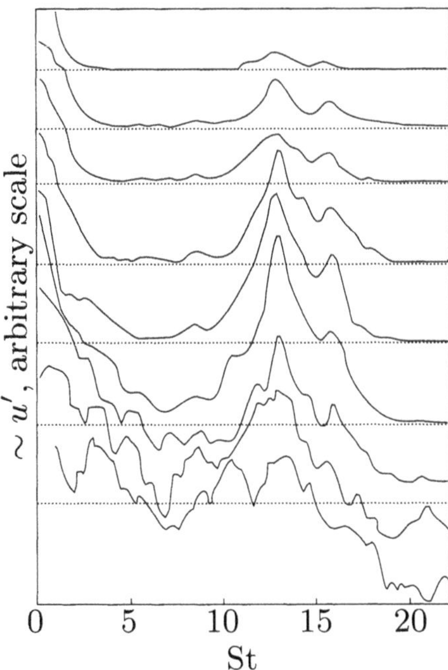

Fig. 6.15. Natural spectra of disturbances in a separation bubble on an aerofoil (Boiko et al. 1989). $\mathrm{Re}_l = 2.6 \times 10^5$; boundary layer separates at $x/l = 0.57$ and reattaches at about $x/l = 0.75$. From *top* to *bottom* $x/l = 0.657, 0.664, 0.671, 0.679, 0.686, 0.693, 0.700, 0.707, 0.714$. Strouhal number $\mathrm{St} = fl/U_\infty$, where l is the chord length. Excitation frequency f_0 corresponds to St = 8.6

of laminar flow breakdown is illustrated in Fig. 6.16 by the spectra of fluctuations measured on an aerofoil in the experiments of Boiko et al. (1989). The data were obtained under a weak controlled excitation within the linear instability range of the flow behind the separation point. Initial growth of the generated instability wave is followed by rapid amplification of low-frequency perturbations centred around the subharmonic $f_0/2$ of the fundamental disturbance. Then modelling an interaction between two-dimensional waves at the fundamental and the subharmonic frequencies, it was found that the low-frequency peak in the spectra of Fig. 6.15 originated from a resonant coupling of the fundamental disturbance with background fluctuations. The resonance width, estimated as $f_0/2$ (Kachanov and Levchenko 1984), results in amplification of a broad range of low-frequency oscillations with subsequent transition to turbulent motion.

The spatial structure of perturbed flow induced by the resonance depends on the wave number spectrum of subharmonic disturbances involved in the interaction. In a flat plate boundary layer, the resonant selection of three-dimensional subharmonic components – keeping phase synchronism with a two-dimensional fundamental wave – takes place, which results in a spanwise periodicity of the disturbances (Craik 1971; Volodin and Zelman 1978; Herbert 1984; Kachanov and Levchenko 1984). On the other hand, in free-shear layers the interactions of two-dimensional and almost two-dimensional waves are significant, so that the generation of subharmonics is not followed by pronounced three-dimensional flow distortion (Browand 1966; Kelly 1967;

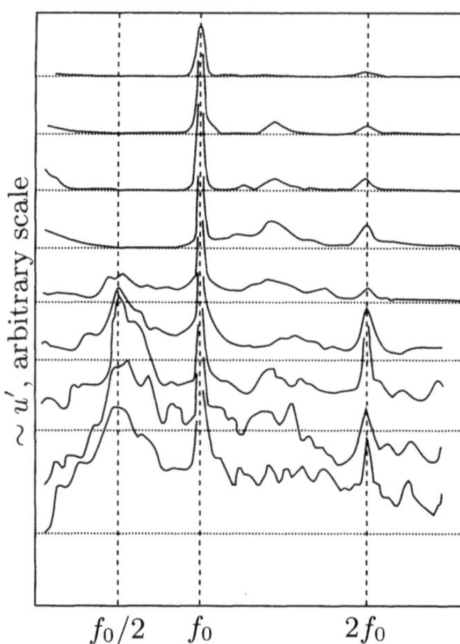

Fig. 6.16. Subharmonic excitation in the separation bubble of Fig. 6.15. From *top* to *bottom* $x/l = 0.657, 0.664, 0.671, 0.679, 0.686, 0.693, 0.700, 0.707, 0.714$. Excitation frequency f_0 corresponds to St = 8.6

Miksad 1972; Pierrehumbert and Widnall 1982; Monkewitz 1988). In separation bubbles with a quite small dispersion of propagating velocities (see Sect. 6.2), the resonant interaction may occur in a wide wave number spectrum of two- and three-dimensional subharmonic disturbances, as was found in the wind tunnel studies of Boiko et al. (1989, 1991a). Resonant amplification of two-dimensional waves was observed by Boiko et al. (1989), while the results obtained in Boiko et al. (1991a) showed that – interacting with a fundamental two-dimensional wave – two- and three-dimensional subharmonic perturbations have comparable growth rates.

The subharmonic excitation in separation regions was investigated theoretically by Nayfeh and Ragab (1987), Zelman and Smorodski (1991) and Masad and Nayfeh (1992a). Nayfeh and Ragab (1987) and Masad and Nayfeh (1992a) performed calculations for the flow behind a hump on a flat plate surface, and the solutions of Zelman and Smorodski (1991) were obtained for inflectional boundary layer profiles, including separated ones. The theoretical data show that at a small mean flow perturbation of the Blasius boundary layer (before separation or in a small separation bubble), the wave number spectrum of generated subharmonics is practically the same as that in the flat plate flow where the most amplified are three-dimensional subharmonic disturbances. However, Nayfeh and Ragab (1987) found that when the hump height increases and the separation region turns larger, the maximum growth moves towards the two-dimensional waves whose amplification rates become comparable with increments of three-dimensional subharmonic oscillations; this correlates with the experimental data of Boiko et al. (1989, 1991a). In the absence of a sharp resonant peak, the spectrum of generated subharmonic perturbations primarily depends on their initial background distribution over wave numbers rather than the process of wave selection. In this respect the transition to turbulence in separation bubbles resembles the free-shear layer transition more than that of a wall-bounded flow.

6.4.2 Some other non-linear mechanisms

Another way of energy distribution over a wide frequency spectrum of perturbations at the non-linear stage of transition consists of wave combinations of the spectral components being amplified in the linear instability region. In wind tunnel studies on laminar–turbulent transition, this scenario has been modeled in a flat plate boundary layer (Kachanov et al. 1980), in the wake behind a thin plate (Sato 1970) and in a free-shear layer (Miksad 1973). In these experiments, under excitation of controlled oscillations at two frequencies f_1 and f_2, combination modes were generated at $mf_2 \pm nf_1$ (Kachanov et al. 1980), $n(f_1 \pm f_2)$ (Sato 1970), and $(nf_2/m) \pm (pf_1/q)$ (Miksad 1973), where m, n, p and q are integers. In this case one expects a complete spectrum of perturbations to appear at incommensurable frequencies of the interacting waves. A similar result for the separation bubble on an aerofoil is shown in Fig. 6.17 as disturbances spectra measured during simultaneous excitation of two instability waves with frequencies f_1 and f_2. After streamwise amplification in the linear instability region, they grow to finite amplitudes and produce the spectral peaks at $mf_2 \pm nf_1$.

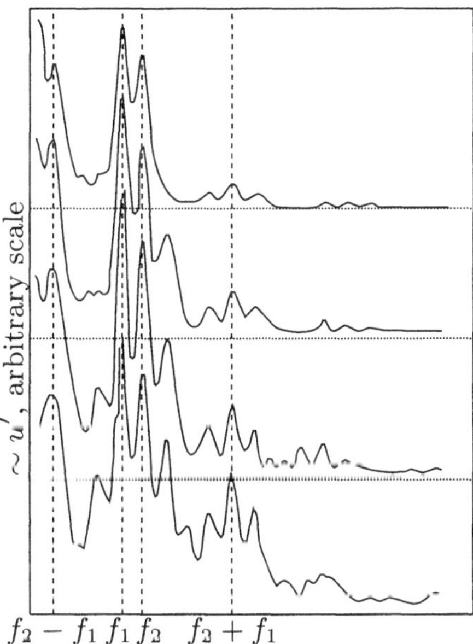

Fig. 6.17. Waves combinations in the separation bubble of Fig. 6.15. From top to bottom at x/l = 0.643, 0.650, 0.657, 0.664. Excitation frequencies f_1 and f_2 correspond to St = 13.4 and 16.8, respectively

Some types of non-linear interactions in separation regions have been reproduced by direct numerical simulation. The results of calculations for boundary layer separation caused by a local variation of external pressure gradient were reported by Rist (1994), Rist and Maucher (1994), Rist et al. (1996) and Maucher et al. (2000). Varying the initial spectra of disturbances over frequencies and wave numbers, they investigated fundamental and subharmonic resonances of two- and three-dimensional waves as well as the interaction of two symmetric three-dimensional waves, so-called oblique breakdown, which was found to be the most effective non-linear mechanism. The reliability of oblique breakdown in separation regions was further supported in the experiments of Ablaev et al. (2000) for the boundary layer separating at a two-dimensional hump on a flat plate.

6.4.3 Effects of the initial spectrum on the transition

The relative contribution of various non-linear mechanisms to the transition process depends on the initial conditions; that is, frequency and wave number spectra which result from a preceding growth in the linear instability. Such examples for open-flow systems are given by wind tunnel data obtained in the Blasius boundary layer by Saric et al. (1984) and Kachanov and Levchenko (1984), and in a plane channel flow by Kozlov and Ramazanov (1983).

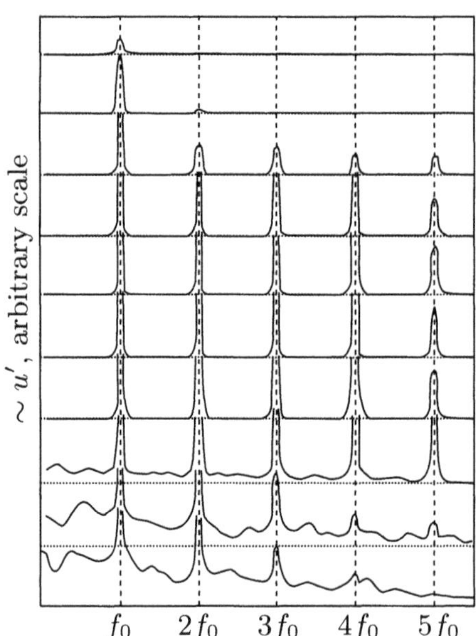

Fig. 6.18. Spectra of disturbances in the separation bubble of Fig. 6.15 under high-amplitude excitation. From *top* to *bottom* at x/l = 0.586, 0.614, 0.643, 0.671, 0.700, 0.729, 0.757, 0.786, 0.814, 0.843. Excitation frequency f_0 corresponds to St = 10.4

The subharmonic route of transition in a separation bubble discussed in Sect. 6.4.1 was modelled in experiments of Boiko et al. (1989) and Boiko et al. (1991b) under low-magnitude harmonic forcing in order to reproduce the natural flow occurring with a fairly smooth initial spectrum. The perturbations of a separation region excited with much larger amplitudes may evolve in completely different manner (Boiko et al. 1989; Dovgal and Boiko 2000). The spectra of disturbances measured by Boiko et al. (1989) in the mid-chord separation bubble on an aerofoil under a 'strong' periodic forcing are shown in Fig. 6.18. Comparing them with similar data for the controlled subharmonic excitation (see Fig. 6.16) and the natural flow (see Fig. 6.15), one can observe that upon increase of the initial amplitude of oscillations the disturbances become more regular and the transition to chaos moves downstream. A possible reason of such an unexpected behaviour (normally, the transition occurs faster at a higher level of perturbations) is the backward influence of the disturbed flow in the separation region; this point is considered in Sect. 6.5.

Under high-amplitude excitation, as well as at its lower magnitudes, growth of the generated perturbations starts from linear instability waves in the upstream part of the separation bubble. Further downstream, a regular non-linear disturbance appears in the frequency spectra as the fundamental wave and its higher harmonics nf_0, $n = 1, 2, \ldots$. At this stage, prior to amplification of other spectral components, the excited two-dimensional perturbation evolves without noticeable three-dimensional distortion. In terms of vortex dynamics, this corresponds to the formation of two-dimensional periodic vortices which shed from the separation bubble and dominate the flow pattern, instead of the laminar–turbulent transition sketched in Fig. 6.1.

As one can see in Fig. 6.18, the 'high-amplitude' case is distinguished by suppression of the subharmonic resonance compare with Fig. 6.16. A similar result was obtained by Zaman (1992). When dealing with acoustic forcing of the leading-edge separation on an aerofoil, he found that at rather strong excitation the disturbances evolved without subharmonic components, which was explained by the proximity of the wing surface preventing the vortex pairing in the separated layer. As regards the interrelation between laminar–turbulent transition and generation of coherent vortices at separation, we notice also the numerical results of Lin and Pauley (1996). Referring to Morkovin (1964), they argue that at a low Reynolds number the effect of a neighbouring wall may be the production of periodic vortices in the separated shear layer without transition to turbulence. We will come back to this issue in Sect. 6.6.

6.5 Backward effects of the disturbed flow in separation regions

In the unstable regime, the flow pattern in a separation bubble – including its laminar part – depends on the transition occurring in the region of separation, or nearby (Chapman et al. 1958). Amplifying instability waves, excited by some sort of external disturbances, modify the transition process and induce perturbations of the reattaching flow which spread all over the bubble affecting its mean and oscillatory characteristics; see Fig. 6.3. In principal, such an upstream influence at separation is a manifestation of viscous–inviscid interaction between the external and the near-wall portions of fluid, which break down the fundamentals of the boundary layer theory.

On the one hand, the effect of transition against the flow direction appears as a mean flow variation close to the separation point produced by excitation of the instability waves, which was reported at the beginning of this chapter. Stimulation of the transition process in the separated layer results in diminution of the circulation region and deformation of the mean velocity profile in laminar part of separation bubble; see Fig. 6.4. Moreover, this may take place even in laminar separation regions with a small amount of instability when the transition occurs far downstream of the reattachment. In the last case the mean flow variation is caused by disturbances growing through the bubble up to non-linear amplitudes (Boiko et al. 1991b); see also Dovgal and Kozlov (1995). However, the mean flow sensitivity to amplifying perturbations does not prevent a linear stability analysis yielding solutions which agree with experimental data and the results of direct numerical simulations; see in Sect. 6.2.

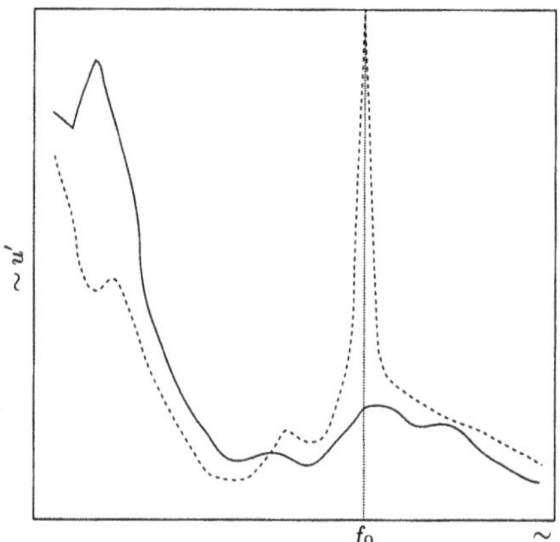

Fig. 6.19. Suppression of background oscillations under excitation of instability waves behind a two-dimensional hump on a flat plate (Boiko et al. 1991b). $\mathrm{Re}_h = 760$, natural (*solid line*) and excited (*dashed line*) spectra

Another backward effect of the transition process consists of non-local interactions between the excited amplifying disturbances and background velocity fluctuations. Figure 6.19 shows frequency spectra of perturbations measured in the laminar part of a separation bubble behind a two-dimensional hump on a flat plate under natural wind tunnel conditions and controlled generation of an instability wave (Boiko et al. 1991b). The excitation is followed by a decrease in the low-frequency irregular oscillations. Similar phenomenon, that is the suppression of background noise, is known from experimental observations of the laminar–turbulent transition in free-shear layers where it is attributed to energy transfer between non-linear spectral components of disturbances (Sato and Kuriki 1961; Zaman and Hussain 1981). In the present case the superposition of the generated perturbation and the continuous spectrum breaks down just behind the point of separation where the local amplitude of the excited wave is only 0.22% of the external flow velocity U_∞, and in all other respects it behaves like a small-amplitude disturbance.

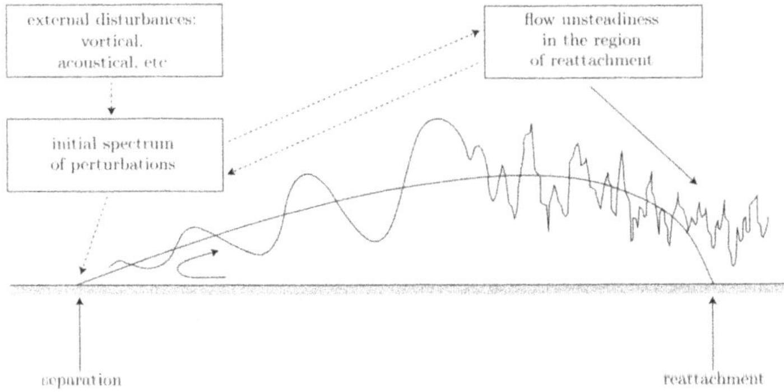

Fig. 6.20. Backward effect of disturbances upon the spectrum of perturbations close to the point of separation

In the experiments of Boiko et al. (1991b) it was shown that such a suppression of the continuous spectrum is caused by the backward influence of the perturbed flow. An interpretation of the above observation is that the initial spectrum of fluctuations near to the separation includes the response of the separation bubble to environmental disturbances as well as to high-amplitude perturbations of the reattaching flow, while the latter are affected by the excitation; see the sketch in Fig. 6.20. This may also be considered as a reduction of low-frequency vertical oscillations of the separated layer induced by the disturbed flow in the region of reattachment.

Non-local interactions between the excited waves and the background disturbances in separation bubbles are of special interest due to a dependence of the transition process on the initial spectrum of perturbations. Despite they are not, in principle, involved in amplification of the small-amplitude disturbances, they are much more important in the non-linear stage of transition. In particular, a key mechanism of the laminar flow breakdown is generation of subharmonic oscillations which are extracted by the resonance from the low-frequency irregular spectrum. One expects that suppression of this might be a reason for the delay in subharmonic generation and the removal of turbulent flow downstream that is observed when the excitation is increased; see Sect. 6.4.

6.6 Laminar–turbulent transition and the origin of coherent vortices

In previous sections of this chapter we focused on the physical model of locally separated flows dominated by the transition to turbulence in the separated shear layer, which is fairly universal and well substantiated by a good deal of exploration results. At the same time, the research data obtained in experiments and calculations reveal another phenomenon besides laminar–turbulent transition, which is initiated by instability of separation bubbles. It consists of the formation of large-scale quasi-periodic two- and three-dimensional vortices shedding into the reattached boundary layer. In this case, the transition to small-scale turbulence plays a minor role whereas the separation region is controlled by the dynamics of the coherent vortices.

Generally their contribution to the flow pattern depends on configuration of the base flow, its mean parameters and environmental perturbations. Vortex shedding is observed at separation behind sharp edges, high steps and on aerofoils, and that induced by a large variation in the external pressure gradient. For experimental data on this subject, see the review papers by Eaton and Johnston (1981) and Kiya (1989); numerical results are reported by Bestek et al. (1989, 1993), Pauley et al. (1990), Tafti and Vanka (1991), Ripley and Pauley (1993), Dallmann et al. (1995) and Lin and Pauley (1996). A tendency that is found in calculations is the onset of vortex shedding upon increases – up to some critical values – of the Reynolds number, step height and pressure gradient (Bestek et al. 1989; Pauley et al. 1990; Dallmann et al. 1995). The effects of external flow disturbances on the vortex motion were investigated in experimental studies by forcing separation regions using some sort of controlled periodic excitation (Roos and Kegelman 1986; Sigurdson and Roshko 1988; Kiya et al. 1991, 1993; Dovgal 1999b); an example given in Sect. 6.4.

Naturally, laminar–turbulent transition in the separated shear layer and the coherent vortices can be distinguished just by their appearances. However, one may expect a much more fundamental reason to consider them as essentially different (but interrelated) instability phenomena. A notion about the origin of vortex shedding, which one can find in a number of research papers, is that the vortices are generated through streamwise growth of the separated-layer disturbances. In this way, the vortex structures are produced at the frequency of the most amplified perturbations in the linear stability problem or at lower frequencies due to vortex pairing and their further amalgamations (Roos and Kegelman 1986; Bestek et al. 1989; Kiya 1989; Pauley et al. 1990; Watmuff 1991, 1999; Hasan 1992; Hein 2000).

On the other hand, it is supposed that the coherent vortices can be initiated by a significantly different instability of separation bubbles; that is, the 'shedding'-type instability as defined by Sigurdson and Roshko (1988). Experimental support for this viewpoint was obtained in the experiments of Dovgal and Sorokin (2001), who examined a separation bubble behind a

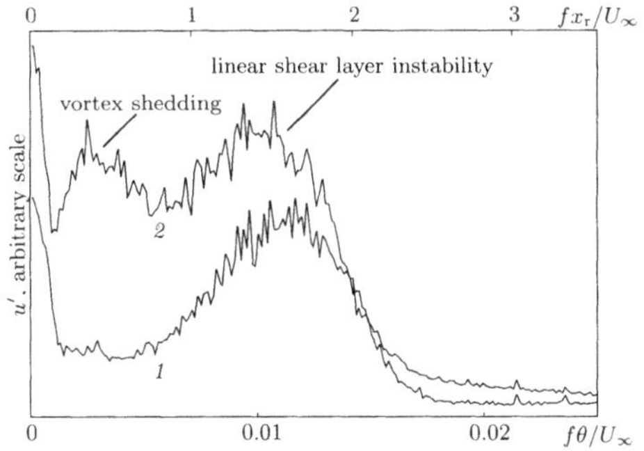

Fig. 6.21. Separated-layer instability and periodic vortices behind a two-dimensional step on a plate (Dovgal and Sorokin 2001), $\mathrm{Re}_h = 1700$. Spectra of disturbances in the upstream part of the bubble (*1*) and in the region of reattachment (*2*). The frequency is normalized by momentum thickness at separation ϑ and the length of reattachment x_r

two-dimensional backstep and observed two different 'modes' of instability associated with the disturbances amplifying in the separated shear layer and the oscillations at the vortex-shedding frequency. At a low level of oncoming perturbations in the wind tunnel and a small Reynolds number, they dealt with the unstable separated flow which, at the same time, was laminar all over the bubble including the region of reattachment. Under these conditions the instability waves of the separated layer and the oscillations at the vortex-shedding frequency, emerging in the reattaching flow, were separated from the frequency spectra of the disturbances (Fig. 6.21). It was found that they occurred independently and did not correlate with each other; thus, the periodic vortices were not produced by amplification of the shear-layer disturbances but resulted from a separation bubble instability to long-wave perturbations.

There is evidence that the vortex-shedding instability indicates that the perturbations of a separation region are regulated not by global rather then local stability properties. Experimental data on periodic vortices indicate a correlation of their kinematics with the length of separation bubbles. That is, the frequency of vortex shedding normalized by the length of reattachment is approximately the same at separation behind backsteps over a wide range of Reynolds numbers (Troutt et al. 1984; Roos and Kegelman 1986; Driver et al. 1987; Hasan 1992) and sharp edges (Kiya 1989), and is close to that of the pressure fluctuations dominating perturbations of the reattaching layer in a variety of flow configurations (Mabey 1972). The results of these observations definitely point to the involvement of a feedback mechanism in the dynamics of coherent vortices, and substantiate the idea that separation bubbles are self-excited systems (Kiya et al. 1993).

The inference that the vortex shedding may be essentially a global dynamic phenomenon which is beyond the accepted model of transitional separation bubbles is further supported by theoretical results. According to Gaster (1992) and Michalke (1993), one expects an absolute instability in separation bubbles which may occur as a precursor of global modes. Results of a global stability analysis applied to the regions of boundary layer separation have been reported by Hammond and Redekopp (1998), Hein et al. (1998),

Bogucki and Redekopp (1999), Theofilis (2000) and Wang et al. (2000). In particular, Hammond and Redekopp (1998) argue that at enlargement of the separation region, a coherent motion synchronized over the entire bubble may occur that overwhelm the convective disturbances generated by external flow perturbations. These theoretical results correlate with the tendency for vortex shedding to occur at an increase of the base-flow parameters, which was found in the numerical simulations by Bestek et al. (1989), Pauley et al. (1990) and Dallmann et al. (1995), and in the wind tunnel data of Dikovskaya et al. (1999), who observed the onset of periodic vortices, passing from a 'small' to a 'large' separation bubble.

6.7 Implication of instability to separation control

The subject of separated-flow instability applies to active control of separation by external periodic forcing. Basically, the method employs the interrelation between the flow pattern in separation regions and their disturbances, so that one can affect mean and non-stationary flow characteristics by the controlled generation of perturbations. In some cases this makes it possible to completely eliminate the separation region at extremely low amplitudes of excitation, thereby minimizing power expenditure of the control process. Obviously to achieve such an effect the imposed oscillations should couple with flow instabilities and energize amplifying disturbances. A variety of control techniques have been examined in this way, including external acoustic excitation and local injection of perturbations by periodic suction, periodic surface heating, vibrations and oscillating flaps.

As the formation of a separation bubble is dominated by the shear-layer transition, one can reduce the separated flow region by stimulating the laminar flow breakdown. Then, with increasing perturbations the momentum transfer in the separated layer becomes larger and the flow tends to reattach early. This was observed in the experiments of Dovgal and Kozlov (1983a) and Zaman and McKinzie (1991), where mid-chord separation bubbles on aerofoils were excited by external acoustic disturbances. In the wind tunnel study of Haggmark (2000), a separation bubble on a plate was diminished exciting the flow by periodic sucking and blowing through a narrow slot in the model surface. The results showed that the bubbles got smaller at quite weak forcing that matched the instability frequencies of the separating layer. A similar effect of periodic forcing upon the mean flow in a separation bubble was found by direct numerical simulations (Rist 1994). Actually, in such a case of separation control the excitation works like other environmental perturbations (for example, free-stream turbulence or surface roughness) which promote the separated-layer transition.

When transition to small-scale turbulence in separation bubbles is shadowed by coherent vortex motion, the controlling disturbances modify the dynamics of vortices; then the method is applicable both to laminar and turbulent separation. Experimental data on this point were obtained by Bhattacharjee et al. (1986) and Roos and Kegelman (1986) for acoustic and oscillating-flap excitation of backstep flows, Sigurdson and Roshko (1988), Kiya et al. (1991) and Kiya et al. (1993) when investigating separation bubbles on blunt circular cylinders during periodic blowing through narrow slots at the leading edge, and Montividas et al. (1992), who examined a separation bubble behind a flap on a flat plate controlled by an oscillatory jet at the surface. It turns out that the excitation in some frequency band reduces the separated flow region, enhancing roll-up and amalgamation of the vortices. A vortex model explaining shortening of the separation bubble under forcing by modification of vortex merging is proposed by Kiya et al. (1993).

A much more pronounced control effect is observed when the oscillatory techniques are applied to globally separated flows. Extensive experiments in separation control on low Reynolds number aerofoils indicate substantial diminution of a separated flow region caused by generation of linear and non-linear disturbances under near- and post-stall conditions; for details and references see the review papers by Gad-el-Hak (1990a), Gad-el-Hak and Bushnell (1991), Fernholz (1993) and Dovgal (1999a). The results obtained in a majority of studies for nominally two-dimensional flows are further supplemented by wind tunnel data on three-dimensional aspects of separation control. Those results are effects of spanwise variation of the controlling disturbances and modification of the global three-dimensional flow structure under the excitation. Experimental results on these points were reported by Kozlov et al. (1993), Dovgal and Grosche (1994), Lushin (1992) and Zanin (1997).

7 Transition prediction and control

In this chapter we consider some applications of the unstable-flow physics for laminar–turbulent transition prediction and control. Basically the purpose of transition prediction is to clarify whether the transition takes place in a flow under consideration and to find (calculate or measure) the Reynolds number of transition, $\mathrm{Re_T}$. If inside the neutral stability curve the disturbance becomes strong enough at its propagation in the streamwise direction, nonlinear mechanisms come to play which lead to the flow turbulization at $\mathrm{Re_T}$. Below we show how the linear stability approach in combination with empirical correlations can be used to predict the location of laminar–turbulent transition with a reasonable accuracy in certain practical situations.

The main purpose of transition control is a delay or acceleration of the laminar flow breakdown. In a laminar flow, the wall friction can be an order of magnitude smaller than that in a turbulent boundary layer (Fig. 7.1). In the case of a vehicle, this means more effective operation with a number of advantages including a reduced fuel consumption. On the other hand, turbulent flows provide high rates of mass and heat transfer and are less prone to separation. Usually, it is easier to turbulize the flow than to keep it in a laminar state, so that in the following sections we will concentrate on the transition delay.

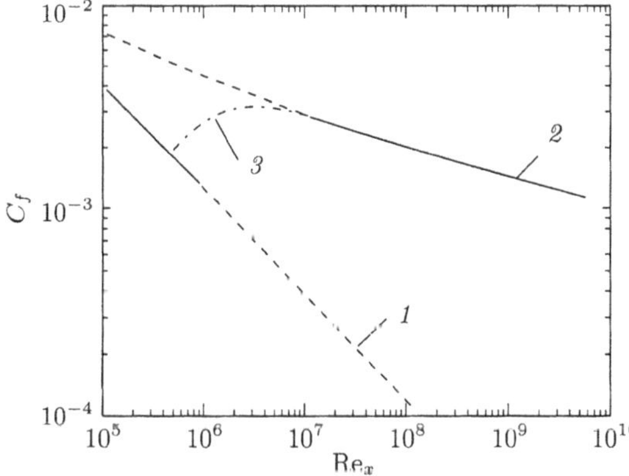

Fig. 7.1. Variation of the wall shear stress coefficient with Reynolds number on a flat plate (Gad-el-Hak 1990b): laminar (*1*), turbulent (*2*) and transitional flows (*3*)

7.1 Transition prediction on the basis of linear stability theory

The laminar–turbulent transition in a boundary layer is a continuous process, starting from excitation of small-amplitude disturbances up to establishment of a developed turbulent flow. However, in applications the notion of 'transition point' or transition Reynolds number $\mathrm{Re_T}$ is generally adopted. Different ways of experimental determination of the transition point are based on observation of variations in the flow structure and its integral characteristics. These include a deviation of the mean velocity distribution from a laminar one; changes of the total pressure distribution (one of the most widespread methods), friction coefficients and heat transfer; and the appearance of the first turbulent spikes and intermittency (see Chap. 4.4.1).

The practical importance of determining the transition position highlighted the problem of its theoretical prediction. A group of the methods is based on the Taylor's model of turbulence initiation by local separations (Dorodnitsyn and Loytsyanskiy 1945; Kozlov 1968; Ostoslavskiy and Svistchev 1975). Another approach employs the model equations of a developed turbulent flow (Glushko 1973; Arnal and Juillen 1977). From a physical point of view, the most justified are the methods based on the concept of hydrodynamic stability. In this case, the prediction of transition point should include three basic elements:

1. Determination of the structure of initial boundary layer disturbances excited by various external perturbations.
2. Calculation of the linear development of small disturbances in a boundary layer.
3. Determination and calculation of the dominant non-linear processes that characterize the beginning of final laminar flow breakdown.

Certain techniques has been developed allowing calculation of the initial amplitudes of Tollmien–Schlichting waves in some practical situations (see Chap. 3). The computations of the linear development of boundary layer disturbances are quite well advanced, and the adequacy of the description of Tollmien–Schlichting wave amplification by the linear theory of hydrodynamic stability has been confirmed experimentally (Chaps. 1, 2). The calculation of the non-linear stage of transition is a more difficult problem, which can be sometimes bypassed in practice. It is known that at low free-stream turbulence level, non-linear processes are usually very fast, and in the major part of a transitional boundary layer (90–95%), the development of small-amplitude disturbances described by the linear stability theory takes place. This enables – in a number of cases – the use of linear theory for prediction of the transition point, neglecting the details of non-linear processes.

According to experimental data and theoretical estimations, the non-linear processes usually begin to play a noticeable role at amplitudes of Tollmien–Schlichting waves of about $u' \approx 1\%$ of free-stream velocity U_∞

(see Chap. 4). Since the non-linear zone is relatively short, it is possible to admit the linear growth of disturbances in calculations a little above their real saturation amplitude and accept a transition point at which the amplitude of a Tollmien–Schlichting wave of a certain frequency (defined during the calculations) reaches the value $u' = 2$–4% of U_∞. In such a way, knowing the initial amplitude–frequency characteristics of Tollmien–Schlichting waves (from an experiment or a solution of a relevant receptivity problem), it is possible to calculate the transition position with reasonable accuracy. This ideas constitute the essence of the so-called e^n-method which received wide acceptance (Smith and Gamberoni 1956; Jaffe et al. 1970; Michel et al. 1985a). It represents a good example of the use of academic results on hydrodynamic stability to purely practical problem of predicting the transition location in convectively unstable flows. Its successful application to various complex situations allows us to look optimistically at the construction of a rational way of the transition prediction at certain knowledge of initial perturbations.

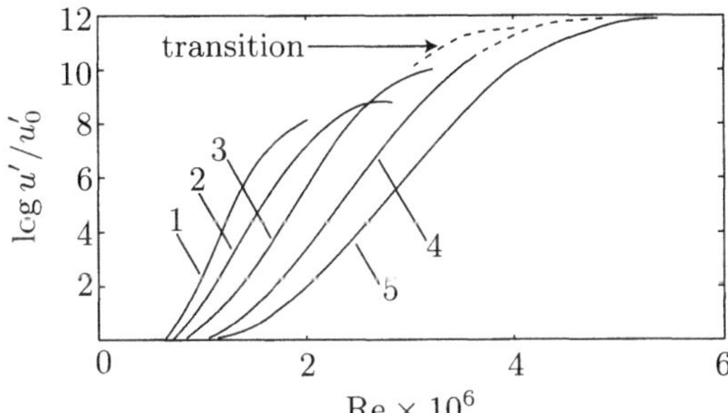

Fig. 7.2. Amplification coefficient for a flat plate (Levchenko et al. 1975): $F \times 10^6 =$ 42.0 (*1*); 34.8 (*2*); 28.7 (*3*); 29.7 (*4*); 20.0 (*5*)

According to the method, the amplification of disturbances of various frequencies is calculated as

$$\bar{A}(\mathrm{Re}) = u'/u'_0 = \mathrm{e}^n, \tag{7.1}$$

so that

$$n = \ln(u'/u'_0) = -\int_{\mathrm{Re}_0}^{\mathrm{Re}} \alpha_i \mathrm{dRe},$$

where index '0' corresponds to branch I of the neutral stability curve. The curves $\bar{A}(\mathrm{Re})$ have maxima when the waves pass branch II of the neutral curve and then decay (Fig. 7.2). The calculated value of $\bar{A}_{\max}$ corresponding to the experimental $\mathrm{Re_T}$ is determined by the exponent n in (7.1).

The effect of external disturbances (an initial amplitude of the wave) is taken in account implicitly in the method by prescribing n for classes of flows with close free stream and boundary conditions. However, the value

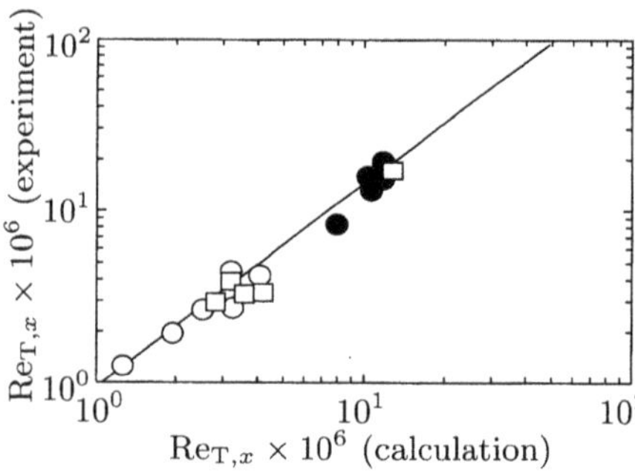

Fig. 7.3. Comparison of the data on transition (Jaffe et al. 1970). The Reynolds number Re_{T} is calculated by $u'/u'_{\min} = \mathrm{e}^{10}$: ○, flat plate; •, aerofoil; □, bodies of revolution

of exponent n depends not only on actual flow conditions but also on the method used for the calculation of amplification. Application of a spatial or time stability theory, accounting or neglecting the flow non-parallelity, proper approximation of mean velocity profiles, etc., can significantly affect the value of n.

In such a way, Jaffe et al. (1970) conducted calculations of spatial increments on the basis of numerical integration of the Orr–Sommerfeld equation, and found that the available experimental data on Re_{T} for wing profiles and bodies of revolution obtained in wind tunnels with a small degree of turbulence and in flight experiments satisfactorily correlate for $n = 10$ (Fig. 7.3). According to earlier calculations based on temporal stability theory, $n = 9$, which gives satisfactory value of Re_{T} in a number of flight experiments.

The successful correlation of experimental data on transition by the e^n-method ($n \approx 9$–10) confirms a determining role of the linear region of disturbance development during the laminar–turbulent transition at low free stream turbulence level. It suggests presence of a universal criterium of laminar flow destruction related to the Tollmien–Schlichting wave amplitude, and indicates that flow disturbances in modern low-turbulence wind tunnels and under flight conditions are approximately identical, though they may have different sources.

For external flow perturbations that are considerably different, the value of n varies. The results of experiments for definition of the transition point on a glider are well generalized by the e^n-method, but with the exponent $n = 15$ (Runiyan and Gerge-Falvy 1979). Such a high value of the exponent is stipulated by the absence of disturbing sources compared with usual flight experiments connected, for example, with propulsion system. To account for the influence of external turbulence on transition in a boundary layer on a flat plate, Mack (1975) obtained the interpolation formula $n = -8.43 - 2.4 \ln A_1$. The value $n = 9$ corresponds to turbulence level $\mathrm{Tu} = 0.07\%$ and $\mathrm{Re}_{\mathrm{T}} = 3.5 \times 10^6$.

The experimental observations of transition in a rotating disc flow and in a swept-wing boundary layer show the presence of quite extended regions of linear-waves development, which enables extension of the e^n-method to three-dimensional boundary layers. The transition to turbulence in a swept-wing boundary layer can occur in the region of negative pressure gradient as a result of crossflow instability, or downstream where there is an amplification of Tollmien–Schlichting waves. Then it is possible to apply the e^n-method to both kinds of instability and to consider that transition occurred if at least one of the criterium is satisfied (Michel et al. 1985a).

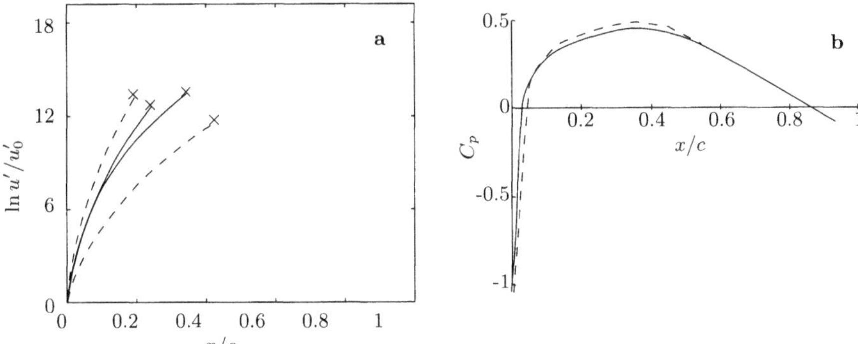

Fig. 7.4. Calculation of disturbance amplification for four frequencies in the boundary layers (Lekoudis 1979) on wings: *dashed line*, wing sweep angle of 30° and *solid line*, wing sweep angle of 40° (**a**); the relevant pressure profiles (**b**); ×, position of transition (Bolz et al. 1960)

The attempts to correlate experimental data on transition on swept wings by the e^n-method are encouraging, though the exponent n appears to be different in various experiments. The experimental data obtained in a wind tunnel by Bolz et al. (1960) were generalized by Lekoudis (1979) at $n = 12$ (Fig. 7.4). At the same time, calculations by a method proposed by Srokowski and Orszag (1977) give $n = 10$. The correlation of results of wind tunnel and flight experiments on the swept-wing transition with an average value of $n = 7.2$ or 9.8 depending on the method used for computation of the spatial wave amplification was carried out by Hefner and Bushnell (1979).

The well-substantiated feasibility of a linear stability theory to locally separated flows is used for the prediction of transition to turbulence initiated by boundary layer separation with the help of the e^n-method (Nayfeh et al. 1988; Cebeci and Egan 1989; Haggmark 2000), and in semi-empirical models of separation bubbles in which the location of the transition point is not prescribed by empirical correlations but is calculated through a stability analysis (Van Ingen 1975, 1991; Drela and Giles 1987; Dini et al. 1992). In the studies of Al-Maaitah et al. (1990a, b), Masad and Nayfeh (1993), Masad and Iyer (1994) and Masad and Malik (1994), the latter approach is extended to compressible flows, separation under non-isothermal conditions and at boundary layer suction.

Further development of the e^n-method of transition prediction is connected with taking into account the background disturbances: their spectral structure and intensity; i.e. with solution of the receptivity problem (Crouch and Ng 2000; Gilev 1985; Saric et al. 1991; Bodonyi and Duck 1992). From an engineering point of view, of special significance is the determination of the conversion coefficient of the external disturbances in the instability waves of a boundary layer and of the initial phase of excited waves for possible application of wave cancelling technique (see Sect. 7.2.3). The control can also be achieved by selection of relevant parameters of possible technological irregularities so as to ensure the minimal factors of receptivity (Kobayashi et al. 1995), or through some non-local effects of a hysteresis type (Kozlov 1985). Besides, from this point of view the study of receptivity effects is also significant to devices that are intended for flow control promoting effective wave transformation (Spalart 1993).

7.2 Basic flow control techniques

Following the basic knowledge on boundary layer transition, the main control methods can be divided into two groups. The first one comprises those which modify the stability characteristics of controlled flow and reduce the growth of perturbations. These methods, coming of the results of linear stability theory, include shaping of a body contour, surface cooling (or heating) in air (or in water), boundary layer suction and wall motion. In the second group are the methods which directly affect linear and non-linear laminar flow disturbances by the adjustment of initial and boundary conditions. The wave number and frequency spectra of instability waves can be modified by reduction of external acoustic, oncoming-flow turbulence and surface vibrations, and by smoothing of the wall and protection from its contamination by a technique of waves cancellation. Transition control by riblets, which is still not well understood, seems to be in the same list as well. The laminar flow can be extended by individual application of the above methods and their combinations as well as by more exotic techniques employing the phenomenon of cavitation and chemical reactions.

From an engineering viewpoint, the methods of transition control are distinguished as passive or active. In the first case, the flow is affected without expending energy in the control process. To drive the latter, an energy-consuming device is used. Normally the passive control does not employ a feedback loop of the transition detection and manipulation, whereas the active one is adaptive and follows changes of the flow structure. Practically, a proper choice of the control method depends on its complexity, the costs of manufacture and maintenance, as well as indirect costs which, for example, are bound with environmental ecology. Often one have to search for a compromise between the above points because of contradictions which may occur.

In what follows we focus on some of main control methods, both accepted and perspective. For more details on transition control see other reviews, for example, by Riley et al. (1988), Carpenter (1990), Gad-el-Hak (1990a, b) and Hefner (1991).

7.2.1 Mean flow adjustment

Shaping and pressure gradient modification. This method consists of body contouring to achieve an external flow pressure distribution that maximizes the extent of the laminar flow. In this respect, a negative streamwise pressure gradient is favourable and a positive one is unfavourable for the transition delay. Moreover, in a decelerating flow the boundary layer may separate from the wall, which is extremely undesirable in terms of the control goals. Thus, a purpose of shaping is to maintain the attached adverse pressure gradient flow as long as possible. Reduction of $\mathrm{Re_T}$ at $\nabla p > 0$ and its growth at $\nabla p < 0$, being the effects of pressure gradient upon boundary layer stability, are well know from transition research on aerofoils (Schlichting and Gersten 2000). Stabilization of a boundary layer to Tollmien–Schlichting waves by a negative pressure gradient and vice versa had already been observed in the transition experiments of Schubauer and Skramstad (1948).

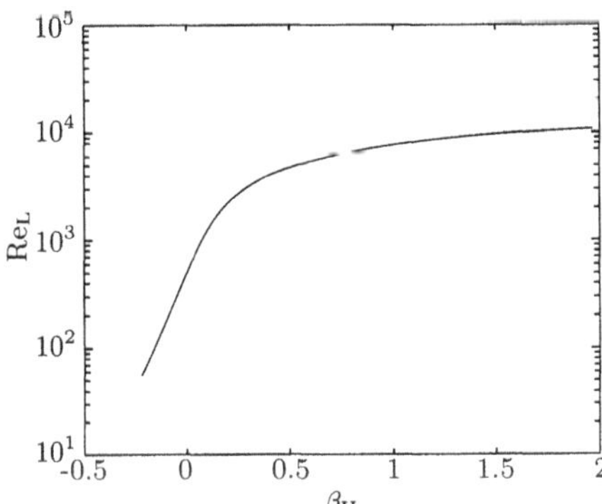

Fig. 7.5. Effect of pressure gradient on the stability of Falkner–Skan profiles (Volodin and Gaponov 1970)

The stability analysis of a pressure gradient flow is carried out in the same way as that of a flat plate boundary layer, assuming a dependence of the mean velocity on the normal coordinate only. Then the pressure gradient affects the stability characteristics through the shape of the velocity distribution $U(y)$. The results of corresponding stability calculations for the Falkner–Skan profiles are shown in Fig. 7.5. As the streamwise pressure grows, the critical Reynolds number is reduced stimulating the transition to turbulence. On the contrary, as the pressure goes down the boundary layer turns more stable promoting the transition delay.

Boundary layer suction. An effective control method is continuous boundary layer suction which enables a drastic increase in the transition Reynolds number (Fig. 7.6). The transition delay by suction is achieved through its double effect on the near-wall flow. Firstly, it reduces the thickness of the boundary layer which becomes less prone to laminar–turbulent transition. Secondly, it makes the mean velocity profiles more stable with a higher critical Reynolds number, as found theoretically by Levchenko and Soloviev (1970) and numerically by Maksimov (1975). Experimentally the stabilization of a flat plate boundary layer to Tollmien–Schlichting waves by suction through a transverse slot on the surface was observed by Kozlov et al. (1978) (Fig. 7.7). In the controlled flow the amplitude of perturbations is diminished across the whole boundary layer in front of and behind the slot. Meanwhile, the small-amplitude disturbances remain linear, i.e. their amplitudes are independent of their initial amplitude.

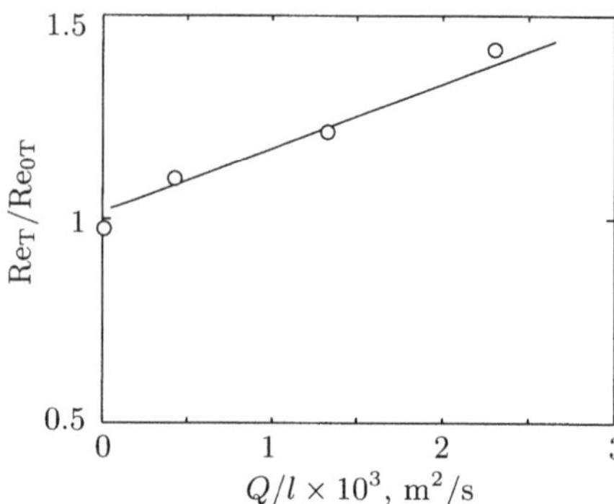

Fig. 7.6. Variation of transition Reynolds number with strength of suction on a flat plate (Kaidalov and Morin 1972)

There are different variants of the method including a distributed suction through porous or perforated surfaces and a localized one through slots or holes. Though it is not a simple problem from a theoretical viewpoint, the transition control by suction nowadays is often just a technical problem. The method is less suitable for underwater applications because of contamination of the suction devices which reduces their effectiveness and, moreover, can destabilize the boundary layer. When maintaining laminar flow an important point is optimization of the suction rate. One expects that its increase makes the boundary layer thinner with the onset of instability occurring well downstream compared to the situation without control. However, this becomes unreasonable in cases where the energy saving due to drag reduction is comparable with its expenditure in the control process. Moreover, in a too-thin laminar boundary layer, the wall shear stress becomes rather high, thereby increasing the drag of a body.

Possible applications of boundary layer suction to flow control are not exhausted by suppression of the instability to Tollmien–Schlichting waves. In a three-dimensional boundary layer the method can be used for stabilization of the crossflow component (Mack 1980; Bippes et al. 1999). Also, some effect can be obtained by controlling non-linear disturbances and secondary instabilities. In the experiments of Arnal et al. (1997), a postponement of the tran-

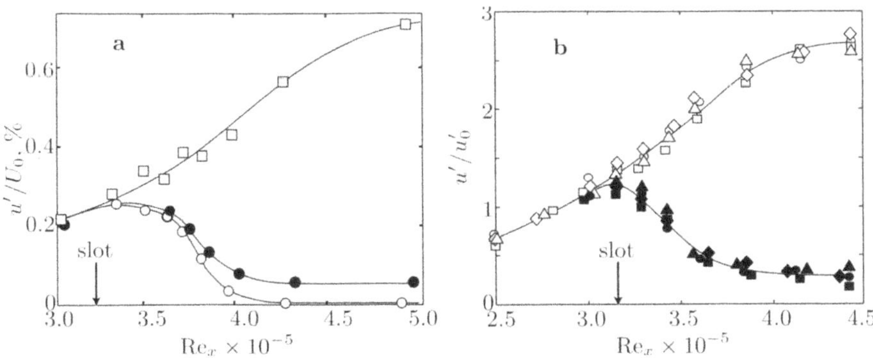

Fig. 7.7. Effect of suction on the transition. **a** streamwise behaviour of the maximum amplitude of boundary layer disturbances with suction through a slot on a flat plate: □, without suction; •, with suction rate $\mathrm{Re}_d = U_s d/\nu = 52$; ○, 104; **b** different initial disturbance amplitudes: without suction $u'_0/U_0 = 0.10$ (○), 0.27 (△), 0.36 (□) and 0.66% (◇). With suction at $\mathrm{Re}_x = 3 \times 10^5$, $\mathrm{Re}_d = 52$ for the same values of u'_0 (*filled symbols*) (d is slot width) (Kozlov et al. 1978)

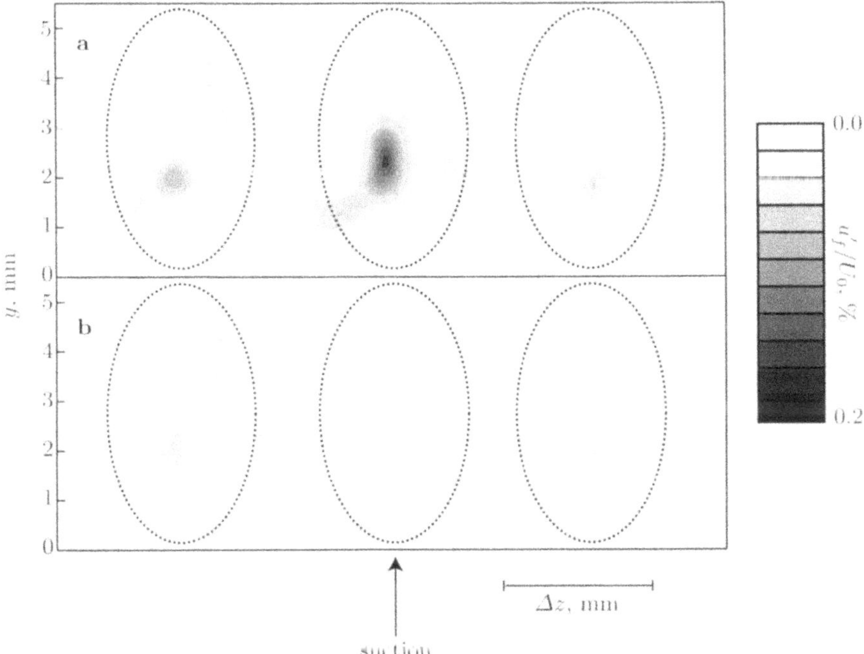

Fig. 7.8. Effect of suction on secondary instability of streamwise stationary vortices in a boundary layer (Boiko et al. 1999). Amplitude of the travelling waves without (**a**) and with (**b**) suction

sition on a swept wing caused by a leading-edge contamination was observed. When investigating a subharmonic boundary layer instability, El-Hady (1991) found a variation of the flow pattern in the non-linear region of transition during a rather strong suction. In the wind tunnel tests of Boiko et al. (1999) and Bakchinov et al. (1999), a boundary layer modulated with streamwise stationary and non-stationary vortices was controlled by suction through a small hole on the surface. As a result, a decay of secondary perturbations propagating along the vortices was observed, which was most pronounced for the waves travelling on the vortex just above the suction hole (Fig. 7.8). Boundary layer suction for secondary-instability control in a swept-wing flow with stationary crossflow vortices was used in the experiments of Egami and Kohama (1999).

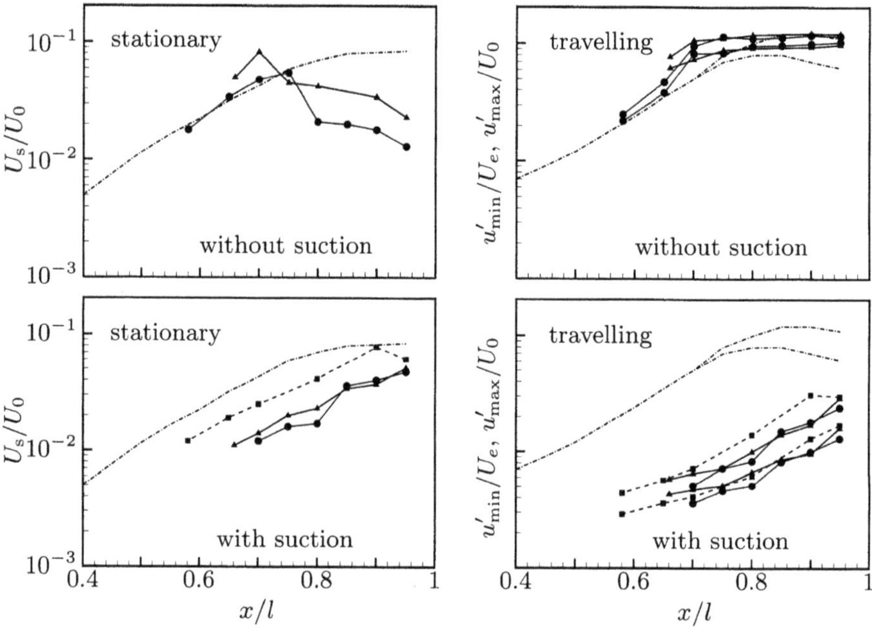

Fig. 7.9. Spatial growth of the stationary (U_s/Q_e) and minimal/maximal (along the wing span) travelling u'_{min}/Q_e, u'_{max}/Q_e disturbance amplitudes (*dashed lines*) for different suction surface geometries: ▲, holes 80 μm; ■, holes 120 μm; •, slots 100 μm; *dash-dotted line*, polished surface (Deyhle and Bippes 1996); chord length l=500 mm (Abegg et al. 2000a)

The experiments with swept-wing models carried out by Bippes et al. (1999) and Abegg et al. (2000b) (see also Abegg et al. (2000a) for a review) were aimed to elucidate not only the physical mechanisms of suction, but also to optimize the suction surface geometry (holes or slots) with respect to the effective control of transition. For a constant porosity of 1% and suction rate of $c_q = 0.1\%$, three different suction panels were used: spanwise slots and perforated sheets with holes of two different diameters. As one can see in Fig. 7.9, the travelling disturbances are more strongly damped than the stationary modes, with the spanwise slots producing better attenuation of the disturbances. Furthermore, the disturbance amplitudes in all cases tested with suction were much lower at the end of the measurement region than those without suction, and even without the suction panels over the polished surface.

Heat transfer. A difference between the temperatures of the wall T_w and the ambient stream T_∞ produces a heat flux which modifies mean velocity profiles and the boundary layer stability through variation of the fluid viscosity. In a gas, which becomes more viscous with increasing temperature, wall heating destabilizes the flow while wall cooling has the opposite effect. The stability characteristics of non-isothermal flows were calculated asymptotically by Lin (1955) and numerically, by Gaponov and Maslov (1971), who determined, in particular, critical Reynolds numbers at low subsonic free-stream velocities (Fig. 7.10a).

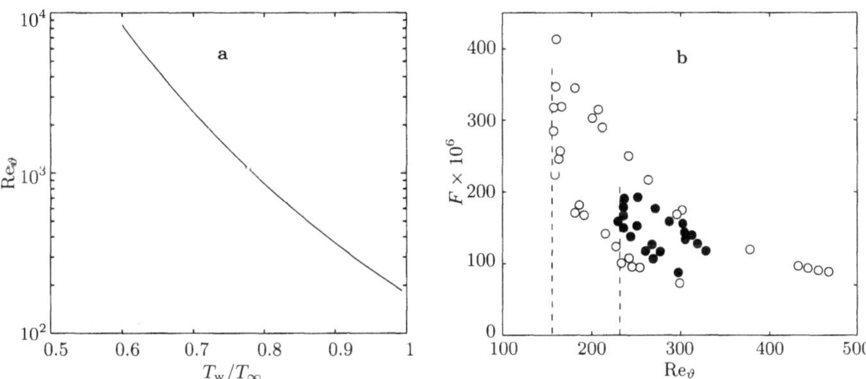

Fig. 7.10. Effect of wall cooling upon flat plate boundary layer stability in air: **a** linear stability calculations of the critical Reynolds number (Gaponov and Maslov 1971); **b** experimental neutral curves of for isothermal flow (○) and wall cooling at $T_w/T_\infty = 0.945$ (•) (Kachanov et al. 1974)

Wind-tunnel experiments on boundary layer stability during wall cooling were carried out by Kachanov et al. (1974). Their results on the neutral stability of a flat plate flow are shown in Fig. 7.10b. Under the control the neutral curve moves towards higher Reynolds numbers over a reduced frequency range of amplifying disturbances. The critical Reynolds number Re_ϑ, where ϑ is the momentum thickness, grows from 155 to 235 (i.e. increasing by a factor of 1.5).

Thus the wall cooling results in an increase of the critical Reynolds number, reduction of the frequency band of amplifying disturbances and their growth rates. Therefore the stability region becomes more extended and the following stages of transition move downstream. We note that when stabilizing the boundary flow to Tollmien–Schlichting waves, the method appears less effective at suppression of other instability modes (Lekoudis 1980; Mack 1980).

While the destabilizing influence of wall heating in a gas flow has been known for quite a long time, the results of linear stability calculations and experimental data indicate that under certain conditions it leads to the opposite effect and can be used for boundary layer stabilization. This becomes possible with a non-uniform surface temperature distribution. In a limited flow section the heat flux is directed from the surface to the gas so that downstream of the heating area a situation which is similar – but not identical – to wall cooling occurs, when the boundary layer develops over the surface which has relatively low temperature. In this way, despite some destabilization in the initial flow region, the subsequent reduction of the growth rates of disturbances produces a nett positive effect.

Among the first theoretical studies focusing on the stabilizing influence of non-uniform wall heating upon boundary layers were those by Lebedev and Fomichev (1987), Kazakov et al. (1985) and Struminskiy et al. (1986). The results of linear stability calculations were later justified in experimental study of Dovgal et al. (1990). The wind tunnel results showed that the local heating of a flat plate near to its leading edge or its downstream sections reduces amplification of the boundary layer oscillations and increases the transition Reynolds number. Similarly, a postponement of the boundary layer transition induced by a three-dimensional roughness element was observed. Moreover, it was found that the method can be used for laminarization of a three-dimensional boundary layer, but only in the case when transition is initiated by Tollmien–Schlichting waves. For control of the crossflow instability, the localized heating of a surface appeared ineffective.

Contrary to a gas, in a flow of liquid the viscosity decreases with temperature growth. Accordingly, the wall heating stabilizes and its cooling destabilizes a boundary layer. In water with high heat conductivity and strong dependence of viscosity on temperature, the method appears very effective for transition delay and drag reduction. As a result, the wall heating by a few degrees can be sufficient to produce a noticeable control effect, which was found in the experiments of Barker and Jennings (1977), Strazisar et al. (1977), Arakeri (1980) and Lauchle and Gurney (1984). In particular, Lauchle and Gurney (1984) observed an almost tenfold increase in the transition Reynolds number at a temperature difference about 25°C.

Stability calculations for boundary layers in water were carried out in a parallel-flow approximation by Wazzan et al. (1970) and Lowell and Reshotko (1974). In the first of these studies the solutions were obtained whilst neglecting temperature and viscosity fluctuations, while the second one took them into account. The results of both investigations coincided well, indicating a weak coupling between vortical and temperature fluctuations, and leading

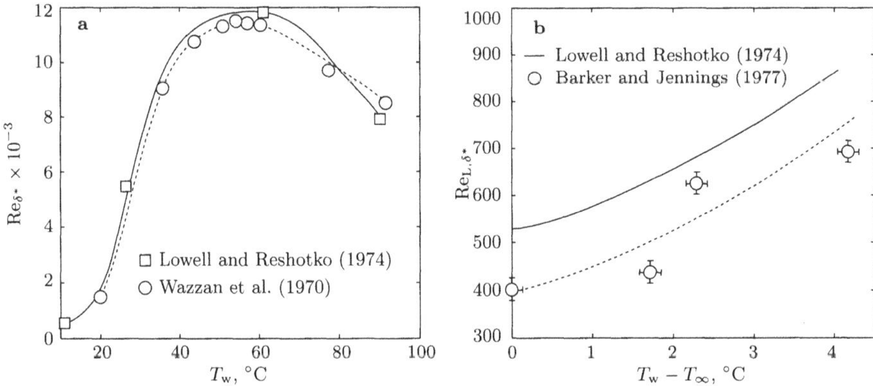

Fig. 7.11. Effect of wall cooling upon the stability of a flat plate boundary layer in water: **a** variation of the critical Reynolds number of a flat plate flow with wall temperature; **b** stabilization of the boundary layer by surface heating. *Solid line*, stability solution; *dashed line*, approximation of experimental data

to the prediction of a pronounced boundary layer stabilization at moderate heating of the wall (Fig. 7.11a). Also, it follows that with an increase of wall temperature the Reynolds number at which the flow becomes destabilized shifts to higher values.

The effect of wall heating upon small-amplitude disturbances of boundary layers were explored experimentally by Barker and Jennings (1977) and Strazisar et al. (1977). The data obtained were in qualitative agreement with the theoretical results (Fig. 7.11b). In the study of Strazisar et al. (1977), the growth rates of boundary layer disturbances were also measured. Within the examined range of base-flow parameters, a weak heating of the wall resulted in an increase of the critical Reynolds number, reduction of the growth rates, and narrowing of the instability ranges over frequency and wave numbers of the oscillations.

7.2.2 Riblets

Streamwise small grooves on a body surface used for boundary layer control are known as riblets. Research data on the natural riblets on shark skin in the mid-1960s resulted in an idea to apply them to the reduction of turbulent friction (Fig. 7.12). In later studies, optimization of the riblet shape and their spanwise spacing and depth was performed (Walsh 1983; Bechert and Bartenwerfer 1989; Choi 1989; Savill 1990; Luchini et al. 1991) (Fig. 7.13). Though the mechanism of how the riblets influence the near-wall flow is not well-understood, this control method is accepted as an effective one for the manipulation of turbulent boundary layers. Engineering applications of ribbed surfaces to flow control are discussed by Coustols and Cousteix (1990), Savill (1990) and Coustols (1991).

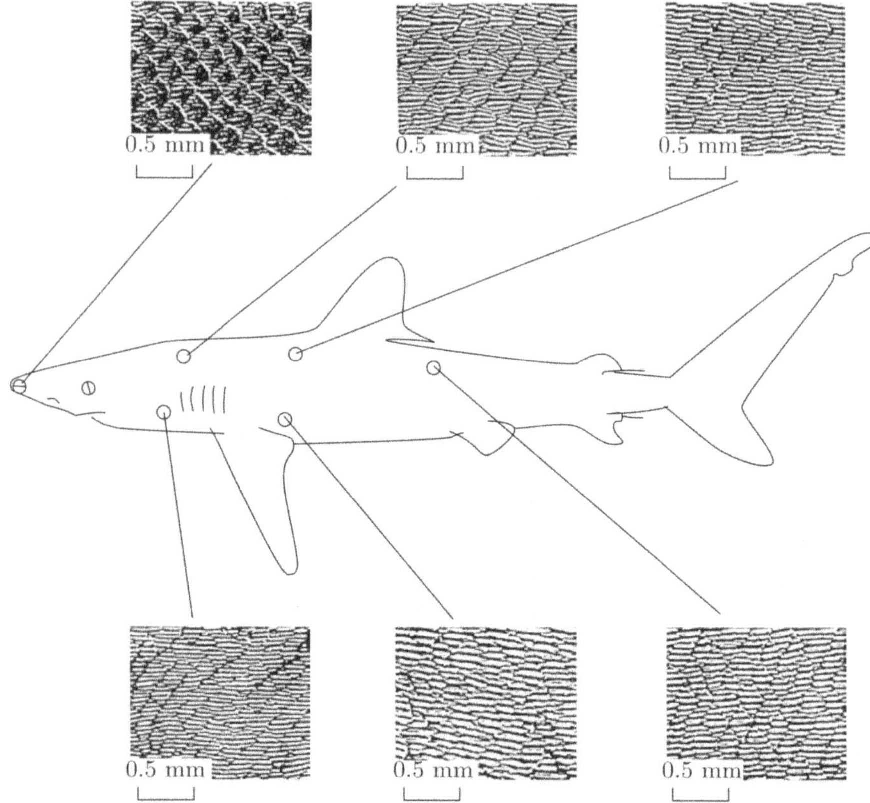

Fig. 7.12. Riblets in nature (Bechert and Bartenwerfer 1989)

The successful control of turbulent flows initiated research work on the method in transitional boundary layers. Belov et al. (1990) found that riblets covering the entire length of a flat plate surface accelerate the laminar–turbulent transition irrespective of their streamwise or crosswise arrangement. On the other hand, in the study of Kozlov et al. (1990), where riblets were also stretched over the whole test surface, it was observed that their influence upon the transition could be positive or negative depending on the external pressure gradient and flow details near the leading edge of the experimental model. A beneficial effect of riblets on turbulent structures originating at the transition on a body of revolution was reported also by Neumann and Dinkelacker (1991). A drag reduction in the transitional region on a ribbed surface was obtained in the calculations of Chu et al. (1992).

It seems that these results are somewhat contradictory because in the experiments the overall effect of riblets on transition was examined without focusing on its different stages and dominant perturbations including linear and non-linear waves, Λ-structures, turbulent spots, etc. More details of tran-

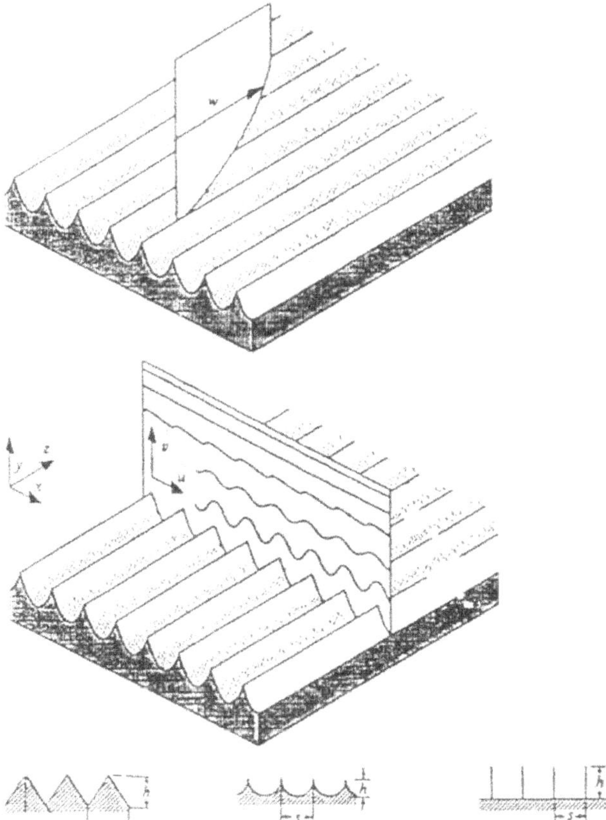

Fig. 7.13. Typical riblet shapes (Luchini et al. 1991)

sition control were elucidated through experimental modelling in a series of studies by Grek et al. (1995, 1996a, b). Wind tunnel results have shown that riblets mounted in a flat plate boundary layer stimulate the transition destabilizing the flow to Tollmien–Schlichting waves, but have an opposite effect when modifying evolution of Λ-structures and stationary vortices. Thus it was found that with this control method one can extend the late stages of laminar–turbulent transition, while accelerating its initial phase.

An enhancement of Tollmien–Schlichting waves by both of streamwise and transversal grooves on a flat plate surface is shown in Fig. 7.14a. The experimental data correlate with a theoretical conclusion about boundary layer destabilization by riblets in the linear instability region (Luchini et al. 1991; Luchini 1995; Ehrenstein 1996). However, in the non-linear region of the transition, the longitudinal riblets reduce the amplitude of perturbations and delay the transformation of an isolated Λ-structure in a turbulent spot (Fig. 7.14b) (at the same time, the crosswise riblets still accelerate the break-

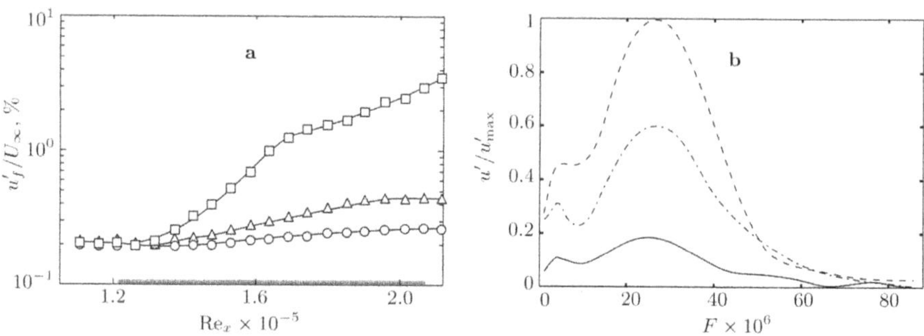

Fig. 7.14. Effect of riblets upon different transitional structures: **a** amplification of a Tollmien–Schlichting wave over riblets on a plate (Grek et al. 1996b); **b** power spectra of Λ-vortices on a plate with riblets (Grek et al. 1996b) at $Re_x = 1.96 \times 10^5$. □ (*solid line*), longitudinal grooves; △ (*dashed line*), transversal grooves; ○ (*dash-dotted line*), smooth surface; *grey region*, riblet range

down of a localized disturbance). At K-type of boundary layer transition when the structures are arranged with a small spacing (Fig. 4.4), the influence of riblets upon the perturbations appeared to be the same (Fig. 7.15). Optimization of the streamwise grooves on their influence upon a Λ-structure was carried out by Grek et al. (1996b), using triangular riblets, which are considered among most promising. The height that produced the maximum effect was determined, which appeared close to that used for control of turbulent flows.

In the experiments of Grek et al. (1996a), riblets were applied to control the flow past a cylindrical roughness element on a flat plate. The result was a narrowing of the wake behind the roughness and delay of the laminar–turbulent transition. The suppression of turbulence, as well as the above effect of riblets on Λ-structures, can be explained by inhibition of a transverse motion in the near-wall region by the streamwise grooves which, in turn, prevents the generation of secondary instabilities.

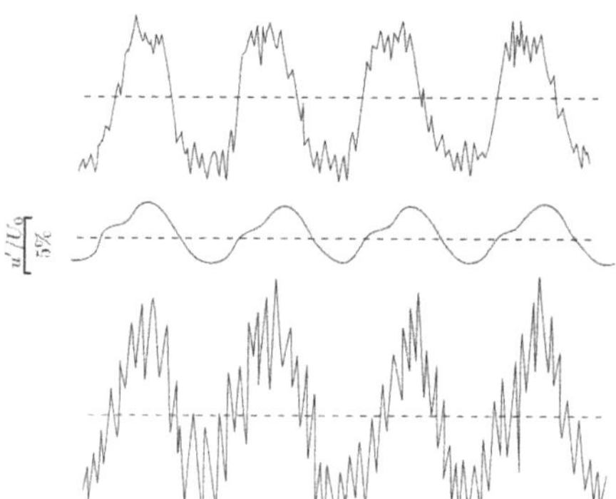

Fig. 7.15. Time traces of the maximum boundary layer disturbances at the K-type of transition on a plate (Grek et al. 1996b). Smooth surface (*top*), longitudinal (*middle*) and transversal (*bottom*) riblets at $Re_x = 2.4 \times 10^5$

One further possible application of riblets is control of a boundary layer modulated by streamwise stationary vortices, where the transition to turbulence can occur as a result of their instability to high-frequency oscillations (see Chap. 5). This problem was explored by Grek et al. (1995) for a flat plate flow perturbed by streamwise vortices which were generated by three-dimensional roughness elements periodically spaced across the test surface. The influence of riblets on mean velocity and natural boundary layer disturbances is shown in Fig. 7.16. In the controlled flow, the transverse mean flow modulation is reduced and the convective perturbations travelling along the vortices are diminished. The same was observed when modelling the high-frequency oscillations excited by external acoustic excitation, when the growth of disturbances on the smooth surface resulted in the transition to turbulence, whereas on the ribbed one the flow remained laminar with a low level of fluctuations (Fig. 7.17). Similar results were obtained in experiments by Boiko et al. (1997c), who dealt with transition to turbulence in an isolated stationary vortex of large amplitude embedded in a swept-wing boundary layer. They found that riblets had a large influence on the strength of vortex and stabilized the flow to secondary convective perturbations.

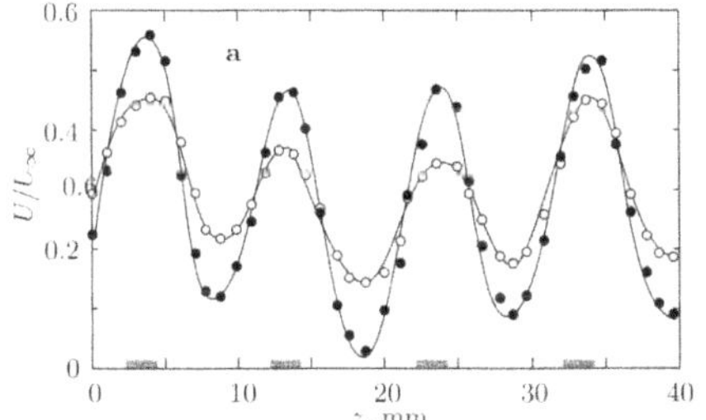

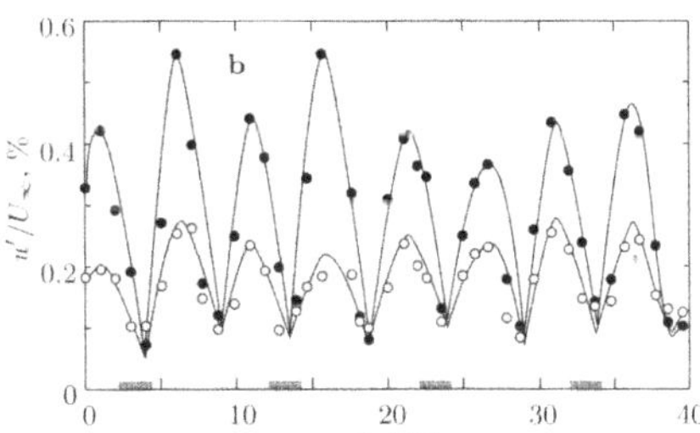

Fig. 7.16. Effect of riblets on a transitional flat plate boundary layer modulated by stationary streamwise vortices (Grek et al. 1995). Transversal variations of the mean flow (**a**) and amplitude of natural disturbances (**b**) on the smooth (•) and grooved (○) surfaces at $\mathrm{Re}_x = 2.5 \times 10^5$. *Grey regions*, location of the vortex generating roughness

7.2.3 Wave cancellation

An approach to transition control which consists of manipulation of the boundary layer disturbances without affecting the base-flow characteristics is suppression of the instability waves by external periodic forcing with an appropriate phase difference from the controlled perturbations. A number of techniques can be used for this, including periodic sucking and blowing (Biringen 1984; Gilev and Kozlov 1987; Danabasoglu et al. 1991; Laurien and Kleiser 1989; Boiko et al. 1999), wall heating and cooling (Liepmann et al

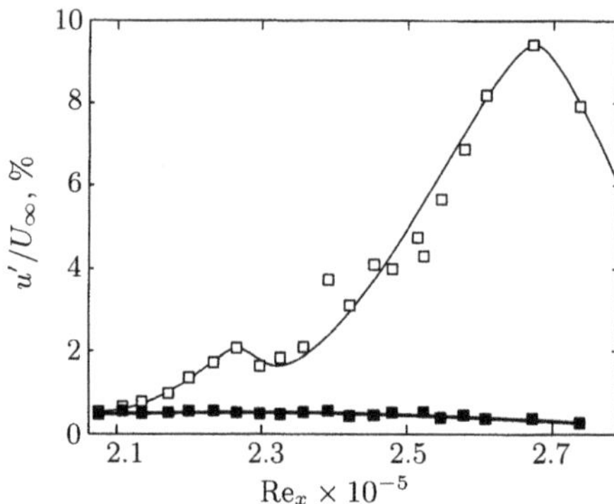

Fig. 7.17. Amplification of travelling disturbances initiated by external acoustics in the boundary layer perturbed by streamwise vortices (Grek et al. 1995). Smooth (*empty symbols*) and grooved (*filled symbols*) surfaces

1982; Liepmann and Nosenchuk 1982; Bayliss et al. 1986), localized surface vibrations (Gedney 1983; Gilev 1985; Gilev and Kozlov 1985; Efremov et al. 1988) and vibrating devices inserted into the boundary layer (Milling 1981; Thomas 1983). In these ways one can reduce the amplitudes of amplifying wavy disturbances and increase the transition Reynolds number. A merit of the method is its low energy consumption.

Among the experimental results on wave cancellation are those of Milling (1981), who excited instability waves using a vibrating ribbon and controlled them by oscillations that were generated with an opposite phase. Liepmann et al. (1982) and Liepmann and Nosenchuk (1982) affected the boundary layer transition in water by periodic heating of the wall. By appropriate choice of the phase of the oscillations they damped a dominant perturbation induced by background disturbances of their facility, increasing the transition Reynolds number by up to 30%. According to these results the same transition delay in water can be obtained using two order of magnitude less electric power for oscillatory heating than for continuous heating. Gilev and Kozlov (1985) used surface vibrations for the same purpose, and Gilev and Kozlov (1987) employed a periodic sucking and blowing through a perforated wall section and a transverse slot, controlling the instability waves excited by a vibrating ribbon. In both cases the boundary layer transition was postponed at opposite phases of the oscillations (Fig. 7.18). In the first of these studies an amplitude of vibrations of several microns was enough to get a substantial transition delay.

Boiko et al. (1999) applied the method for suppression of secondary instability in a boundary layer modulated by stationary streamwise vortices. In their experiments, the travelling waves developing on the vortices were modelled by external acoustic excitation and the controlling disturbances were generated by periodic sucking and blowing through a hole on the test surface. As a result, they managed to reduce the amplitude of secondary perturbations within the vortex which was above the source of oscillations (Fig. 7.19). Further extension of the idea to postpone the onset of turbulence by injection of controlled boundary layer disturbances was reported

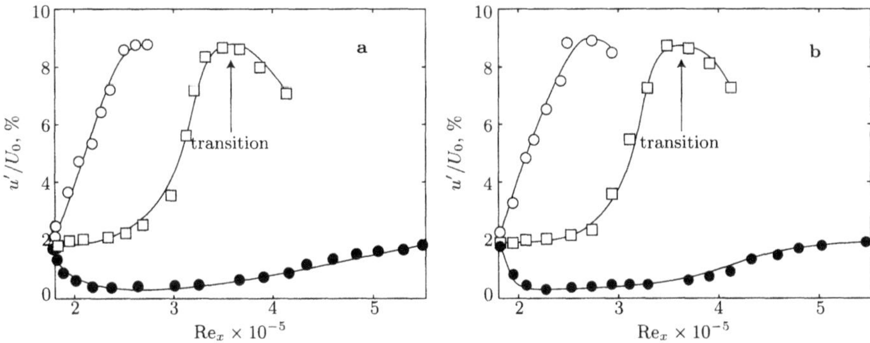

Fig. 7.18. Cancellation of two-dimensional flat plate Tollmien–Schlichting waves: **a** by surface vibrations (Gilev and Kozlov 1985); **b** by periodic suction and blowing through a set of holes (Gilev and Kozlov 1987). Transition induced by a vibrating ribbon (□) is affected by excitation with in-phase (∘) and anti-phase (•) oscillations

by Bakchinov et al. (2000). When turning to bypass transition they delayed the laminar flow breakdown, affecting three-dimensional localized non-linear boundary layer disturbances by pulses of sucking and blowing.

Besides experimental activities, the feasibility of the waves cancellation technique to transition control is supported by theoretical studies (Maslennikova and Zelman 1986; Efremov et al. 1988) and numerical simulations (Biringen 1984; Bayliss et al. 1986; Laurien and Kleiser 1989; Danabasoglu et al. 1991; Gmelin et al. 2000). The calculation results show that the method can be used for diminution of the two-dimensional linear instability waves and, sometimes, three-dimensional and non-linear perturbations as well.

7.2.4 Engineering application of the wave cancellation technique

A problem concerning engineering applications of the above results obtained by modelling of disturbances interactions is that during natural transition to turbulence a broad spectral band of oscillations normally grows. If so, one expects the method to be used for manipulation of some dominant disturbances. For this purpose, along with selection of an effective excitation technique, a control system has to be developed as a circuit of flow detectors and generators of the controlling oscillations. Such a system is not very complex when the flow geometry prescribes a dominance of two-dimensional waves and the control result can be obtained by suppression of the largest spectral component or that with the highest amplification rate (Joslin et al. 1995; Baumann et al. 2000; Evert et al. 2000). In the general case, however, many waves with similar amplitudes are involved in the transition process so that a control system employing a large number of detectors and generators is needed (Opfer and Ronneberger 2001; Opfer et al. 2001).

It is believed that identification of the fine structure of boundary layer perturbations and their selective control can be achieved through progress in

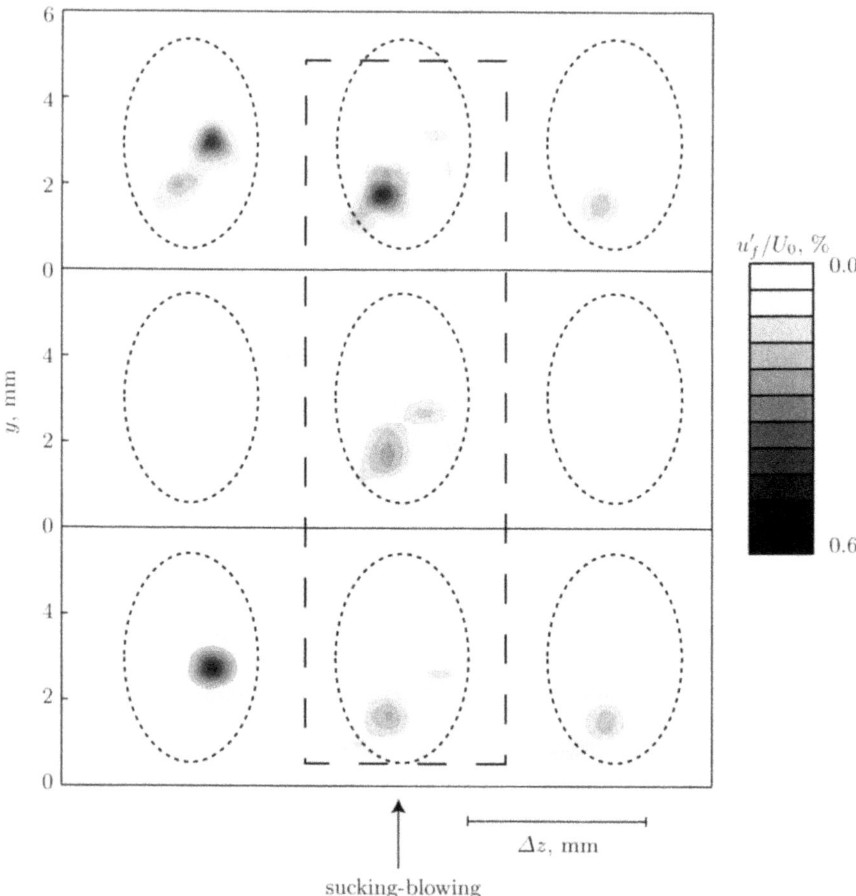

Fig. 7.19. Wave cancellation in a boundary layer with embedded streamwise vortices (Boiko et al. 1999). Travelling waves excited in the vortices by external acoustic (**a**), periodic sucking and blowing (**b**) and its anti-phase excitation

micro-electro-mechanical systems (MEMS). Being applied to the transition control, the micromachine manufacture of extremely small mechanical parts and devices makes it possible to create tiny sensors and actuators with a high frequency response (Ho 1994; Ho and Tai 1994). By integrating them using microelectronics, one can interactively control the boundary layer transition by resolving small scales of perturbations (Fig. 7.20). To realize this concept of control a number of problems have to be dealt with concerning the micromechanics of solids and fluids as well as organization of the distributed control circuit. Different versions of the MEMS application to flow control are considered by Tsao et al. (1994) and Liu et al. (1995). Many details on this topic are discussed in Ho and Tai (1996, 1998) and Löfdahl and Gad-el-Hak (1999).

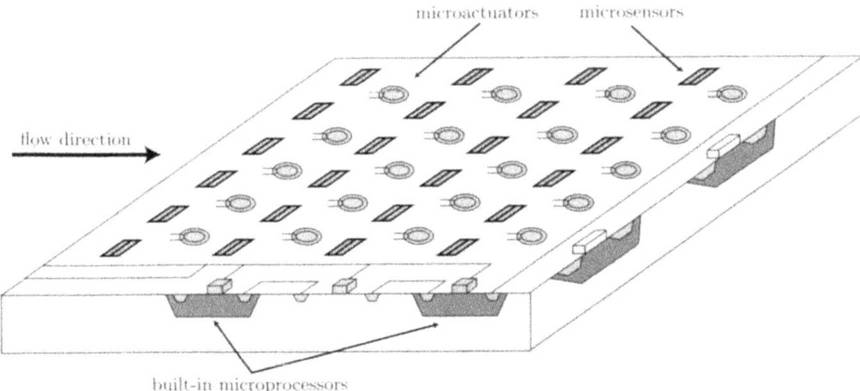

Fig. 7.20. Active control of boundary layer by MEMS; based on the figure by (Ho 1995)

References

Abegg, C., Bippes, H., Boiko, A., Krishnan, V., Lerche, T., Pöthke, A., Wu, Y. and Dallmann, U. (2000a). Transitional flow physics and plow control for swept wings: Experiments on boundary-layer receptivity, instability excitation and HLF-technology. In Proc. CEAS/DragNet European Drag Reduction Conf. (pp. 675–680). Potsdam.

Abegg, C., Bippes, H. and Janke, E. (2000b). Stabilization of boundary-layer flows subject to crossflow instability with the aid of suction. In H.F. Fasel and W.S. Saric (Eds.), Laminar-Turbulent Transition. Berlin Heidelberg New York: Springer.

Ablaev, A.R., Grek, G.R., Dovgal, A.V., Katasonov, M.M. and Kozlov, V.V. (2000). Transition to turbulence in a separation bubble caused by interaction of two oblique waves. Thermophys. Aeromech., 7(3), 353–360.

Aihara, Y. (1962). Transition in incompressible boundary-layer along a concave wall. Bull. Aeronaut. Res. Inst., 3(4), 195–240.

Aizin, L.B. and Polyakov, N.F. (1979). Generation of Tollmien–Schlichting wave by sound at an isolated surface roughness (Preprint 17–79). Novosibirsk: RAS. Sib. Branch Inst. Theoret. Appl. Mech. (In Russian.)

Alfredsson, P.H., Bakchinov, A.A., Kozlov, V.V. and Matsubara, M. (1996). Laminar–turbulent transition at a high level of a free stream turbulence. In P.W. Duck and P.Hall (Eds.), Nonlinear Instability and Transition in Three-Dimensional Boundary Layers (pp. 423–436). Dordrecht: Kluwer.

Al-Maaitah, A.A., Nayfeh, A.H. and Ragab, S.A. (1990a). Effect of cooling on the stability of compressible subsonic flows over smooth humps and backward-facing steps. Phys. Fluids A, 2(3), 381–389.

Al-Maaitah, A.A., Nayfeh, A.H. and Ragab, S.A. (1990b). Effect of suction on the stability of compressible subsonic flows over backward-facing steps. AIAA J., 28(11), 1916–1924.

Amini, J. and Lespinard, G. (1982). Experimental study of an "incipient spot" in a transitional boundary layer. Phys. Fluids, 25(10), 1743–1750.

Andersson, P., Berggren, M. and Henningson, D.S. (1999). Optimal disturbances and bypass transition in boundary layers. Phys. Fluids, 11(1), 134–150.

Arakeri, V.H. (1980). Delay of turbulent transition by surface heating in water. In R.Eppler and H.Fasel (Eds.), Laminar-Turbulent Transition (pp. 341–350). Berlin Heidelberg New York: Springer.

Arena, A.V. and Mueller, T.J. (1980). Laminar separation, transition, and turbulent reattachment near the leading edge of airfoils. AIAA J., 18(7), 747–753.

Arnal, D. (1992). Transition description and prediction. In O.Pironneau, W.Rodi, I.L. Ryhming, A.M. Savill and T.V. Truong (Eds.), Numerical Simulation of Unsteady Flows and Transition to Turbulence (pp. 303–316). Cambridge: Cambridge University Press.

Arnal, D. and Juillen, J.C. (1977). Étude expérimentale et théorique de la transition de la couche limite. Rech. Aérospat., 2, 75–88.

Arnal, D. and Juillen, J.C. (1978). Contribution expérimentale à l'étude de la réceptivité d'une couche limite laminaire, à la turbulence de l'écoulement général (CERT RT 1/5018 AYD). ONERA.

Arnal, D., Casalis, G. and Juillen, J.C. (1990). Experimental and theoretical analysis of natural transition on 'infifnite' swept wing. In D.Arnal and R.Michel (Eds.), Laminar-Turbulent Transition (pp. 311–325). Berlin Heidelberg New York: Springer.

Arnal, D., Juillen, J.C., Reneaux, J. and Gasparian, G. (1997). Effect of wall suction on leading edge contamination. Aerosp. Sci. Technol., 8, 505–517.

Asai, M. and Kaneko, M. (1998). Experimental investigation of the receptivity of separated flow. In Proc. Third Internat. Conf. Fluid Mechanics (pp. 231–237). Beijing: Beijing Institute of Technology.

Ashpis, D.E. and Reshotko, E. (1990). The vibrating ribbon problem revisited. J. Fluid Mech., 213, 531–547.

Azad, R.S. and Doell, B. (1990). Behaviour of separation bubble with different roughness elements at the leading edge of a flat plate. In A.Gyr (Ed.), Structure of Turbulence and Drag Reduction (pp. 85–92). Berlin Heidelberg New York: Springer.

Bakchinov, A.A., Grek, G.R. and Kozlov, V.V. (1994a). Development of instability waves between artificial turbulent regions. Thermophys. Aeromech., 1(1), 45–50.

Bakchinov, A.A., Grek, G.R. and Kozlov, V.V. (1994b). Experimental study of localized disturbances in a laminar boundary layer. Thermophys. Aeromech., 1(1), 45–52. ((In Russian).)

Bakchinov, A.A., Grek, H.R., Klingmann, B.G.B. and Kozlov, V.V. (1995). Transition experiments in a boundary layer with embedded streamwise vortices. Phys. Fluids A, 7(4), 820–832.

Bakchinov, A.A., Grek, G.R., Katasonov, M.M. and Kozlov, V.V. (1997). Experimental investigation of the structure and development characteristics of the localized disturbances in the flat plate boundary layer

(Preprint 1–97). Novosibirsk: RAS. Sib. Branch Inst. Theoret. Appl. Mech. (In Russian.)

Bakchinov, A.A., Grek, G.R., Katasonov, M.M. and Kozlov, V.V. (1998a). Experimental investigation of the interaction of longitudinal streaky structures with a high-frequency disturbance. Fluid Dyn., 33(5), 667–675.

Bakchinov, A.A., Westin, K.J.A., Kozlov, V.V. and Alfredsson, P.H. (1998b). Experiments on localized disturbances in a flat plate boundary layer. Part 2. Interaction between localized disturbances abd TS-waves. European J. Mech. B Fluids, 17(6), 847–873.

Bakchinov, A.A., Katasonov, M.M., Alfredsson, P.H. and Kozlov, V.V. (1999). Control of boundary layer transition at high FST by localized suction. In G.E.A. Meier and P.R. Viswanath (Eds.), Mechanics of passive and active flow control (pp. 159–164). Dordrecht: Kluwer.

Bakchinov, A.A., Katasonov, M.M., Alfredsson, P.H. and Kozlov, V.V. (2000). Control of streaky structures by localized blowing and suction. In H.F. Fasel and W.S. Saric (Eds.), Laminar-Turbulent Transition (pp. 161–166). Berlin Heidelberg New York: Springer.

Bake, S., Kachanov, Y.S. and Fernholz, H.H. (1996). Subharmonic K-regime of boundary-layer breakdown. In R.A.W.M. Henkes and J.L. Van Ingen (Eds.), Transitional Boundary Layers in Aerodynamics, Proc. Colloquium (pp. 81–88). Royal Netherlands Academy of Arts and Sciences, Amsterdam: Elsevier.

Balachandrar, S., Streett, C.L. and Malik, M.R. (1992). Secondary instability in rotating-disk flow. J. Fluid Mech., 242, 323–347.

Balakumar, P., Hall, P. and Malik, M.R. (1990). On the receptivity and non-parallel stability of traveling disturbances in rotating disk flow (ICASE Rep. 90–89).

Barker, S.J. and Jennings, C. (1977). The effect of wall heating upon transition in water boundary layer. In AGARD-CP–224 Laminar–Turbulent Transition. (Paper No. 19.)

Barrow, J., Barnes, F.H. and Ross, M.A.S. (1984). The structure of a turbulent spot in Blasius flow. J. Fluid Mech., 149, 319–337.

Bassom, A.P. and Hall, P. (1991). Concerning the interaction of non-stationary crossflow vortices in a three-dimensional boundary layer. Q. J. Mech. Appl. Math., 44, 147–172.

Bassom, A.P. and Hall, P. (1994). The receptivity problem for $O(1)$ wavelength Görtler vortices. Proc. R. Soc. Lond. A, 446, 449–516.

Bassom, A.P. and Seddougui, S.O. (1995). Receptivity mechanisms for Görtler vortex modes. Theoret. Comput. Fluid Dyn., 7, 317–339.

Batchelor, G.K. (1964). Axial flow in trailing line vortices. J. Fluid Mech., 20, 645–658.

Baumann, M., Sturzebecher, D. and Nitsche, W. (2000). Active control of TS-instabilities on an unswept wing. In H.F. Fasel and W.S. Saric

(Eds.), Laminar-Turbulent Transition (pp. 155–160). Berlin Heidelberg New York: Springer.

Bayliss, A., Maestrello, L., Parikh, P. and Turkel, E. (1986). Numerical simulation of boundary-layer excitation by surface heating/cooling. AIAA J., 24(7), 1095–1101.

Bayly, B.J. and Orszag, S.A. (1988). Instability mechanisms in shear-flow transition. Ann. Rev. Fluid Mech., 20, 359–391.

Bech, K.H., Henningson, D.S. and Henkes, R.A.W.M. (1998). Linear and nonlinear development of localized disturbances in zero and adverse pressure gradient boundary-layers. Phys. Fluids, 10(6), 1405–1418.

Bechert, D.W. and Bartenwerfer, M. (1989). The viscous flow on surfaces with longitudinal ribs. J. Fluid Mech., 206, 105–129.

Belov, I.A., Enutin, G.V. and Litvinov, V.N. (1990). Effect of streamwise and transverse ribs on a flat plate upon laminar–turbulent transition. Uchen. Zap. TsAGI, 21(6), 107–111. (In Russian.)

Benney, D.J. and Bergeron, E.F. (1968). A new class of nonlinear waves in parallel flows. Stud. Appl. Math., 48(3), 181–204.

Benney, D.J. and Gustavsson, L.H. (1981). A new mechanism for linear and nonlinear hydrodynamic stability. Stud. Appl. Math., 64, 185–209.

Benney, D.J. and Rosenblat, S. (1964). Stability of spatially varying and time dependent flows. Phys. Fluids, 7(8), 1385.

Berlin, S. and Henningson, D.S. (1994). A study of boundary layer receptivity to disturbances in the free stream [TRITA-MEK 94–14]. In D.S. Henningson (Ed.), Bypass Transition – Proceedings from a Mini-Workshop (pp. 43–49). Stockholm.

Berlin, S., Lundbladh, A. and Henningson, D.S. (1994). Spatial simulations of oblique transition in a boundary layer. Phys. Fluids A, 6(6), 1949–1951.

Berlin, S., Wiegel, M. and Henningson, D.S. (1999). Numerical and experimental investigations of oblique boundary layer transition. J. Fluid Mech., 393, 23–57.

Bertolotti, F.P. (1997). Response of the Blasius boundary layer to free-strem vorticity. Phys. Fluids, 9(8), 2286–2299.

Bertolotti, F.P. (2000). Receptivity of three-dimensional boundary layers to localised wall roughness and suction. Phys. Fluids, 12(7), 1799–1809.

Bertolotti, F.P. and Kendall, J.M. (1997). Response of the boundary layer to controlled free-stream vortices of axial form (AIAA Paper 97–2018).

Bertolotti, F.P., Herbert, T. and Spallart, P.R. (1992). Linear and nonlinear stability of the Blasius boundary layer. J. Fluid Mech., 242, 441–474.

Bestek, H., Gruber, K. and Fasel, H. (1989). Self-excited unsteadiness of laminar separation bubbles caused by natural transition. In The Prediction and Exploitation of Separated Flows (pp. 14.1–14.16). London: R. Aeronaut. Soc.

Bestek, H., Gruber, K. and Fasel, H. (1993). Direct numerical simulation of unsteady separated boundary-layer flows over smooth backward-facing steps. In K.Gersten (Ed.), Physics of Separated Flows – Numerical, Experimental, and Theoretical Aspects (Vol. 40, pp. 73–80). Braunschweig: Vieweg.

Betchov, R. (1960). On the mechanism of turbulent transition. Phys. Fluids, 3(6), 1026–1027.

Betchov, R. and Criminale, W.O. (1967). Stability of parallel flows. New York: Academic.

Bhattacharjee, S., Scheelke, B. and Troutt, T.R. (1986). Modification of vortex interaction in a reattaching separated flow. AIAA J., 24(4), 623–629.

Bippes, H. (1977). Experimente zur Entwicklung der freien Wirbel hinter einem Rechteckflügel. Acta Mech., 26, 223–245.

Bippes, H. (1990). Instability features appearing on swept wing configurations. In D.Arnal and R.Michel (Eds.), Laminar-Turbulent Transition (pp. 419–430). Berlin Heidelberg New York: Springer.

Bippes, H. (1999). Basic experiments on transition in three-dimensional boundary layers dominated by crossflow instability. Progr. Aerosp. Sci., 35(4), 363–412.

Bippes, H., Wiegel, M. and Bertolotti, F. (1999). Experiments on the control of crossflow instability with the aid of suction through perforated walls. In G.E.A. Meier and P.R. Viswanath (Eds.), Mechanics of passive and active flow control (pp. 165–170). Dordrecht: Kluwer.

Biringen, S. (1984). Active control of transition by periodic suction-blowing. Phys. Fluids, 27(6), 1345–1347.

Blackwelder, R.F. (1983). Analogies between transitional and turbulent boundary layers. Phys. Fluids, 26(10), 2807–2815.

Blackwelder, R.F. and Swearingen, J.D. (1990). The role of inflectional velocity profiles in wall bounded flows. In S.J. Kline and N.H. Afgan (Eds.), Near-Wall Turbulence. Proc. 1988 Zorian Zorac Memorial Conference (pp. 268–288). Hemisphere.

Blair, M.F. (1992). Boundary-layer transition in accelerating flows with intense freestream turbulence: Part 1 – Disturbances upstream of transition onset. Trans. ASME Ser. I: J. Fluids Eng., 114, 313–321.

Bodonyi, R.J. and Duck, P.W. (1992). Boundary layer receptivity to a wall suction control of Tollmien–Schlichting waves. Phys. Fluids A, 4(6), 1206–1214.

Bodonyi, R.J., Welch, W.J.C., Duck, P.W. and Tadjfar, M. (1989). A numerical study of the interaction between unsteady free-stream disturbances and localized variations in surface geometry. J. Fluid Mech., 209, 285–308.

Bogucki, D.J. and Redekopp, L.G. (1999). A mechanism for sediment resuspension by internal solitary waves. Geophys. Res. Lett., 26(9), 1317–1320.

Boiko, A.V. (2000). Receptivity of boundary layers to free streamaxial vortices (IB 223–2000 A10). Göttingen: German Aerospace Center (DLR) – Institute for Fluid Mechanics.

Boiko, A.V., Dovgal, A.V., Kozlov, V.V., Simonov, O.A. and Scherbakov, V.A. (1988). Laminar flow separation at 2D bumps in a boundary layer (Preprint 7–88). Novosibirsk: RAS. Sib. Branch Inst. Theoret. Appl. Mech. (In Russian.)

Boiko, A.V., Dovgal, A.V. and Kozlov, V.V. (1989). Nonlinear interactions between perturbations in transition to turbulence in the zone of laminar boundary-layer separation. Soviet J. Appl. Phys., 3(2), 46–52.

Boiko, A.V., Dovgal, A.V., Kozlov, V.V. and Scherbakov, V.A. (1990). Instability and receptivity of a boundary layer close to 2D surface imperfections. Izv. Sib. Otd. Akad. Nauk SSSR, Ser. Tekhn. Nauk[1], 1, 50–56.

Boiko, A.V., Dovgal, A.V., Simonov, O.A. and Scherbakov, V.A. (1991a). Effects of laminar–turbulent transition in separation bubbles. In V.V. Kozlov and A.V. Dovgal (Eds.), Separated Flows and Jets (pp. 565–572). Berlin Heidelberg New York: Springer.

Boiko, A.V., Dovgal, A.V. and Scherbakov, V.A. (1991b). An effect of small-amplitude disturbances on a laminar separation bubble (Preprint 5–91). Novosibirsk: RAS. Sib. Branch Inst. Theoret. Appl. Mech. (In Russian.)

Boiko, A.V. and Dovgal, A.V. (1992). Instability of locally separated flows to small-amplitude disturbances. Sib. Fiz.-Techn. Zh., 34(3), 19–24. (In Russian.)

Boiko, A.V., Westin, K.J.A., Klingmann, K.G.B., Kozlov, V.V. and Alfredsson, P.H. (1994). Experiments in a boundary layer subjected to free stream turbulence. Part 2. The role of TS-waves in the transition process. J. Fluid Mech., 281, 219–245.

Boiko, A.V., Kozlov, V.V., Syzrantsev, V.V. and Scherbakov, V.A. (1995a). Experimental investigation of high frequency secondary disturbances in swept wing boundary layer. J. Appl. Mech. Tehn. Phys., 36(3), 74–83. (In Russian.)

Boiko, A.V., Kozlov, V.V., Syzrantsev, V.V. and Scherbakov, V.A. (1995b). Experimental investigation of the transition process at an isolated stationary disturbance in swept wing boundary layer. J. Appl. Mech. Tehn. Phys., 36(1), 72–84. (In Russian.)

Boiko, A.V., Chung, Y.M. and Sung, H.J. (1997a). Spatial simulation of the instability of channel flow with local suction/blowing. Phys. Fluids, 9(11), 3258–3266.

Boiko, A.V., Kozlov, V.V., Syzrantsev, V.V. and Scherbakov, V.A. (1997b). A study of the influence of internal structure of a streamwise vortex

[1] Translated in Siberian Phys. Techn. J. (former Sov. J. Appl. Phys.)

on the development of traveling disturbances inside it. Thermophys. Aeromech., 4(4), 343–354.

Boiko, A.V., Kozlov, V.V., Syzrantsev, V.V. and Scherbakov, V.A. (1997c). Transition control by riblets in swept wing boundary layer with embedded streamwise vortex. European J. Mech. B Fluids, 16(4), 465–482.

Boiko, A.V., Kozlov, V.V., Syzrantsev, V.V. and Shcherbakov, V.A. (1999). Active control over secondary instability in a three-dimensional boundary layer. Thermophys. Aeromech., 6(2), 167–178.

Boiko, A.V., Kozlov, V.V., Sova, V.A. and Shcherbakov, V.A. (2000). Generation of streamwise structures in a boundary layer of a swept wing and their secondary instability. Thermophys. Aeromech., 7(1), 25–35.

Bolz, F.W., Kenyon, G.C. and Allen, C.Q. (1960). Effects of sweep angle on the boundary layer stability characteristics of an untapered wing at low speeds (NASA TN D-338).

Borodulin, V.I. and Kachanov, Y.S. (1994). Experimental study of nonlinear stages of a boundary layer breakdown. In S.P. Lin, W.R.C. Phillips and D.T. Valentine (Eds.), Nonlinear Instability of Nonparallel Flows (pp. 69–80). Berlin Heidelberg New York: Springer.

Bottaro, A. and Klingmann, B.G.B. (1996). On the linear breakdown of Görtler vortices. European J. Mech. B Fluids, 15(3), 301–330.

Bottaro, A. and Luchini, P. (1999). Görtler vortices: Are they amenable to local eigenvalue analysis? European J. Mech. B Fluids, 18(1), 47–65.

Bouthier, M. (1971). Stabilité linéaire des écoulements presque parallèles des échelles multiples. Compt. Rendus Acad. Sci., Sèr. A, 273, 1101–1104.

Bouthier, M. (1972). Stabilité linéaire des écoulements presque parallèles. Part 1. J. Mécanique, 11(4), 599–621.

Bouthier, M. (1973). Stabilité linéaire des écoulements presque parallèles. Part 2. La couche limite de Blasius. J. Mécanique, 12(1), 75–95.

Brendel, M. and Mueller, T.J. (1988). Boundary-layer measurements on an airfoil at low Reynolds numbers. J. Aircraft, 25(7), 612–617.

Breuer, K.S. and Haritonidis, J.H. (1990). The evolution of a localized disturbance in a laminar boundary layer. Part 1. Weak disturbances. J. Fluid Mech., 220, 569–594.

Breuer, K.S. and Kuraishi, T. (1994). Transient growth in two- and three-dimensional boundary layers. Phys. Fluids A, 6(6), 1983–1993.

Breuer, K.S. and Landahl, M.T. (1990). The evolution of a localized disturbance in a laminar boundary layer. Part 2. Strong disturbances. J. Fluid Mech., 220, 595–621.

Breuer, K.S., Cohen, J. and Haritonidis, J.H. (1997). The late stages of transition induced by a low-amplitude wave packet in a laminar boundary layer. J. Fluid Mech., 340, 395–411.

Brevdo, L. (1995). Convectively unstable wave packets in the Blasius boundary layer. Z. Angew. Math. Mech., 75(6), 423–476.

Briggs, R.J. (1964). Electron-stream interactions in plasmas. Cambridge, Mass.: MIT.

Briley, W.R. (1971). A numerical study of laminar separation bubble using the Navier–Stokes equations. J. Fluid Mech., 47, 713–736.

Briley, W.R. and McDonald, H. (1975). Numerical prediction of incompressible separation bubbles. J. Fluid Mech., 69, 631–656.

Browand, F.K. (1966). An experimental investigation of the instability of an incompressible separated shear layer. J. Fluid Mech., 26, 281–307.

Burnel, S. and Cougat, P. (1972). Étude du développement des instabilités naturelles dans la couche limite laminaire en écoulement incompressible. Compt. Rendus Acad. Sci., Sèr. A, 274(16), 1251–1254.

Bushnell, D.M. (1989). Notes on initial disturbance fields for the transition problem (Tech. Rep. TM-1051). NASP.

Butler, K.M. and Farrell, B.F. (1992). Three-dimensional perturbations in viscous shear flow. Phys. Fluids A, 4(8), 1627–1650.

Butler, K.M. and Farrell, B.F. (1993). Optimal perturbations and streak spacing in wall-bounded shear flows. Phys. Fluids A, 5, 774–777.

Cantwell, D., Coles, D. and Dimotakis, P. (1978). Structure and entrainment in the plane of symmetry of a turbulent spot. J. Fluid Mech., 87, 641–671.

Carpenter, P.W. (1990). Status of transition delay using compliant walls. In D.M. Bushnell and J.N. Heffner (Eds.), Viscous Drag Reduction in Boundary Layers (pp. 79–113). New York: AIAA.

Case, K.M. (1960). Stability of inviscid plane Couette flow. Phys. Fluids, 3(2), 143–148.

Cebeci, T. and Egan, D.A. (1989). Prediction of transition due to isolated roughness. AIAA J., 27(7), 870–875.

Chang, P.K. (1970). Separation of flow. Oxford: Pergamon.

Chapman, D.R., Kuehn, D.M. and Larson, H.K. (1958). Investigation of separated flows in supersonic and subsonic streams with emphasis on the effect of transition (NACA Rep. 1356).

Chernoray, V.G., Bakchinov, A.A., Kozlov, V.V. and Löfdahl, L. (2000a). Experimental study of K-regime of breakdown in straight and swept wing boundary layers. In Proc. Tenth Internat. Conf. Methods Aerophysical Research (Vol. 1, pp. 65–70). Novosibirsk: RAS. Sib. Branch Inst. Theoret. Appl. Mech.

Chernorai, V.G., Grek, G.R., Katasonov, M.M. and Kozlov, V.V. (2000b). Generation of the localized disturbances by the vibrating surface. Thermophys. Aeromech., 7(3), 329–339.

Choi, D.H. and Kang, D.J. (1991). Calculation of separation bubbles using a partially parabolized Navier–Stokes procedure. AIAA J., 29(8), 1267.

Choi, K.-S. (1989). Near-wall structure of a turbulent boundary layer with riblets. J. Fluid Mech., 208, 417–458.

Chomaz, J.M. (1992). Absolute and convective instabilities in nonlinear systems. Phys. Rev. Lett., 69, 1931–1934.

Chomaz, J.M. and Perrier, M. (1991). Nature of the Görtler instability in a forced experiment. In The Geometry of Turbulence (pp. 23–32). New York: Plenum.

Choudhari, M. (1993). Boundary-layer receptivity due to distributed surface imperfections of a deterministic or random nature. Theoret. Comput. Fluid Dyn., 4, 101–117.

Choudhari, M. (1998). Receptivity. In R.W. Johnson (Ed.), The Handbook of Fluid Dynamics (pp. 13.25–13.40). Berlin Heidelberg New York: Springer.

Choudhari, M. and Streett, C.L. (1990). Boundary layer receptivity phenomena in three-dimensional and high-speed boundary layers (AIAA Paper 90–5258).

Choudhari, M. and Streett, C.L. (1992). A finite Reynolds-number approach for the prediction of boundary-layer receptivity in localized regions. Phys. Fluids A, 4(11), 2495–2514.

Chu, D., Henderson, R. and Karniadakis, G.E. (1992). Parallel spectral-element-Fourier simulation of turbulent flow over riblet-mounted surfaces. Theoret. Comput. Fluid Dyn., 3(4), 219–229.

Cohen, J., Breuer, K.S. and Haritonidis, J.H. (1991). On the evolution of a wave packet in a laminar boundary layer. Part 1. Weak disturbances. J. Fluid Mech., 225, 575–606.

Collier, F.S. and Malik, V.R. (1987). Stationary disturbances in three-dimensional boundary layers over concave surfaces (AIAA Paper 87–1412).

Collis, S.S. and Lele, S.K. (1999). Receptivity to surface roughness near a swept wing leading edge. J. Fluid Mech., 380, 141–168.

Cooke, J.C. (1950). The boundary layer of a class of infinite yawed cylinders. Proc. Camb. Philos. Soc., 46, 645–648.

Corbett, P. and Bottaro, A. (2000). Optimal perturbations for boundary layers subject to stream-wise pressure gradient. Phys. Fluids, 12(1), 120–130.

Corbett, P. and Bottaro, A. (2001). Optimal linear growth in swept boundary layers. J. Fluid Mech., 435, 1–23.

Corke, T.C. and Mangano, R.A. (1989). Resonant growth of three-dimensional modes in transitioning Boundary layers. J. Fluid Mech., 209, 93–150.

Cossu, C. and Chomaz, J.-M. (1997). Global measures of local convective instabilities. Phys. Rev. Lett., 78(23), 4387–4390.

Cousteix, J. and Pailhas, G. (1979). Étude exploratoire d'un processus de transition laminaire-turbulent au voisinage du décollement d'une couche laminaire. Rech. Aérospat., 1979(3), 213–218.

Coustols, E. (1991). Performance of internal manipulators in subsonic three-dimensional flows. In K.-S. Choi (Ed.), Recent Developments in Turbulence Management (pp. 43–64). Dordrecht: Kluwer.

Coustols, E. and Cousteix, J. (1990). Experimental investigation of turbulent boundary layers manipulated with internal devices: Riblets. In A.Gyr (Ed.), Structure of Turbulence and Drag Reduction (pp. 577–584). Berlin Heidelberg New York: Springer.

Craik, A.D.D. (1971). Non-linear resonant instability in boundary layers. J. Fluid Mech., 50, 393–413.

Craik, A.D.D. (1991). The continuous spectrum of the Orr–Sommerfeld equation: Note on a paper of Grosch & Salwen. J. Fluid Mech., 226, 565–571.

Crimi, P. and Reeves, B.L. (1976). Analysis of leading-edge separation bubbles on airfoils. AIAA J., 14(11), 1548–1555.

Crouch, J.D. (1992a). Localized receptivity of boundary layers. Phys. Fluids A, 4(7), 1408–1414.

Crouch, J.D. (1992b). Non-localized receptivity of boundary layers. J. Fluid Mech., 244, 567–581.

Crouch, J.D. (1993a). Receptivity and the evolution of boundary-layer instabilities over short-scale waviness. Phys. Fluids A, 5(3), 561–567.

Crouch, J.D. (1993b). Receptivity of three-dimensional boundary layers (AIAA Paper 93–0074).

Crouch, J.D. (1994a). Distributed excitation of Tollmien–Schlichting waves by vortical free-stream disturbances. Phys. Fluids A, 6(1), 217–223.

Crouch, J.D. (1994b). Theoretical study on the receptivity of boundary layers (AIAA Paper 94–2223).

Crouch, J.D. (1998). Theory of instability and transition. In R.W. Johnson (Ed.), The Handbook of Fluid Dynamics (pp. 13.12–13.25). Berlin Heidelberg New York: Springer.

Crouch, J.D. and Bertolotti, F.P. (1992). Nonlocalized receptivity of boundary layers to three-dimensional disturbances (AIAA Paper 92–0740).

Crouch, J.D., Gaponenko, V.R., Ivanov, A.V. and Kachanov, Y.S. (1997). Theoretical and experimental comparisons for the stability and receptivity of swept-wing boundary layers. Bull. Amer. Phys. Soc., 42, 2174.

Crouch, J.D. and Herbert, T. (1993). Nonlinear evolution of secondary instabilities in boundary-layer transition. Theoret. Comput. Fluid Dyn., 4, 151–175.

Crouch, J.D. and Ng, L.L. (2000). Variable N-factor method for transition prediction in three-dimensional boundary layers. AIAA J., 38(2), 211–216.

Crouch, J.D. and Spalart, P.R. (1995). A study of nonlinear and nonparallel effects on the localized receptivity of boundary layers. J. Fluid Mech., 290, 29–37.

Crouch, J.D., Gaponenko, V.R., Ivanov, A.V. and Kachanov, Y.S. (1998). A method of experimental determination of the linear receptivity coefficients of a 3D boundary layer subjected to microscopic surface non-uniformities. Verification of theory. In Proc. Ninth Internat. Conf. on Methods of Aerophysical Research (Vol. 2, pp. 30–35). Novosibirsk.

Dagenhart, J.R., Saric, W.S., Hoos, J.A. and Mousseux, M.C. (1990). Experiments on swept-wing boundary layers. In D.Arnal and R.Michel (Eds.), Laminar-Turbulent Transition (pp. 369–380). Berlin Heidelberg New York: Springer.

Dallmann, U., Herberg, T., Gebing, H., Su, W.H. and Zhang, H.Q. (1995). Flow field diagnostics: Topological flow changes and spatio-temporal flow structure (AIAA Paper 95–0791).

Danabasoglu, G., Biringen, S. and Streett, C.L. (1991). Spatial simulation of instability control by periodic suction blowing. Phys. Fluids A, 3(9), 2138–2147.

Davies, S.J. and White, C.M. (1928). An experimental study of the flow of water in pipe of rectangular section. Proc. R. Soc. Lond. A, 19, 92–107.

Davis, R.L., Carter, J.E. and Reshotko, E. (1987). Analysis of transitional separation bubbles on infinite swept wings. AIAA J., 25(3), 421–428.

DeBruin, A.C. (1990). The effect of a single cylindrical roughness element on boundary layer transition in a favourable pressure gradient. In D.Arnal and R.Michel (Eds.), Laminar-Turbulent Transition (pp. 645–655). Berlin Heidelberg New York: Springer.

Denier, J.P., Hall, P. and Seddougui, S.O. (1991). On the receptivity problem for Görtler vortices: Vortex motions induced by roughness. Philos. Trans. R. Soc. Lond. A, 335, 51–85.

Deyhle, H. and Bippes, H. (1996). Disturbance growth in an unstable three-dimensional boundary layer and its dependence on environmental conditions. J. Fluid Mech., 316, 73–113.

Dikovskaya, N.D., Dovgal, A.V. and Sorokin, A.M. (1999). Transition to turbulence and formation of regular vortices in the boundary layer separation region. Thermophys. Aeromech., 6(1), 25–33.

Dini, P. and Maughmer, M.D. (1990). A locally interactive laminar separation bubble model (AIAA Paper 90–0570).

Dini, P., Selig, M.S. and Maughmer, M.D. (1992). Simplified linear stability transition prediction method for separated boundary layers. AIAA J., 30(8), 1953–1961.

DiPrima, R.C. and Habetler, G.J. (1969). A completeness theorem for nonself-adjoint eigenvalue problems in hydrodynamic stability. Arch. Rat. Mech. Anal., 34, 218.

Dorodnitsyn, A.A. and Loytsyanskiy, L.G. (1945). On the theory of laminar–turbulent transition. Prikl. Mat. Mekh.[2], 9(4), 269–284.

[2] Translated in Appl. Math. Mech.

Dovgal, A.V. (1985). Evolution of vortical disturbances in flow with laminar separation. In V.V. Kozlov (Ed.), Laminar-Turbulent Transition (pp. 359–366). Berlin Heidelberg New York: Springer.

Dovgal, A.V. (1999a). Flow control by periodic forcing: acoustic excitation and local injection of artificial disturbances. In G.E.A. Meier and P.R. Viswanath (Eds.), Mechanics of passive and active flow control (pp. 291–298). Dordrecht: Kluwer.

Dovgal, A.V. (1999b). Unsteadiness of a laminar separation bubble under external acoustic excitation. Russian J. Eng. Thermophys., 9(4), 285–296.

Dovgal, A.V. and Boiko, A.V. (2000). Effect of harmonic excitation on instability of laminar separation bubble on an airfoil. In H.F. Fasel and W.S. Saric (Eds.), Laminar-Turbulent Transition (pp. 675–680). Berlin Heidelberg New York: Springer.

Dovgal, A.V. and Grosche, F.-R. (1994). Internal-sound technique for control of laminar separation: Optimization of the control device (DLR-IB 223-94 A 20). Göttingen.

Dovgal, A.V. and Kozlov, V.V. (1983a). Influence of acoustic perturbations on the flow structure in a boundary layer with adverse pressure gradient. Fluid Dyn., 18(2), 205–209.

Dovgal, A.V. and Kozlov, V.V. (1983b). Instability of flow separating from a wall with an inflexion. Dokl. Akad. Nauk[3], 270(6), 1356–1358.

Dovgal, A.V. and Kozlov, V.V. (1984). Stability of separated flow in a dihedral corner. Fluid Mech. – Soviet Res., 13(1), 137–143.

Dovgal, A.V. and Kozlov, V.V. (1990). Hydrodynamic instability and receptivity of small scale separation regions. In D.Arnal and R.Michel (Eds.), Laminar-Turbulent Transition (pp. 523–531). Berlin Heidelberg New York: Springer.

Dovgal, A.V. and Kozlov, V.V. (1995). On nonlinearity of transitional boundary-layer flows. Philos. Trans. R. Soc. Lond. A, 352, 473–482.

Dovgal, A.V., Kozlov, V.V. and Levchenko, V.Y. (1980). Experimental investigation into the reacrtion of a boundary layer to external periodic disturbances. Fluid Dyn., 15(4), 602–606.

Dovgal, A.V. and Zanin, B.Y. (1982). Effect of free-stream turbulence on disturbances evolving at boundary layer separation. In V.Y. Levchenko (Ed.), Instability of Sub- and Supersonic Flows (pp. 77–84). Novosibirsk: RAS. Sib. Branch Inst. Theoret. Appl. Mech. (In Russian.)

Dovgal, A.V., Kozlov, V.V. and Simonov, O.A. (1987). Experiments on hydrodynamic instability of boundary layers with separation. In F.T. Smith and S.N. Brown (Eds.), Boundary-Layer Separation (pp. 109–130). Berlin Heidelberg New York: Springer.

[3] Translated in Phys. Dokl.

Dovgal, A.V., Kozlov, V.V. and Simonov, O.A. (1988a). Development of a spatial wave packet of disturbances in a swept-wing boundary layer. Izv. Sib. Otd. Akad. Nauk SSSR, Ser. Tekhn. Nauk[4], 3(11), 43–47.

Dovgal, A.V., Kozlov, V.V. and Simonov, O.A. (1988b). Stability of three-dimensional flow generated upon separation from a wall with an inflexion. Soviet J. Appl. Phys., 2(4), 18–24.

Dovgal, A.V., Kozlov, V.V. and Simonov, O.A. (1989). Acoustic excitation of a swept-wing boundary layer. Uchen. Zap. TsAGI, 20(6), 21–26. (In Russian.)

Dovgal, A.V., Levchenko, V.Y. and Timofeev, V.A. (1990). Boundary layer control by a local heating of the wall. In D.Arnal and R.Michel (Eds.), Laminar-Turbulent Transition (pp. 113–121). Berlin Heidelberg New York: Springer.

Dovgal, A.V., Kozlov, V.V. and Michalke, A. (1995). Contribution to the instability of laminar separating flows along axisymmetric bodies. Part 2: Experiment and comparison with theory. European J. Mech. B Fluids, 14(3), 351–365.

Dovgal, A.V., Kozlov, V.V. and Michalke, A. (1996). On disturbances excited by a point source in an axisymmetric laminar separation bubble. European J. Mech. B Fluids, 15(4), 651–664.

Dovgal, A.V. and Sorokin, A.M. (2001). Instability of a laminar separation bubble to vortex shedding. Thermophys. Aeromech., 8(2), 189–197.

Drazin, P.G. and Reid, W.H. (1981). Hydrodynamic stability. Cambridge: Cambridge University Press.

Drela, M. and Giles, M.B. (1987). Viscous-inviscid analysis of transonic and low Reynolds number airfoils. AIAA J., 25(10), 1347–1355.

Driver, D.M., Seegmiller, H.L. and Marvin, J.G. (1987). Time-dependent behavior of a reattaching shear layer. AIAA J., 25(7), 914–919.

Dryden, H.L. (1934). Boundary layer flow near flat plates. In Proc. Fourth Internat. Congress for Appl. Mech. (p. 175). Cambridge.

Dryden, H.L. (1936). Airflow in the boundary layer near a plate (NACA TR 562).

Dryganets, S.V., Kachanov, Y.S., Levchenko, V.Y. and Ramazanov, M.P. (1990). Resonant flow randomization in K-regime of boundary layer transition. J. Appl. Mech. Tehn. Phys., 31(2), 239–249.

Eaton, J.K. and Johnston, J.P. (1981). A review of research on subsonic turbulent flow reattachment. AIAA J., 19(9), 1093–1100.

Efremov, O.A., Ryzhov, O.S. and Terent'ev, E.D. (1988). Suppression of unstable oscillations in a boundary layer. Fluid Dyn., 22(2), 183–189.

Egami, Y. and Kohama, Y. (1999). Control of crossflow instability field by selective suction system. In G.E.A. Meier and P.R. Viswanath (Eds.), Mechanics of passive and active flow control (pp. 171–176). Dordrecht: Kluwer.

[4] Translated in Siberian Phys. Techn. J. (former Sov. J. Appl. Phys.)

Ehrenstein, U. (1996). On the linear stability of channel flow over riblets. Phys. Fluids, 8(11), 3194–3196.

Ehrenstein, U. and Koch, W. (1991). Three-dimensional wavelike eqilibrium states in plane Poiseuille flow. J. Fluid Mech., 228, 111–148.

Elder, J.W. (1960). An experimental investigation on turbulent spot and breakdown to turbulence. J. Fluid Mech., 9, 235–246.

El-Hady, N.M. (1991). Control of the vortical structure in the early stage of transition in boundary layers. In R.W. Barnwell and M.Y. Hussaini (Eds.), Natural Laminar Flow and Laminar Flow Control (pp. 426–442). Berlin Heidelberg New York: Springer.

Elli, S. and Van Dam, C.P. (1991). The influence of a laminar separation bubble on boundary-layer stability (AIAA Paper 91–3294–CP).

Ellingsen, T. and Palm, E. (1975). Stability of linear flow. Phys. Fluids, 18(4), 487–478.

Elofsson, P.A. and Alfredsson, P.H. (2000). An experimental investigation of oblique transition in a Blasius boundary layer flow. European J. Mech. B Fluids, 19(5), 615-636.

Elofsson, P.A. and Lundbladh, A. (1994). Ribbon induced oblique transition in plane Poiseuille flow [TRITA-MEK 94–14]. In D.S. Henningson (Ed.), Bypass Transition – Proceedings from a Mini-Workshop (pp. 29–41). Stockholm.

Emmons, H.W. (1951). The laminar–turbulent transition in a boundary layer – Part 1. J. Aeronaut. Sci., 18(7), 490–498.

Evert, F., Ronneberger, D. and Grosche, F.-R. (2000). Application of linear and nonlinear adaptive filters for the compensation of disturbances in the laminar boundary layer. ZAMM, 80(1), 85–88.

Farrell, B.F. and Ioannou, P.J. (1993). Optimal excitation of three dimensional perturbations in viscous constant shear flow. Phys. Fluids A, 5(6), 1390–1400.

Fasel, H. (1976). Investigation of the stability of boundary layers by a finite-difference model of the Navier–Stokes equations. J. Fluid Mech., 78, 355–383.

Fasel, H. (1977). Reaktion von zweidimensionalen, laminaren, inkompressiblen Grenzschichten auf periodische Störungen in der Außenströmung. Z. Angew. Math. Mech., 57, T180–T183.

Fasel, H. (1990). Numerical simulation of instability and transition in boundary layer flows. In D.Arnal and R.Michel (Eds.), Laminar-Turbulent Transition (pp. 587–598). Berlin Heidelberg New York: Springer.

Fasel, H. and Konzelmann, U. (1990). Non-parallel stability of a flat-plate boundary layer using the complete Navier–Stokes equations. J. Fluid Mech., 221, 311–347.

Fedorov, A.V. (1984). Excitation of Tollmien–Schlichting waves in a boundary layer by a periodic external source located on the body surface. Fluid Dyn., 19(6), 888-893.

Fedorov, A.V. (1988). Excitation of crossflow instability in swept wing boundary layer. J. Appl. Mech. Tehn. Phys., 29(5), 46–52. (In Russian.)

Fernholz, H.H. (1993). Management and control of turbulent shear flows. Z. Angew. Math. Mech., 73(11), 287–300.

Fischer, T.M. and Dallmann, U. (1991). Primary and secondary stability analysis of a three-dimensional boundary-layer flow. Phys. Fluids A, 3(10), 2378–2391.

Fischer, T.M., Hein, S. and Dallmann, U. (1993). A theoretical approach for describing secondary instability features in three-dimensional boundary layer flows (AIAA Paper 93–0080).

Fjørtoft, R. (1950). Amplification of integral theorems in deriving criteria for instability for laminar flows and the baroclinic circular vortex. Geofus. Publ., 17(6), 1–52.

Floryan, J.M. (1991). On the Görtler instability of boundary layers (TR 1120T). Japan: National Aerospace Laboratory.

Floryan, J.M. and Saric, W.S. (1982). Stability of Görtler vortices in boundary layers. AIAA J., 20(3), 316–324.

Floryan, J.M. and Saric, W.S. (1984). Wavelength selection and growth of Görtler vortices. AIAA J., 22(11), 1529–1538.

Fujimura, K. (1991). Method of centre manifold and multiple scales in the theory of weakly nonlinear stability of fluid motions. Proc. R. Soc. Lond. A, 434, 719–733.

Gad-el-Hak, M. (1990a). Control of low-speed airfoil aerodynamics. AIAA J., 28(9), 1537–1552.

Gad-el-Hak, M. (1990b). Transition control. In M.Y. Hussaini and R.G. Voight (Eds.), Instability and Transition (pp. 319–354). Berlin Heidelberg New York: Springer.

Gad-el-Hak, M. and Bushnell, D. (1991). Status and outlook of flow separation control (AIAA Paper 91–0037).

Galdi, G.P. and Padula, M. (1990). A new approach to energy theory in the stability of fluid motion. Arch. Rat. Mech. Anal., 110, 187–286.

Gaponenko, V.R., Ivanov, A.V. and Kachanov, Y.S. (1995). Experimental study of cross-flow instability of a swept-wing boundary layer with respect to travelling waves. In R.Kobayashi (Ed.), Laminar-Turbulent Transition (pp. 373–380). Berlin Heidelberg New York: Springer.

Gaponenko, V.R., Ivanov, A.V. and Kachanov, Y.S. (1996). Experimental study of 3D boundary-layer receptivity to surface vibration. In P.W. Duck and P Hall (Eds.), Nonlinear Instability and Transition in Three-Dimensional Boundary Layers (pp. 389–398). Dordrecht: Kluwer.

Gaponov, S.A. and Maslov, A.A. (1971). Stability of compressible boundary layer at subsonic flow velocities. Izv. Sib. Otd. Akad. Nauk SSSR, Ser. Tekhn. Nauk[5], 1(3), 24–27.

[5] Translated in Siberian Phys. Techn. J. (former Sov. J. Appl. Phys.)

Gaponov, S.A. and Maslov, A.A. (1980). Development of disturbances in compressible flows. Novosibirsk: Nauka. (In Russian.)

Gaster, M. (1962). A note on the relation between temporally-increasing and spatially-increasing disturbances in hydrodynamic stability. J. Fluid Mech., 14, 222–224.

Gaster, M. (1965). On the generation of spatially growing waves in a boundary layer. J. Fluid Mech., 22, 433–441.

Gaster, M. (1967). The structure and behavior of separation bubbles (ARC R&M 3595).

Gaster, M. (1968). The development of three-dimensional wave-packets in a boundary layer. J. Fluid Mech., 32, 173–184.

Gaster, M. (1974). On the effects of boundary-layer growth on flow stability. J. Fluid Mech., 66, 465–480.

Gaster, M. (1975). A theoretical model of a wave-packet in the boundary layer on a flat plate. Proc. R. Soc. Lond. A, 347, 271–289.

Gaster, M. (1979). The propagation of linear wave packet in laminar boundary layers – asymptotic theory for non-conservative wave systems (AIAA Paper 79–1492).

Gaster, M. (1990). The nonlinear phase of wave growth leading to chaos and breakdown to turbulence in a boundary layer as an example of an open system. Proc. R. Soc. Lond. A, 430, 3–24.

Gaster, M. (1992). Stability of velocity profiles with reverse flow. In M.Y. Hussaini, A.Kumar and C.L. Streett (Eds.), Instability, Transition, and Turbulence (pp. 212–215). Berlin Heidelberg New York: Springer.

Gaster, M. and Grant, T. (1975). An experimental investigation of the formation and development of a wave-packet in a laminar boundary layer. Proc. R. Soc. Lond. A, 347, 253–269.

Gaster, M. and Gupta, T.K.S. (1993). The generation of disturbances in a boundary layer by wall perturbations: The vibrating ribbon revisited once more. In D.E. Ashpis, T.B. Gatski and R.Hirsh (Eds.), Instabilities and Turbulence in Engineering Flows (pp. 31–50). Dordrecht: Kluwer.

Gates, E.M. (1980). Observations of transition on some axisymmetric bodies. In R.Eppler and H.Fasel (Eds.), Laminar-Turbulent Transition (pp. 351–363). Berlin Heidelberg New York: Springer.

Gatski, T.B. and Grosch, C. (1987). Numerical experiments in boundary-layer receptivity. In D.L. Dwoyer and M.Y. Hussaini (Eds.), Stability of Time Dependent and Spatially Varying Flows (pp. 82–96). Berlin Heidelberg New York: Springer.

Gedney, C.J. (1983). The cancellation of a sound-excited Tollmien–Schlichting wave with plane vibration. Phys. Fluids, 26(5), 1158–1160.

Gertsenstein, S.Y. (1966). Effect of a single roughness element on turbulence onset. Fluid Dyn., 1(2), 113–115.

Gilev, V.M. (1985). Tollmien–Schlichting waves excitation on the vibrator and laminar–turbulent transition control. In V.V. Kozlov (Ed.),

Laminar-Turbulent Transition (pp. 243–248). Berlin Heidelberg New York: Springer.
Gilev, V.M. and Kozlov, V.V. (1984). Excitation of Tollmien–Schlichting waves in a boundary layer on a vibrating surface. Zh. Prikl. Mekh. Techn. Fiz., 6, 73–77. (In Russian.)
Gilev, V.M. and Kozlov, V.V. (1985). Use of small localized wall vibrations for control of transition in the boundary-layer. Fluid Mech. – Soviet Res., 14(6), 50–54.
Gilev, V.M. and Kozlov, V.V. (1987). Effect of alternating injection and suction on transition in the boundary layer. Fluid Mech. – Soviet Res., 16(3), 61–68.
Gilev, V.M., Dovgal, A.V., Kachanov, Y.S. and Kozlov, V.V. (1988). Development of three-dimensional disturbances in a boundary layer with pressure gradient. Fluid Dyn., 23(3), 393–399.
Glushko, G.S. (1973). Transition to the turbulent flow mode in the boundary layer of a flat plate for different free-stream turbulence scales. Fluid Dyn., 7(3), 424–425.
Gmelin, C., Rist, U. and Wagner, S. (2000). DNS of active control of disturbances in a Blasius boundary layer. In H.F. Fasel and W.S. Saric (Eds.), Laminar-Turbulent Transition (pp. 149–154). Berlin Heidelberg New York: Springer.
Goldshtik, M.A. and Shtern, V.N. (1977). Hydrodynamic stability and turbulence. Novosibirsk: Nauka. (In Russian.)
Goldstein, M.E. (1983). The evolution of Tollmien–Schlichting waves near a leading edge. J. Fluid Mech., 127, 59–81.
Goldstein, M.E. (1984). Generation of instability waves in flows separating from smooth surfaces. J. Fluid Mech., 145, 71–94.
Goldstein, M.E. (1985). Scattering of acoustic waves into Tollmien–Schlichting waves by small streamwise variations in surface geometry. J. Fluid Mech., 154, 509–529.
Goldstein, M.E. and Choi, S.W. (1989). Nonlinear evolution of interacting oblique waves on two-dimensional shear layers. J. Fluid Mech., 207, 97–120. (See also Corrigendum, J.Fluid Mech., 216, 659–663.)
Goldstein, M.E. and Hultgren, L.S. (1989). Boundary layer receptivity to long wave free-stream disturbances. Ann. Rev. Fluid Mech., 21, 137–166.
Goldstein, M.E. and Lee, S.S. (1992). Fully coupled resonant-triad interaction in an adverse-pressure-gradient boundary layer. J. Fluid Mech., 245, 523–551.
Goldstein, M.E., Leib, S.J. and Cowley, S.J. (1987). Generation of Tollmien–Schlichting waves on interactive marginally separated flows. J. Fluid Mech., 181, 485–517.
Goldstein, M.E., Leib, S.J. and Cowley, S.J. (1992). Distortion of a flat-plate boundary layer by free-stream vorticity normal to the plate. J. Fluid Mech., 237, 231–260.

Görtler, H. (1940). Über eine dreidimensionale Inistabilität laminaren Grenzschichten an konkaven Wänden. Nachr. Ges. Wiss. Göttingen, N.F., 2(1), 1–26.
Gray, W.E. (1952). The effect of wing sweep on laminar flow (RAE TM Aero 255).
Greenspan, H.P. and Benney, D.J. (1963). On shear-layer instability, breakdown and transition. J. Fluid Mech., 15, 133–153.
Gregory, N. and Walker, W.S. (1956). The effect on transition of isolated surface excrescenses in the boundary-layer (ARC R&M 2779).
Gregory, N., Stuart, J.T. and Walker, W.S. (1955). On the stability of three-dimensional boundary layers with application to the flow due to a rotating disk. Philos. Trans. R. Soc. Lond. A, 248, 155–199.
Grek, G.R. and Kozlov, V.V. (1992). Interaction of Tollmien–Schlichting waves and localized disturbances. Sib. Fiz.-Techn. Zh., 34, 68–76. (In Russian.)
Grek, G.R., Kozlov, V.V. and Ramazanov, M.P. (1985). Three types of disturbances from the point source in the boundary layer. In V.V. Kozlov (Ed.), Laminar-Turbulent Transition (pp. 267–272). Berlin Heidelberg New York: Springer.
Grek, G.R., Kozlov, V.V. and Ramazanov, M.P. (1987). Laminar–turbulent transition at high free stream turbulence (Preprint 8–87). Novosibirsk: RAS. Sib. Branch Inst. Theoret. Appl. Mech. (In Russian.)
Grek, G.R., Kozlov, V.V. and Ramazanov, M.P. (1988). Laminar-turbulent transition in a presence of a high level of free-stream turbulence. Fluid Dyn., 23(6), 829–834.
Grek, G.R., Kozlov, V.V. and Ramazanov, M.P. (1989). Laminar–turbulent transition at high free stream turbulence in gradient flow. Izv. Sib. Otd. Akad. Nauk SSSR, Ser. Tekhn. Nauk[6], 3, 66–70.
Grek, G.R., Kozlov, V.V. and Ramazanov, M.P. (1990). Receptivity and stability of the boundary layer at a high turbulence level. In D.Arnal and R.Michel (Eds.), Laminar-Turbulent Transition (pp. 511–521). Berlin Heidelberg New York: Springer.
Grek, G.R., Kozlov, V.V. and Ramazanov, M.P. (1991a). Laminar–turbulent transition at high free stream turbulence: Review. Izv. Sib. Otd. Akad. Nauk SSSR, Ser. Tekhn. Nauk[6], 106–138.
Grek, G.R., Kozlov, V.V. and Ramazanov, M.P. (1991b). Investigation of boundary layer stability in a gradient flow with a high degree of free-stream turbulence. Fluid Dyn., 25(2), 207–213.
Grek, G.R., Kozlov, V.V., Titarenko, S.V. and Klingmann, B.G.B. (1995). The influence of riblets on a boundary layer with embedded streamwise vortices. Phys. Fluids A, 7(10), 2504–2506.
Grek, H.R., Dey, J., Kozlov, V.V., Ramazanov, M.P. and Tuchto, O.N. (1991c). Experimental analysis of the process of the formation of turbu-

[6] Translated in Siberian Phys. Techn. J. (former Sov. J. Appl. Phys.)

lence in the boundary layer at higher degree of turbulence of windstream (Tech. Rep. 91–FM–2). Bangalor: Indian Institute of Science.

Grek, G.R., Kozlov, V.V. and Titarenko, S.V. (1996a). Effects of riblets on vortex development in the wake behind a single roughness element in the laminar boundary layer on a flat plate. Rech. Aérospat., 1996(1), 1–9.

Grek, G.R., Kozlov, V.V. and Titarenko, S.V. (1996b). An experimental study on the influence of riblets on transition. J. Fluid Mech., 315, 31–49.

Grek, G.R., Kozlov, V.V., Katasonov, M.M. and Chernorai, V.G. (2000). Experimental study of a λ-structure and its transformation into the turbulent spot. Current Sci., 79(6), 781–789.

Grosch, C.E. and Salwen, H. (1978). The continuous spectrum of the Orr–Sommerfeld equation. Part 1. The spectrum and eigenfunctions. J. Fluid Mech., 87, 33–54.

Gruber, K., Bestek, H. and Fasel, H. (1987). Interaction between a Tollmien–Schlichting wave and a laminar separation bubble (AIAA Paper 87–1256).

Gulyaev, A.N., Kozlov, V.E., Kuznetsov, V.R., Mineev, B.I. and Sekundov, A.N. (1989). Interaction of a laminar boundary layer with external turbulence. Fluid Dyn., 24(5), 700–710.

Guo, Y. and Finlay, W.H. (1994). Wavenumber selection and irregularity of spatially developing nonlinear Dean and Görtler vortices. J. Fluid Mech., 264, 1–40.

Gustavsson, L.H. (1979). Initial-value problem for boundary layer flows. Phys. Fluids, 22(9), 1602–1605.

Gustavsson, L.H. (1991). Energy growth of three-dimensional disturbances in plane Poiseuille flow. J. Fluid Mech., 224, 241–260.

Haberman, R. (1972). Critical layers in parallel flows. Stud. Appl. Math., 51(2), 139–161.

Haggmark, C.P. (2000). Investigations of disturbances developing in a laminar separation bubble flow. Doctoral thesis, Department of Mechanics, Royal Institute of Technology, Stockholm.

Haggmark, C.P., Bakchinov, A.A. and Alfredsson, P.H. (1997). Transition in a two-dimensional laminar boundary-layer separation bubble. In EUROMECH–359. Stability and Transition of Boundary-Layer Flows. Collection of Abstracts. University of Stuttgart. (Paper No. 32.)

Haggmark, C.P., Bakchinov, A.A. and Alfredsson, P.H. (2000). Experiments on a two-dimensional laminar separation bubble. Philos. Trans. R. Soc. Lond. A, 358, 3193–3205.

Hall, P. (1982). Taylor–Görtler vortices in fully developed or boundary-layer flows: Linear theory. J. Fluid Mech., 124, 475–494.

Hall, P. (1983). The linear development of Görtler vortices in growing boundary layers. J. Fluid Mech., 130, 41–58.

Hall, P. (1988). The nonlinear development of Görtler vortices in growing boundary layers. J. Fluid Mech., 193, 243–266.
Hall, P. (1990). Görtler vortices in growing boundary layers: The leading edge receptivity problem, linear growth and the nonlinear breakdown stage. Mathematika, 37, 151–189.
Hamilton, J.M. and Abernathy, F. (1994). Streamwise vortices and transition to turbulence. J. Fluid Mech., 264, 185–212.
Hämmerlin, G. (1955). Über das Eigenwertproblem der dredimensionalen Instabilität laminarer Grenzschichten an konkaven Wänden. J. Rat. Mech. Anal., 4(2), 279–321.
Hammond, D.A. and Redekopp, L.G. (1998). Local and global instability properties of separation bubbles. European J. Mech. B Fluids, 17(2), 145–164.
Hanifi, A., Schmid, P.J. and Henningson, D.S. (1994). Transient growth in compressible boundary layer flow [TRITA-MEK 94–14]. In D.S. Henningson (Ed.), Bypass Transition – Proceedings from a Mini-Workshop (pp. 51–65). Stockholm.
Hasan, M. A.Z. (1992). The flow over a backward-facing step under controlled perturbation: Laminar separation. J. Fluid Mech., 238, 73–96.
Healey, J. (1995). On the neutral curve of the flat plate boundary layer: Comparison between experiment, Orr–Sommerfeld theory and asymptotic theory. J. Fluid Mech., 288, 59–73.
Hefner, J.N. (1991). Laminar flow control: Introduction and overview. In R.W. Barnwell and M.Y. Hussaini (Eds.), Natural Laminar Flow and Laminar Flow Control (pp. 1–21). Berlin Heidelberg New York: Springer.
Hefner, J.N. and Bushnell, D.M. (1979). Application of stability theory to laminar flow control. In Twelfth AIAA Fluid and Plasma Dynamics Conf. Williamsburg, Virginia.
Hein, S. (2000). Linear and nonlinear nonlocal instability analysis for two-dimensional laminar separation bubbles. In H.F. Fasel and W.S. Saric (Eds.), Laminar-Turbulent Transition (pp. 681–686). Berlin Heidelberg New York: Springer.
Hein, S., Theofilis, V. and Dallmann, U. (1998). Unsteadiness and three-dimensionality of steady two-dimensional laminar separation bubbles as result of linear instability mechanisms (DLR-IB 223–98 A 39). Göttingen.
Heinrich, R.A., Choudhari, M. and Kerschen, E.J. (1988). A comparison of boundary layer receptivity mechanisms (AIAA Paper 88–3758).
Heisenberg, W. (1924). Über Stabilität und Turbulenz von Flüssigkeitsstömmen. Ann. Phys., 74, 577–627. (Translated as NACA TM 1291, 1951.)

Henningson, D. (1996). Comment on "Transition in shear flows. Nonlinear normality versus nonnormal linearity" [Phys. Fluids. 7, 3060 (1995)]. Phys. Fluids, 8(8), 2257–2258.

Henningson, D.S. (1988). The inviscid initial value problem for a piecewise linear mean flow. Stud. Appl. Math., 78(1), 31–56.

Henningson, D.S. (1991). An eigenfunction expansion of localized disturbances. In A.V. Johansson and P.H. Alfredsson (Eds.), Advances in Turbulence 3 (pp. 162–169). Berlin Heidelberg New York: Springer.

Henningson, D.S. (1995). Bypass transition and linear growth mechanisms. In R.Benzi (Ed.), Advances in turbulence (pp. 190–204). Dordrecht: Kluwer.

Henningson, D.S. and Reddy, S.C. (1994). On the role of linear mechanisms in transition to turbulence. Phys. Fluids A, 6(3), 1396–1398.

Henningson, D.S. and Schmid, P.J. (1994). A note on measures on disturbance size in spatially evolving flows. Phys. Fluids A, 6(8), 2862–2864.

Henningson, D.S., Johansson, A.V. and Lunbladh, A. (1990). On the evolution of localized disturbances in laminar shear flows. In D.Arnal and R.Michel (Eds.), Laminar-Turbulent Transition (pp. 279–284). Berlin Heidelberg New York: Springer.

Henningson, D.S., Lunbladh, A. and Johansson, A.V. (1993). A mechanism for bypass transition from localized disturbances in wall-bounded shear flows. J. Fluid Mech., 250, 169–207.

Henningson, D.S., Gustavsson, L.H. and Breuer, K.S. (1994). Localized disturbances in parallel shear flows. Appl. Sci. Res., 53, 51–97.

Herbert, T. (1976). On the stability of the boundary layer along a concave wall. Arch. Mech., 28(1), 5–6.

Herbert, T. (1983a). On perturbation methods in nonlinear stability theory. J. Fluid Mech., 126, 167–186.

Herbert, T. (1983b). Secondary instability of plane channel flow to subharmonic three-dimensional disturbances. Phys. Fluids, 26(4), 871–874.

Herbert, T. (1984). Analysis of the subharmonic route to transition in boundary layers (AIAA Paper 84-0009).

Herbert, T. (1988). Secondary instability of boundary layers. Ann. Rev. Fluid Mech., 20, 487–526.

Herbert, T. (1997). Parabolized stability equations. Ann. Rev. Fluid Mech., 29, 245–283.

Herbert, T. and Bertolotti, F.P. (1987). Stability analysis of nonparallel boundary layers. Bull. Amer. Phys. Soc., 32(10), 2079.

Herbert, T. and Lin, N. (1993). Studies of boundary-layer receptivity with parabolized stability equations (AIAA Paper 93-3053).

Hill, D.C. (1995). Adjoint systems and their role in the receptivity problem for boundary layer. J. Fluid Mech., 292, 183–204.

Ho, C.-M. (1994). Interaction between fluid dynamics and new technology. In N.W.M. Ko, H.E. Fiedler and B.H.K. Lee (Eds.), First Internat. Conf.

Flow Interaction cum Exhibition/Lectures on Interaction of Science & Art (SCART'94) (pp. 1–8). Hong Kong.
Ho, C.-M. (1995). Distributed micro transducers for aerodynamic applications. In N.W.M. Ko, H.E. Fiedler and B.H.K. Lee (Eds.), Seminar Series on MicroElectroMechanical Systems (MEMS). Dept. of Mech. Eng., The Hong Kong University of Science & Technology.
Ho, C.-M. and Tai, Y.-C. (1994). MEMS: Science and technology. In P.R. Bandyopadhyay, K.S. Breuer and C.J. Blechinger (Eds.), Application of Microfabrication to Fluid Mechanics (Vol. FED–197, pp. 39–48). ASME.
Ho, C.-M. and Tai, Y.-C. (1996). Review: MEMS and its applications for flow control. Trans. ASME Ser. I: J. Fluids Eng., 118(9), 437–447.
Ho, C.-M. and Tai, Y.-C. (1998). Micro-electro-mechanical systems (MEMS) and fluid flows. Ann. Rev. Fluid Mech., 30, 579–612.
Holmes, B.J., Obara, C.J., Martin, G.L. and Dormack, C.S. (1986). Manufacturing tolerances for natural laminar flow air frame surfaces (NASA CP 2413).
Horton, H.P. (1967). A semi-empirical theory for the growth and bursting of laminar separation bubbles (Aeronaut. Research Council CP 1073).
Howard, L.N. (1961). Note on a paper by John. W.Miles. J. Fluid Mech., 10, 509–512.
Huai, X., Joslin, R.D. and Piomelli, U. (1997). Large-eddy simulation of boundary layer transition on swept wings (AIAA Paper 97–0750).
Huerre, P. and Monkewitz, P.A. (1985). Absolute and convective instabilities in free shear layers. J. Fluid Mech., 159, 151–168.
Huerre, P. and Monkewitz, P.A. (1990). Local and global instabilities in spatially developing flows. Ann. Rev. Fluid Mech., 22, 473–537.
Huerre, P. and Rossi, M. (1998). Hydrodynamic instabilities in open flow. In C.Godrèche and P.Manneville (Eds.), Hydrodynamics and Nonlinear Instabilities (pp. 81–294). Cambridge: Cambridge University Press.
Hultgren, L.S. and Gustavsson, L.H. (1981). Algebraic growth of disturbances in a laminar boundary layer. Phys. Fluids, 24(6), 1000–1004.
Iordanskiy, S.V. and Kulikovskiy, A.G. (1965). On absolute stability of some plane parallel flows at high Reynolds numbers. Zh. Èksper. Teoret. Fiz., 49(10), 1326. (In Russian.)
Itoh, N. (1974). Spatial growth of finite wave disturbances in parallel and nearly parallel flows. Part 2. The numerical results for the flat plate boundary layer. Trans. Japan Soc. Aero. Space Sci., 17, 175.
Itoh, N. (1986). The origin and subsequent development in space of Tollmien–Schlichting waves in a boundary layer. Fluid Dyn. Res., 1, 119–130.
Ivanov, A.V., Kachanov, Y.S. and Koptsev, D.B. (1994). An experimental investigation of instability wave excitation in three-dimensional boundary layer at acoustic wave scattering on a vibrator. Thermophys. Aeromech., 4(4), 359–372.

Ivanov, A.V., Kachanov, Y.S., Obolentseva, T.G. and Michalke, A. (1998). Receptivity of the Blasius boundary layer to surface vibrations. Comparison of theory and experiment. In Proc. Ninth Internat. Conf. on Methods of Aerophysical Research (Vol. 1, pp. 93–98). Novosibirsk.

Jaffe, N.A., Okamura, T.T. and Smith, A.M.O. (1970). Determination of spatial amplification factors and their application to prediction transition. AIAA J., 8(2), 301–308.

Janke, E. (2000). On the secondary instability of three-dimensional boundary layers. Theoret. Comput. Fluid Dyn., 14, 167–194.

Jordisson, R. (1970). The flat plate boundary layer. Part 1. Numerical integration of the Orr–Sommerfeld equation. J. Fluid Mech., 43, 801–811.

Joseph, D.D. (1976). Stability of Fluid Motion (Vol. 1). Berlin Heidelberg New York: Springer.

Joslin, R.D. (1995). Evolution of stationary crossflow vortices in boundary layers on swept wings. AIAA J., 33(7), 1279–1285.

Joslin, R.D. and Streett, C.L. (1994). The role of stationary crossflow vortices in boundary-layer transition on swept wings. Phys. Fluids, 6(10), 3442–3453.

Joslin, R.D., Nicolaides, R.A., Erlebacher, G., Hussaini, M.Y. and Gunzburger, M.D. (1995). Active control of boundary-layer instabilities: Use of sensors and spectral controller. AIAA J., 33(8), 1521–1523.

Kachanov, Y.S. (1985). Development of spatial wave packets in boundary layer. In V.V. Kozlov (Ed.), Laminar-Turbulent Transition (pp. 115–123). Berlin Heidelberg New York: Springer.

Kachanov, Y.S. (1987). On the resonant nature of the breakdown of a laminar boundary layer. J. Fluid Mech., 184, 43–74.

Kachanov, Y.S. (1994). Physical mechanisms of laminar-boundary-layer transition. Ann. Rev. Fluid Mech., 26, 411–482.

Kachanov, Y.S. (1996). Generation, development and interaction of instability modes in swept-wing boundary layer. In P.W. Duck and P.Hall (Eds.), Nonlinear Instability and Transition in Three-Dimensional Boundary Layers (pp. 115–132). Dordrecht: Kluwer.

Kachanov, Y.S. (2000). Three-dimensional receptivity of boundary layers. European J. Mech. B Fluids, 19(5), 723–744.

Kachanov, Y.S., Kozlov, V.V. and Levchenko, V.Y. (1974). Experimental study on the effect of wall cooling upon laminar boundary-layer stability. Izv. Sib. Otd. Akad. Nauk SSSR, Ser. Tekhn. Nauk[7], 9(8), 75–79.

Kachanov, Y.S., Kozlov, V.V. and Levchenko, V.Y. (1975). Generation and development of small amplitude disturbances in laminar boundary layer in presence of acoustic field. Izv. Sib. Otd. Akad. Nauk SSSR, Ser. Tekhn. Nauk[7], 3(13), 18–26.

[7] Translated in Siberian Phys. Techn. J. (former Sov. J. Appl. Phys.)

Kachanov, Y.S. and Levchenko, V.Y. (1984). The resonant interaction of disturbances at laminar–turbulent transition in a boundary layer. J. Fluid Mech., 138, 209–247.

Kachanov, Y.S. and Michalke, A. (1994). Three-dimensional instability of flat-plate boundary layers: Theory and experiment. European J. Mech. B Fluids, 13(4), 401–422.

Kachanov, Y.S., Ryzhov, O.S. and Smith, F.T. (1993). Formation of solitons in transitional boundary layers: Theory and experiment. J. Fluid Mech., 251, 273–297.

Kachanov, Y.S. and Tararykin, O.I. (1990). The experimental investigation of stability and receptivity of a swept-wing flow. In D.Arnal and R.Michel (Eds.), Laminar-Turbulent Transition (pp. 499–509). Berlin Heidelberg New York: Springer.

Kachanov, Y.S. and Tararykin, O.I. (1991). The development of 3-D separated flows and their influence on laminar–turbulent transition. In V.V. Kozlov and A.V. Dovgal (Eds.), Separated Flows and Jets (pp. 737–740). Berlin Heidelberg New York: Springer.

Kachanov, Y.S., Kozlov, V.V. and Levchenko, V.Y. (1978). Nonlinear development of a wave in a boundary layer. Fluid Dyn., 12(3), 383–390.

Kachanov, Y.S., Kozlov, V.V. and Levchenko, V.Y. (1979a). The development of small-amplitude oscillations in a laminar boundary layer. Fluid Mech. – Soviet Res., 8(2), 152–156.

Kachanov, Y.S., Kozlov, V.V. and Levchenko, V.Y. (1979b). Occurence of Tollmien–Schlichting waves in the boundary layer under the effect of external perturbations. Fluid Dyn., 13(5), 85–94.

Kachanov, Y.S., Kozlov, V.V., Levchenko, V.Y. and Maksimov, V.P. (1979c). Transformation of external disturbances into the boundary layer waves. In Proc. Sixth Intern. Conf. on Numerical Methods in Fluid Dyn. (pp. 299–307). Berlin Heidelberg New York: Springer.

Kachanov, Y.S., Kozlov, V.V. and Levchenko, V.Y. (1980). Experiments on nonlinear interaction of waves in boundary layer. In R.Eppler and H.Fasel (Eds.), Laminar-Turbulent Transition (pp. 135–152). Berlin Heidelberg New York: Springer.

Kachanov, Y.S., Kozlov, V.V. and Levchenko, V.Y. (1982). Origin of turbulence in boundary layer. Novosibirsk: Nauka.

Kachanov, Y.S., Gaponenko, V.R. and Ivanov, A.V. (1997). Experimental study of swept-wing boundary-layer receptivity to stationary and non-stationary surface non-uniformities. In Stability and Transition of Boundary-Layer Flows. Collection of Abstracts (p. Paper No. 5). Universität Stuttgart, Germany.

Kaidalov, G.N. and Morin, O.V. (1972). Effect of suction upon boundary layer transition. In A.M. Kharitonov (Ed.), Aerophysical Studies (p. 129). Novosibirsk: RAS. Sib. Branch Inst. Theoret. Appl. Mech. (In Russian.)

Kazakov, A.V., Kogan, M.N. and Kuparev, V.A. (1985). On increase of subsonic boundary-layer stability at surface heating close the leading edge. Dokl. Akad. Nauk[8], 283(2), 333–335.

Kelly, R.E. (1967). On the stability of an inviscid shear layer which is periodic in space and time. J. Fluid Mech., 27, 657–689.

Kendall, J.M. (1985). Experimental study of disturbances produced in a pre-transitional laminar boundary layer by weak free stream turbulence (AIAA Paper 85–1695).

Kendall, J.M. (1990). Boundary layer receptivity to freestream turbulence (AIAA Paper 90–1504).

Kendall, J.M. (1991). Studies on laminar boundary layer receptivity to freestream turbulence near a leading edge. In D.C. Reda, H.L. Reed and R.Kobayashi (Eds.), Boundary Layer Stability and Transition to Turbulence (Vol. FED–114, pp. 23–30). ASME.

Kerschen, E.J. (1991). Linear and non-linear receptivity to vortical freestream disturbances. In D.C. Reda, H.L. Reed and R.Kobayashi (Eds.), Boundary Layer Stability and Transition to Turbulence (Vol. FED–114). ASME.

King, R.A. and Breuer, K.S. (2001). Acoustic receptivity and evolution of two-dimensional and oblique disturbances in a blasius boundary layer. J. Fluid Mech., 432, 69–90.

Kiya, M. (1989). Separation bubbles. In P.Germain, M.Piau and D.Caillerie (Eds.), Theoretical and Applied Mechanics (pp. 173–191). Elsevier.

Kiya, M., Mochizuki, O., Tanaka, H. and Tsukasaki, T. (1991). Control of a turbulent leading-edge separation bubble. In V.V. Kozlov and A.V. Dovgal (Eds.), Separated Flows and Jets (pp. 647–656). Berlin Heidelberg New York: Springer.

Kiya, M., Shimizu, M., Mochizuki, O., Ido, Y. and Tezuka, H. (1993). Active forcing of an axisymmetric leading-edge turbulent separation bubble (AIAA Paper 93–3245).

Klebanoff, P.S. and Tidstrom, K.D. (1959). Evolution of amplified waves leading to transition in a boundary layer with zero pressure gradient (NACA TN D-195).

Klebanoff, P.S. and Tidstrom, K.D. (1972). Mechanism by which a two-dimensional roughness element induces boundary-layer transition. Phys. Fluids, 15(7), 1173–1188.

Klebanoff, P.S., Tidstrom, K.D. and Sargent, L.M. (1962). The three-dimensional nature of boundary-layer instability. J. Fluid Mech., 12, 1–34.

Kleiser, L. and Zang, T.A. (1991). Numerical simulation of transition in wall-bounded shear flows. Ann. Rev. Fluid Mech., 23, 495–537.

Klingmann, B.G.B. (1991). Laminar–turbulent transition in plane Poiseuille flow. Doctoral dissertation, Department of Gasdynamics, Stockholm.

[8] Translated in Phys. Dokl.

Klingmann, B.G.B. (1992). On transition due to three-dimensional disturbances in plane Poiseuille flow. J. Fluid Mech., 240, 167–195.

Klingmann, B.G.B., Boiko, A.V., Westin, K.J.A., Kozlov, V.V. and Alfredsson, P.H. (1993). Experiments on the stability of Tollmien–Schlichting waves. European J. Mech. B Fluids, 12(4), 493–514.

Knapp, C.F. and Roache, P.J. (1968). A combined visual and hot-wire anemometer investigation of boundary layer transition. AIAA J., 6(1), 29–36.

Kobayashi, R., Kohama, Y. and Kurosawa, M. (1983). Boundary-layer transition on a rotating cone in axial flow. J. Fluid Mech., 127, 341–352.

Kobayashi, R., Kohama, Y., Arai, T. and Ukaku, M. (1987). The boundary-layer transition on rotating cones in axial flow with free-stream turbulence. Japan. Soc. Mech. Eng. Internat. J., 30(261), 423–429.

Kobayashi, R., Fukunishi, Y. and Kato, T. (1995). Laminar-flow control of boundary layers utilizing acoustic receptivity. In Proceedings of The 6th Asian Congress of Fluid Mechanics (pp. 629–633). Singapore.

Koch, W., Bertolotti, F.P., Stolte, A. and Hein, S. (2000). Nonlinear equilibrium solutions in a three-dimensional boundary layer and their secondary instability. J. Fluid Mech., 406, 131–174.

Kohama, Y. (1987). Some expectation on the mechanism of cross-flow instability in a swept-wing flow. Acta Mech., 66, 21–38.

Kohama, Y., Saric, W.S. and Hoos, J.A. (1991). A high-frequency, secondary instability of crossflow vortices that leads to transition. In Proc. Boundary-Layer Transition and Control (pp. 4.1–4.13). Cambridge: R. Aeronaut. Soc.

Kohama, Y., Onodera, T. and Egami, Y. (1996). Design and control of crossflow instability field. In P.W. Duck and P.Hall (Eds.), Nonlinear Instability and Transition in Three-Dimensional Boundary Layers (pp. 147–156). Dordrecht: Kluwer.

Konzelmann, U. (1990). Numerische Untersuchungen zur räumlichen Entwicklung dreidimensionalen Wellenpakete in einer Plattengrenzschichtsströmung. Doctoral dissertation, Universität Stuttgart.

Kosorygin, V.S. and Polyakov, N.P. (1990). Laminar boundary layers in turbulent flows. In D.Arnal and R.Michel (Eds.), Laminar-Turbulent Transition (pp. 573–578). Berlin Heidelberg New York: Springer.

Kovasznay, L.S., Komoda, H. and Vasudeva, B.R. (1962). Detailed flow field in transition. In Proc. Heat Transfer and Fluid Mech. Institute (pp. 1–26). Stanford: Stanford University.

Kozlov, L.F. (1968). Optimal suction of the boundary layer taking account of initial turbulence and surface roughness. J. Fluid Mech., 31, 53–64.

Kozlov, V.V. (1985). Interrelation of the flow separation and stability. In V.V. Kozlov (Ed.), Laminar-Turbulent Transition (pp. 349–358). Berlin Heidelberg New York: Springer.

Kozlov, L.F. and Babenko, V.V. (1978). Experimental investigations of boundary layer. Kiev: Naukova dumka. (In Russian.)

Kozlov, V.V. and Ramazanov, S.P. (1982). An experimental investigation of the stability of Poiseuille flow. Fluid Mech. – Soviet Res., 11(4), 41–46.

Kozlov, V.V. and Ramazanov, S.P. (1983). Development of finite-amplitude perturbations in Poiseuille flow. Fluid Dyn., 18(1), 30–33.

Kozlov, V.V. and Ryzhov, O.S. (1990). Receptivity of boundary layers: Asymptotic theory and experiment. Proc. R. Soc. Lond. A, 429, 341–373.

Kozlov, V.V., Levchenko, V.Y. and Scherbakov, V.A. (1978). Development of boundary layer disturbances under suction through a slot. Uchen. Zap. TsAGI, 9(2), 99–105. (In Russian.)

Kozlov, V.E., Kuznetsov, V.R., Mineev, B.I. and Sekundov, A.N. (1990). The influence of free-stream turbulence and surface ribbing on the characteristics of a transitional boundary layer. In S.J. Kline and N.H. Afgan (Eds.), Near-Wall Turbulence. Proc. 1988 Zorian Zorac Memorial Conference (pp. 172–189). Hemisphere.

Kozlov, V.V., Grosche, F.-R., Dovgal, A.V., Bippes, H., Kuhn, A. and Stiewitt, H. (1993). Control of leading-edge separation by acoustic excitation (DLR-IB 222-93 A 10). Göttingen.

Kozlov, V.V., Levchenko, V.Y., Sova, V.A. and Scherbakov, V.A. (1999). Influence of acoustic field on flow structure and laminar–turbulent transition at swept wing. In Modern Problems of Aerohydromechanics (Vol. 2, pp. 145–155). Moscow: Inst. Prikl. Mehan. RAS. (In Russian.)

Kulikovskiy, A.G. (1966). On stability of uniform states. Prikl. Mat. Mekh.[9], 30, 148–153.

Kwon, O.K. and Pletcher, R.H. (1979). Prediction of incompressible separated boundary layer including viscous-inviscid interaction. Trans. ASME Ser. I: J. Fluids Eng., 101(4), 466–472.

Kyriakides, N.K., Kastrinakis, E.G., Nychas, S.G. and Goulas, A. (1999). A bypass wake induced laminar/turbulent transition. European J. Mech. B Fluids, 18(6), 1049–1065.

Landahl, M.L. (1972). Wave mechanism of breakdown. J. Fluid Mech., 56, 775–802.

Landahl, M.L. (1980). A note on an algebraic instability of inviscid parallel shear flows. J. Fluid Mech., 98, 243–251.

Landau, L.D. and Lifshitz, E.M. (1959). Fluid mechanics. Oxford: Pergamon.

Landau, L.D. and Lifshitz, E.M. (1982). Mechanics (Vol. 1, 3rd ed.). Oxford: Pergamon.

Lauchle, G.C. and Gurney, G.B. (1984). Laminar boundary-layer transition on a heated underwater body. J. Fluid Mech., 144, 79–101.

Laurien, E. and Kleiser, L. (1989). Numerical simulation of boundary-layer transition and transition control. J. Fluid Mech., 199, 403–440.

[9] Translated in Appl. Math. Mech.

Lebedev, Y.B. and Fomichev, V.M. (1987). Stability of the fluid boundary layer with nonuniform surface temperature distribution. J. Eng. Phys. Thermophys., 52(6), 948–953.

Leblanc, P., Blackwelder, R. and Liebeck, R. (1991). Experimental results on laminar separation on two airfoils at low Reynolds numbers. In Twenty-ninth Aerospace Sciences Meeting. Reno, U.S.A.

Leehey, P. and Shapiro, P. (1980). Leading edge effect in laminar boundary layer excitation by sound. In R.Eppler and H.Fasel (Eds.), Laminar-Turbulent Transition (pp. 321–331). Berlin Heidelberg New York: Springer.

Lekoudis, S.G. (1979). Stability of three-dimensional compressible boundary layers over wings with suction (AIAA Paper 79–0265).

Lekoudis, S.G. (1980). Stability of the boundary layer on a swept wing with wall cooling. AIAA J., 18(9), 1029–1035.

Levchenko, V.Y. and Soloviev, A.S. (1970). On boundary layer stability at pressure gradient and suction. Izv. Sib. Otd. Akad. Nauk SSSR, Ser. Tekhn. Nauk[10], 2(8), 65–67.

Levchenko, V.Y., Volodin, A.G. and Gaponov, S.A. (1975). Stability characteristics of boundary layers. Novosibirsk: Nauka. (In Russian.)

Li, F. and Malik, M.R. (1995). Fundamental and subharmonic secondary instabilities of Görtler vortices. J. Fluid Mech., 297, 77–100.

Libly, P.A. and Fox, H. (1964). Some perturbation solutions in laminar boundary-layer theory. J. Fluid Mech., 17, 433–449.

Liepmann, H.W., Brown, G.L. and Nosenchuk, D.M. (1982). Control of laminar instability waves using a new technique. J. Fluid Mech., 118, 187–200.

Liepmann, H.W. and Nosenchuk, D.M. (1982). Active control of laminar–turbulent transition. J. Fluid Mech., 118, 201–204.

Lin, C.C. (1955). The theory of hydrodynamic stability. Cambridge: Cambridge University Press.

Lin, C.C. (1961). Some mathematical problems in the theory of the stability of parallel flows. J. Fluid Mech., 10, 430–438.

Lin, J.C.M. and Pauley, L.L. (1996). Low-Reynolds-number separation on an airfoil. AIAA J., 34(8), 1570–1577.

Lindors, M. and Laine, S. (1976). Numerical solution of the Navier–Stokes equations for unsteady boundary-layer flows past a wave-like bulge on a flat plate. Lect. Notes Phys., 59, 293–299.

Lingwood, R.J. (1995). Absolute instability of the boundary layer on a rotating disk. J. Fluid Mech., 299, 17–33.

Liu, C., Tsao, T. and Tai, Y.-C. (1995). Out-of-plane permalloy magnetic actuators for delta-wing control. In Proc. IEEE Micro Electro Mechanical Systems Workshop (MEMS '95), IEEE Catalog No. 95CH35754 (pp. 7–12). Amsterdam, Netherland.

[10] Translated in Siberian Phys. Techn. J. (former Sov. J. Appl. Phys.)

Löfdahl, L. and Gad-el-Hak, M. (1999). MEMS applications in turbulence and flow control. Progr. Aerosp. Sci., 35, 101–203.

Lowell, R.L. and Reshotko, E. (1974). Numerical study of the stability of a heated water boundary layer (FTAS/TR 73-93). Cleveland: Case Western Reserve University.

Luchini, P. (1995). Asymptotic analysis of laminar boundary-layer flow over finely grooved surfaces. European J. Mech. B Fluids, 14(2), 169–195.

Luchini, P. (1996). Reynolds-number-independent instability of the boundary layer over a flat surface. J. Fluid Mech., 327, 101–115.

Luchini, P. (2000). Reynolds-number-independent instability of the boundary layer over a flat surface: Optimal perturbations. J. Fluid Mech., 404, 289–309.

Luchini, P. and Bottaro, A. (1998). Görtler vortices: A backward in time approach to the receptivity problem. J. Fluid Mech., 363, 1–23.

Luchini, P., Manzo, M. and Pozzi, A. (1991). Resistance of a grooved surface to parallel flow and cross-flow. J. Fluid Mech., 228, 87–109.

Lushin, V.N. (1992). A finite-span wing flow under acoustic excitation. Sib. Fiz.-Techn. Zh., 4, 64–68. (In Russian.)

Mabey, D.G. (1972). Analysis and correlation of data on pressure fluctuations in separated flow. J. Aircraft, 9(9), 642–645.

Mack, L.M. (1975). A numerical method for the prediction of high speed boundary layer transition using linear theory (NASA SP-347).

Mack, L.M. (1976). A numerical study of the temporal eigenvalue spectrum of the Blasius boundary layer. J. Fluid Mech., 73, 497–520.

Mack, L.M. (1980). On the stabilization of three-dimensional boundary layers by suction and cooling. In R.Eppler and H.Fasel (Eds.), Laminar-Turbulent Transition (pp. 223–238). Berlin Heidelberg New York: Springer.

Mack, L.M. (1982). Compressible boundary-layer stability calculations for sweptback wings with suction. AIAA J., 20(3), 363–369.

Mack, L.M. (1984). Boundary layer stability theory: Special course on stability and transition of laminar flow (AGARD Report 709).

Maksimov, V.P. (1975). Numerical investigation of the effect of local variations in viscous incompressible gas flow. In Problems of Gasdynamics (pp. 118–121). Novosibirsk: RAS. Sib. Branch Inst. Theoret. Appl. Mech. (In Russian.)

Maksimov, V.P. (1979). Origin of Tollmien–Schlichting waves in oscillating boundary layers. In V.Y. Levchenko (Ed.), Development of Disturbances in Boundary Layer (pp. 68–75). Novosibirsk: RAS. Sib. Branch Inst. Theoret. Appl. Mech. (In Russian.)

Malik, M.R. and Hussaini, M.Y. (1990). Numerical simulation of interactions between Görtler vortices and Tollmien–Schlichting waves. J. Fluid Mech., 210, 183–199.

Malik, M.R., Li, F. and Chang, C.-L. (1994). Crossflow disturbances in three-dimensional boundary layers: Nonlinear development, wave interaction and secondary instability. J. Fluid Mech., 268, 1–36.

Malik, M.R. and Orszag, S.A. (1980). Comparison of methods for prediction of transition by stability analysis. AIAA J., 18(12), 1485–1489.

Marchuk, G.I. (1977). Methods of numerical mathematics. Moscow: Nauka. (In Russian.)

Masad, J.A. and Iyer, V. (1994). Transition prediction and control in subsonic flow over a hump. Phys. Fluids, 6(1), 313–327.

Masad, J.A. and Malik, M.R. (1994). On the link between flow separation and transition onset (AIAA Paper 94–2370).

Masad, J.A. and Nayfeh, A.H. (1992a). Effect of a bulge on the subharmonic instability of subsonic boundary layers. AIAA J., 30(7), 1731–1737.

Masad, J.A. and Nayfeh, A.H. (1992b). Stability of separating boundary layers. In Fourth Internat. Conf. Fluid Mechanics (Vol. 1, pp. 261–278). Alexandria.

Masad, J.A. and Nayfeh, A.H. (1993). The influence of imperfections on the stability of subsonic boundary layers. In D.E. Ashpis, T.B. Gatski and R.Hirsh (Eds.), Instabilities and Turbulence in Engineering Flows (pp. 65–82). Dordrecht: Kluwer.

Maslennikova, I.I. and Zelman, M.B. (1986). On development of nonlinear disturbances at active boundary-layer transition control (Preprint 16–86). Novosibirsk: RAS. Sib. Branch Inst. Theoret. Appl. Mech. (In Russian.)

Maslowe, S.A. (1985). Shear flow instabilities and transition. In H.L. Swinney and J.P. Golub (Eds.), Hydrodynamic Instabilities and the Transition to Turbulence (Vol. 45, 2nd ed., pp. 181–228). Berlin Heidelberg New York: Springer.

Maslowe, S.A. and Spiteri, R.J. (2001). The continuous spectrum for a boundary layer in a streamwise pressure gradient. Phys. Fluids, 13(5), 1294–1299.

Matsubara, M. and Alfredsson, P.H. (2001). Disturbance growth in boundary layers subjected to free-stream turbulence. J. Fluid Mech., 430, 149–168.

Matsubara, M., V.Kozlov, V., Alfredsson, P.H., Bakchinov, A.A. and Westin, K.J.A. (1996). On flat plate boundary layer perturbations at high free stream turbulence level. In Proc. Eighth Internat. Conf. on Methods of Aerophysical Research (Vol. 1, pp. 174–179). Novosibirsk.

Maucher, U., Rist, U. and Wagner, S. (1994). Direct numerical simulation of airfoil separation bubbles. In S.Wagner, E.H. Hirschel, J.Périaux and R.Piva (Eds.), Computational fluid dynamics '94 (pp. 471–477). New York: Wiley.

Maucher, U., Rist, U. and Wagner, S. (2000). Secondary disturbance amplification and transition in laminar separation bubbles. In H.F. Fasel and

W.S. Saric (Eds.), Laminar-Turbulent Transition (pp. 657–662). Berlin Heidelberg New York: Springer.

McGhee, R.J., Jones, G.S. and Jouty, R. (1988). Performance characteristics from wind-tunnel tests of a low-Reynolds-number airfoil (AIAA Paper 88–0607).

Meyer, D.G.W., Rist, U., Borodulin, V.I., Gaponenko, V.R., Kachanov, Y.S., Lian, Q.X. and Lee, C.B. (2000). Late-stage transitional boundary-layer structures. Direct numerical simulation and experiment. In H.F. Fasel and W.S. Saric (Eds.), Laminar-Turbulent Transition (pp. 167–172). Berlin Heidelberg New York: Springer.

Meyer, F. and Kleiser, L. (1990). Numerical simulation of transition due to crossflow instability. In D.Arnal and R.Michel (Eds.), Laminar-Turbulent Transition (pp. 609–619). Berlin Heidelberg New York: Springer.

Michalke, A. (1990). On the inviscid instability of wall-bounded velocity profiles close to separation. Z. Flugwiss. Weltraumforsch., 14, 24–31.

Michalke, A. (1991). On the instability of wall-boundary layers close to separation. In V.V. Kozlov and A.V. Dovgal (Eds.), Separated Flows and Jets (pp. 557–564). Berlin Heidelberg New York: Springer.

Michalke, A. (1993). On the receptivity of a compressible two-dimensional vortex sheet close to a wall for various types of excitation. European J. Mech. B Fluids, 12(4), 421–445.

Michalke, A. (1995). Receptivity of axisymmetric boundary layers due to excitation by a Dirac point source at the wall. European J. Mech. B Fluids, 14(4), 373–393.

Michalke, A. (1997). Excitation of small disturbances by a Dirac line source at the wall and their growth in a decelerated laminar boundary layer. European J. Mech. B Fluids, 16(1), 17–37.

Michalke, A. and Al-Maaitah, A.A. (1992). On the receptivity of the unstable wall boundary layer along a surface hump excited by a 2-D Dirac source at the wall. European J. Mech. B Fluids, 11(5), 521–542.

Michalke, A., Kozlov, V.V. and Dovgal, A.V. (1995). Contribution to the instability of laminar separating flows along axisymmetric bodies. Part 1: Theory. European J. Mech. B Fluids, 14(3), 333–350.

Michel, R., Arnal, D. and Coustols, E. (1985a). Stability calculations and transition criteria on two- and three-dimensional flows. In V.V. Kozlov (Ed.), Laminar-Turbulent Transition (pp. 455–461). Berlin Heidelberg New York: Springer.

Michel, R., Arnal, D., Coustols, E. and Juillen, J.C. (1985b). Experimental and theoretical studies of boundary layer transition on a swept infinite wing. In V.V. Kozlov (Ed.), Laminar-Turbulent Transition (pp. 553–560). Berlin Heidelberg New York: Springer.

Miksad, R.W. (1972). Experiments on the non-linear stages of free-shear layer transition. J. Fluid Mech., 56, 695–719.

Miksad, R.W. (1973). Experiments on the nonlinear interactions in the transition of a free shear layer. J. Fluid Mech., 59, 1–21.

Milling, R.W. (1981). Tollmien–Schlichting wave cancellation. Phys. Fluids, 24(5), 979–981.

Monkewitz, P. (1988). Subharmonic resonance, pairing and shredding in the mixing layer. J. Fluid Mech., 188, 223–252.

Monkewitz, P.A. (1990). The role of absolute and convective instability in predicting the behavior of fluid systems. European J. Mech. B Fluids, 9(5), 395–413.

Monkewitz, P.A. and Huerre, P. (1982). Influence of the velocity ratio on the spatial instability of mixing layers. Phys. Fluids, 25(7), 1137–1143.

Montividas, R.E., Acharya, M. and Metwally, M.H. (1992). Reactive control of an unsteady separating flow. AIAA J., 30(4), 1133–1134.

Morkovin, M.V. (1964). Flow around circular cylinder – a kaleidoscope of challenging flow phenomena. In A.G. Hansen (Ed.), Fully Separated Flows (pp. 102–118). New York: Amer. Soc. Mech. Eng.

Morkovin, M.V. (1968). Critical evaluation of transition from laminar to turbulent shear layers with emphasis of hypersonically traveling bodies (AFFDL TR 68–149).

Morkovin, M.V. (1984). Bypass transition to turbulence and research desiderata. In NASA CP–2386 Transition in Turbines (pp. 161–204).

Moston, J., Stewart, P.A. and Cowley, S.J. (2000). On the nonlinear growth of two-dimensional Tollmien–Schlichting waves in a flat-plate boundary layer. J. Fluid Mech., 425, 259–300.

Mueller, T.J. and Batill, S.M. (1982). Experimental studies of separation on a two-dimensional airfoil at low Reynolds numbers. AIAA J., 20(4), 457–463.

Müller, B. and Bippes, H. (1988). Experimental study of instability modes in a three-dimensional boundary layer. In AGARD-CP–438 Fluid Dynamics of Three-Dimensional Turbulent Shear Flows and Transition (pp. 13.1–13.15). Chismet, Turkey.

Murdock, J.W. (1977). A numerical study of nonlinear effects on boundary layer stability. AIAA J., 15(8), 1167–1173.

Murdock, J.W. (1980). The generation of Tollmien–Schlichting waves by a sound wave. Proc. R. Soc. Lond. A, 372, 517–534.

Murdock, J.W. and Stewartson, K. (1977). Spectra of Orr–Sommerfeld equation. Phys. Fluids, 20(9), 1404–1411.

Nagata, M. (1990). Three-dimensional finite-amplitude solutions in plane Couette flow: Bifurcation from infinity. J. Fluid Mech., 217, 519–527.

Narasimha, R., Subramanian, C. and Narayanan, M.A. (1982). Turbulent spot growth in favourable pressure gradients (Tech. Rep. 82–FM–12). Bangalore, India.

Narayanan, M.A. and Narayana, T. (1967). Some studies on transition from laminar to turbulent flow in a two-dimensional channel. Z. Angew. Math. Phys., 18, 642–650.

Nayfeh, A.H. (1973). Perturbation methods. New York: Wiley-Interscience.

Nayfeh, A.H. (1980). Stability of three-dimensional boundary layers. AIAA J., 18(4), 406–416.

Nayfeh, A.H. (1987). nonlinear stability of boundary layers (AIAA Paper 87–0044).

Nayfeh, A.H. and Padhye, A. (1979). Relation between temporal and spatial stability in three-dimensional flows. AIAA J., 17(10), 1084–1090.

Nayfeh, A.H. and Ragab, S.A. (1987). Effect of a bulge on the secondary instability of boundary layers (AIAA Paper 87–0045).

Nayfeh, A.H., Saric, W.S. and Mook, D.T. (1974). Stability of non-parallel flows. Arch. Mech., 26(3), 401–406.

Nayfeh, A.H., Ragab, S.A. and Al-Maaitah, A.A. (1988). Effects of bulges on the stability of boundary layers. Phys. Fluids, 31(4), 796–806.

Nayfeh, A.H., Ragab, S.A. and Masad, J.A. (1990). Effect of a bulge on the subharmonic instability of boundary layers. Phys. Fluids A, 2(6), 937–948.

Neumann, D. and Dinkelacker, A. (1991). Drag mechanisms on V-grooved surfaces on a body of revolution in axial flow. Appl. Sci. Res., 48, 101–114.

Ng, L.L. and Crouch, J.D. (1999). Roughness-induced receptivity to crossflow vortices on a swept wing. Phys. Fluids, 11(2), 432–438.

Nishioka, M. and Morkovin, M.V. (1986). Boundary-layer receptivity to unsteady pressure gradients: Experiments and overview. J. Fluid Mech., 171, 219–261.

Nishioka, M., Iida, S. and Ichikawa, Y. (1975). An experimental investigation of the stability of plane Poiseuille flow. J. Fluid Mech., 72, 731–751.

Nishioka, M., Asai, M. and Iida, S. (1980). An experimental investigation of the secondary instability. In R.Eppler and H.Fasel (Eds.), Laminar-Turbulent Transition (pp. 37–46). Berlin Heidelberg New York: Springer.

Nitschke-Kowsky, P. and Bippes, H. (1988). Instability and transition of a three-dimensional boundary layer on a swept flat plate. Phys. Fluids, 31(4), 786–795.

Oertel, J. and Stank, R. (1999). Dynamics of localized disturbances in transonic wing boundary layers (AIAA Paper 99–0551).

O'Meara, M.M. and Mueller, T.J. (1987). Laminar separation bubble characteristics on an airfoil at low Reynolds numbers. AIAA J., 25(8), 1033–1041.

Opfer, H. and Ronneberger, D. (2001). Propagation and cancellation of 3D disturbances in a flat plate's boundary layer. In GAMM, Book of abstracts (p. 102). Zürich.

Opfer, H., Ronneberger, D. and Dallmann, U. (2001). Demands on a 3D wave cancellation system in a 2D laminar boundary layer. In EUROMECH Symposium 423. Stuttgart.

Orr, W.M. (1907a). The stability or instability of the steady motions of a perfect liquid and of a viscous liquid. Part 1: A perfect liquid. Proc. R. Irish Acad. A, 27, 9–68.

Orr, W.M. (1907b). The stability or instability of the steady motions of a perfect liquid and of a viscous liquid. Part 2: A viscous liquid. Proc. R. Irish Acad. A, 27, 69–138.

Orszag, S.A. (1971). Accurate solution of the Orr–Sommerfeld stability equation. J. Fluid Mech., 50, 689–703.

Orszag, S.A. and Patera, A.T. (1983). Secondary instability of wall-bounded shear flows. J. Fluid Mech., 128, 347–385.

Ostoslavskiy, I.S. and Svistchev, G.P. (1975). Calculation of laminar–turbulent transition point. Trudy TsAGI, 1725, 1–11. (In Russian.)

Park, D.S. and Huerre, P. (1995). Primary and secondary instabilities of the asymptotic suction boundary layer on a curved plate. J. Fluid Mech., 283, 249–272.

Patel, V.C. and Head, M.R. (1969). Some observations on skin friction and velocity profiles in fully developed pipe and channel flows. J. Fluid Mech., 38, 181–201.

Pauley, L.P., Moin, P. and Reynolds, W.C. (1990). The structure of two-dimensional separation. J. Fluid Mech., 220, 397–411.

Perraud, J. (1998). Linear stability of the incompressible boundary layer over 2D steps and gaps. In EUROMECH Colloquium 380/ERCOFTAC SIG 33 Conf., Book of Abstracts. Göttingen. (Paper 13.)

Pierrehumbert, R.T. and Widnall, S.E. (1982). The two- and three-dimensional instabilities of a spatially periodic shear layer. J. Fluid Mech., 114, 59–82.

Pironneau, O., Rodi, W., Ryhming, I.L., Savill, A.M. and Truong, T.V. (Eds.). (1992). Numerical simulation of unsteady flows and transition to turbulence. Cambridge: Cambridge University Press.

Poll, D.I.A. (1979). Transition in the infinite swept attachment line boundary layer. Aeronaut. Quart., 30, 607–629.

Poll, D.I.A. (1985). Some observations on the transition process on the windward face of a long yawed cylinder. J. Fluid Mech., 150, 329–356.

Polyakov, N.F. (1975). Induction of hydrodynamic waves in laminar boundary layer by streamwise acoustic field. In Proc. Symp. on Physics of Acousto-Hydrodynamic Phenomena (pp. 216–223). Moscow: Nauka. (In Russian.)

Polyakov, N.F. (1979). Laminar boundary layer at 'natural' transition to turbulent flow. In V.Y. Levchenko (Ed.), Development of Disturbances in Boundary Layer (pp. 23–67). Novosibirsk: RAS. Sib. Branch Inst. Theoret. Appl. Mech. (In Russian.)

Prandtl, L. (1921). Bemerkungen über die Entstehung der Turbulenz. Z. Angew. Math. Mech., 1, 431–436.
Prandtl, L. (1922). Bemerkungen über die Entstehung der Turbulenz. Phys. Z., 23, 19–25.
Prandtl, L. (1935). The mechanics of viscous fluids. In W.F. Durand (Ed.), Aerodynamic Theory (Vol. 3). Berlin Heidelberg New York: Springer.
Radeztsky, R.H., Reibert, M.S. and Saric, W.S. (1994). Development of stationary crossflow vortices on a swept wing (AIAA Paper 94–2373).
Radeztsky, R.H., Reibert, M.S. and Saric, W.S. (1999). Effect of isolated micron-sized roughness on transition in swept-wing flows. AIAA J., 37(11), 1370–1377.
Rai, M.M. and Moin, P. (1991). Direct simulation of transition and turbulence in a spatially evolving boundary layer (AIAA Paper 91–1607–CP).
Ramazanov, M.P. (1985). Development of finite-amplitude disturbances in Poiseuille flow. In V.V. Kozlov (Ed.), Laminar-Turbulent Transition (pp. 183–190). Berlin Heidelberg New York: Springer.
Rannacher, J. (1969). Untersuchung von geraden ebenen Flugelgittern im kritischen Reynoldszahlbereich. Kurzfassung in Maschinenbautechnik, 18, 2–10.
Rayleigh, J.W.S. (1880). On the stability, or instability, of certain fluid motions. Proc. Lond. Math. Soc., 9, 57–70.
Reddy, S.C. and Henningson, D.S. (1993). Energy growth in viscous channel flows. J. Fluid Mech., 252, 209–238.
Reddy, S.C., Schmid, P.J. and Henningson, D.S. (1993). Pseudospectra of Orr–Sommerfeld operator. SIAM J. Appl. Math., 53(1), 15–47.
Reed, H.L. (1987). Wave interaction in swept-wing flows. Phys. Fluids, 30(11), 3419–3426.
Reed, H.L. and Saric, W.S. (1989). Stability of three-dimensional boundary layers. Ann. Rev. Fluid Mech., 21, 235–284.
Reibert, M.S., Saric, W.S., Carrillo, R.B. and Chapman, K.L. (1996). Experiments in nonlinear saturation of stationary crossflow vortices in a swept-wing boundary layer (AIAA Paper 96–0184).
Reshotko, E. (1976). Boundary layer stability and transition. Ann. Rev. Fluid Mech., 8, 311–349.
Reynolds, O. (1883). An experimental investigation of the circumstances which determine whether the motion of water shall be direct or sinuous, and of the law of resistance in parallel channels. Philos. Trans. R. Soc. Lond. A, 174, 935–982.
Riley, J.J., Gad-el-Hak, M. and Metcalfe, R.W. (1988). Compliant coatings. Ann. Rev. Fluid Mech., 20, 393–420.
Ripley, M.D. and Pauley, L.L. (1993). The unsteady structure of two-dimensional steady laminar separation. Phys. Fluids A, 5(12), 3099–3106.

Rist, U. (1994). Nonlinear effects of 2D and 3D disturbances on laminar separation bubbles. In S.P. Lin, W.R.C. Phillips and D.T. Valentine (Eds.), Nonlinear Instability of Nonparallel Flows (pp. 324–333). Berlin Heidelberg New York: Springer.

Rist, U. and Fasel, H.F. (1995). Direct numerical simulation of controlled transition in a flat-plate boundary layer. J. Fluid Mech., 298, 211–248.

Rist, U. and Maucher, U. (1994). Direct numerical simulation of 2D and 3D instability waves in a laminar separation bubble. In AGARD-CP-551 Application of Direct and Large eddy Simulation to Transition and Turbulence (pp. 34.1–34.7). Chania, Crete, Greece.

Rist, U., Maucher, U. and Wagner, S. (1996). Direct numerical simulation of some fundamental problems related to transition in laminar separation bubbles. In J.-A. Désidéri, C.Hirsch, P.L. Tallec, M.Pandolfi and J.Périaux (Eds.), Computational Fluid Dynamics '96 (pp. 319–325). New York: Wiley.

Rist, U., Müller, K. and Wagner, S. (1998). Visualisation of late-stage transitional structures in numerical data using vortex identification and feature extraction. In G.M. Carlomango and I.Grant (Eds.), Eighth International Symposium on Flow Visualisation. Sorrento: Published and distributed on CD by the editors. (Paper No. 109.)

Roach, P.E. and Brierley, D.H. (1992). The influence of a turbulent free-stream on zero pressure gradient transitional boundary layer development. Part 1: Test cases T3A and T3B. In O.Pironneau, W.Rodi, I.L. Ryhming, A.M. Savill and T.V. Truong (Eds.), Numerical Simulation of Unsteady Flows and Transition to Turbulence (pp. 319–347). Cambridge: Cambridge University Press.

Roberts, W.B. (1980). Calculation of laminar separation bubbles and their effect on airfoil performance. AIAA J., 18(1), 25–31.

Roget, C., Brazier, J.P., Cousteix, J. and Mauss, J. (1998). A contribution to the physical analysis of separated flows past three-dimensional humps. European J. Mech. B Fluids, 17(3), 307–329.

Rogler, H.L. and Reshotko, E. (1975). Disturbances in a boundary layer introduced by a low intensity array of vortices. SIAM J. Appl. Math., 28(2), 431–462.

Roos, F.W. and Kegelman, J.T. (1986). Control of coherent structures in reattaching laminar and turbulent shear layers. AIAA J., 24(12), 1956–1963.

Ross, J.A., Barnes, F.H., Burns, J.G. and Ross, M.A.S. (1970). The flat plate boundary layer. Part 3. Comparison of theory with experiment. J. Fluid Mech., 43, 819–832.

Rotenberry, J.M. (1993). Finite amplitude steady waves in the Blasius boundary layer. Phys. Fluids A, 5(7), 1840–1842.

Rothmayer, A.P. and Smith, F.T. (1998). High Reynolds number asymptotic theories. In R.W. Johnson (Ed.), The Handbook of Fluid Dynamics (pp. 23.1–25.26). Berlin Heidelberg New York: Springer.

Ruban, A.I. (1985). On the generation of Tollmien–Schlichting waves by sound. Fluid Dyn., 19(5), 709–716.

Ruban, A.I. (1991). Propagation of wave packets in the boundary layer on a curved surface. Fluid Dyn., 25(2), 213–221.

Runiyan, L.J. and Gerge-Falvy, D. (1979). Amplification factors corresponding to transition on an unswept wing in free flight and on a swept wing in wind tunnel (AIAA Paper 79–0267).

Ryzhov, O.S. and Terent'ev, E.D. (1998). Streamwise absolute instability of a three-dimensional boundary layer at high reynolds numbers. J. Fluid Mech., 373, 111–153.

Saffman, P.G. (1962). On the stability of a laminar flow to a dusty gas. J. Fluid Mech., 13, 120–128.

Salwen, H. and Grosch, C.E. (1981). The continuous spectrum of the Orr–Sommerfeld equation. Part 2. Eigenfunction expansion. J. Fluid Mech., 104, 445–465.

Sandham, N.D. and Kleiser, L. (1992). The late stages of transition to turbulence in channel flow. J. Fluid Mech., 245, 319–348.

Saric, W.S. (1990). Low-speed experiments: Requirements for stability measurements. In M.Y. Hussaini and R.G. Voight (Eds.), Instability and Transition (Vol. 1, pp. 162–172). Berlin Heidelberg New York: Springer.

Saric, W.S. (1994a). Low-speed boundary-layer transition experiments. In T.C. Corke, G.Erlebacher and M.Y. Hussaini (Eds.), Transition, Experiments, Theory & Computations (pp. 1–113). Oxford: Oxford University Press.

Saric, W.S. (1994b). Görtler vortices. Ann. Rev. Fluid Mech., 26, 379–409.

Saric, W.S. and Nayfeh, A.H. (1977). Nonparallel stability of boundary layers with pressure gradients and suction. In AGARD-CP–224 Laminar–Turbulent Transition (pp. 6.1–6.21). Copenhagen.

Saric, W.S. and Yeates, L.G. (1985). Generation of crossflow vortices in a three-dimensional flat plate flow. In V.V. Kozlov (Ed.), Laminar-Turbulent Transition (pp. 429–437). Berlin Heidelberg New York: Springer.

Saric, W.S., Kozlov, V.V. and Levchenko, V.Y. (1984). Forced and unforced subharmonic resonance in boundary-layer transition (AIAA Paper 84–0007).

Saric, W.S., Hoos, J.A. and Radeztsky, R.H. (1991). Boundary-layer receptivity of sound with roughness. In D.C. Reda, H.L. Reed and R.Kobayashi (Eds.), Boundary Layer Stability and Transition to Turbulence (Vol. FED–114, pp. 17–22). ASME.

Sarik, W., Reed, H. and Kerschen, E. (1994). Leading edge receptivity to sound: Experiments, DNS, and theory (AIAA Paper 94–2222).

Sato, H. (1959). Further investigation on the transition of two-dimensional separated layers at subsonic speed. J. Phys. Soc. Japan, 14(12), 1797–1810.

Sato, H. (1970). An experimental study of non-linear interaction of velocity fluctuations in the transition region of a two-dimensional wake. J. Fluid Mech., 44, 741–765.

Sato, H. and Kuriki, K. (1961). The mechanism of transition in the wake of a thin flat plate placed parallel to a uniform flow. J. Fluid Mech., 11, 321–352.

Sattinger, D.H. (1970). The mathematical problem of hydrodynamic stability. J. Math. Mech., 20(9), 797–817.

Savenkov, I.V. (1991). Instability of the boundary layer on a curved surface. Fluid Dyn., 25(1), 151–154.

Savill, A.M. (1990). Drag reduction by passive devices – a review of some recent developments. In A.Gyr (Ed.), Structure of Turbulence and Drag Reduction (pp. 429–465). Berlin Heidelberg New York: Springer.

Sboev, D.S., Grek, G.R. and Kozlov, V.V. (1999a). Experimental study of boundary layer receptivity to localized perturbations of the external flow. Thermophys. Aeromech., 6(1), 1–13.

Sboev, D.S., Grek, G.R. and Kozlov, V.V. (1999b). On the features of the inner structure of the 'streaky structures'. Thermophys. Aeromech., 6(3), 359–370.

Sboev, D.S., Grek, G.R. and Kozlov, V.V. (2000). Experimental study of swept wing boundary layer receptivity to free stream localized disturbances. Thermophys. Aeromech., 7(4), 469–480.

Schensted, I.V. (1960). Contributions to the theory of hydrodynamic stability. Doctoral dissertation, University of Michigan.

Schlichting, H. (1932). Über die Stabilität der Couette-strömung. Ann. D. Phys., 5(14), 905–936.

Schlichting, H. (1933). Zur Entstehung der Turbulenz bei der Plattenströmung. In Math. Phys. Klasse (pp. 181–208). Nachr. Ges. Wiss. Göttingen.

Schlichting, H. (1935). Amplitudenverteilung und Energiebilanz der kleinen Störungen bei der Plattenströmung. In Math. Phys. Klasse, Fachgruppe I (Vol. 1, pp. 47–78). Nachr. Ges. Wiss. Göttingen.

Schlichting, H. and Gersten, K. (2000). Boundary layer theory (8th ed.). Berlin Heidelberg New York: Springer.

Schmid, P.J. and Henningson, D.S. (1994). Optimal energy density growth in Hagen–Poiseuille flow. J. Fluid Mech., 277, 197–225.

Schmid, P.J. and Henningson, D.S. (2000). Stability and transition in shear flows. Berlin Heidelberg New York: Springer.

Schmid, P.J. and Kytomaa, H.K. (1994). Transient and asymptotic stability of granular shear flow. J. Fluid Mech., 264, 255–275.

Schubauer, G.B. and Klebanoff, P.S. (1956). Contributions on the mechanics of boundary layer transition (NACA TR 1289).

Schubauer, G.B. and Skramstad, H.K. (1948). Laminar-boundary layer oscillations and transition on a flat plate (NACA TN 909).

Sen, P.K. and Venkateswarly, D. (1983). On the stability of plane Poiseuille flow to finite-amplitude disturbances, considering the higher-order Landau coefficients. J. Fluid Mech., J. Fluid Mech., 179–206.

Shapiro, P.J. (1977). The influence of sound upon laminar boundary layer instability (Tech. Rep. 83458–83560–1). Acoustic and Vibration Lab.: Mass. Inst. Technol., Cambridge. (See also NTIS AD–A046057.)

Sherlin, G.C. (1960). Behaviour of isolated disturbances superimposed on laminar flow in a rectangular pipe. J. Res. N. B. S., Sec. A, Phys. and Chem., 64A, 281–289.

Sidorenko, N.V. and Erofeev, E.A. (1985). Interaction of vortex disturbances and boundary layer on blunt bodies. In V.V. Kozlov (Ed.), Laminar-Turbulent Transition (pp. 261–266). Berlin Heidelberg New York: Springer.

Sigurdson, L.W. and Roshko, A. (1988). The structure and control of a turbulent reattaching flow. In H.W. Liepmann and R.Narasimha (Eds.), Turbulent management and relaminarization (pp. 497–514). Berlin Heidelberg New York: Springer.

Sinha, S.N., Gupta, A.K. and Oberai, M.M. (1981). Laminar separating flow over backsteps and cavities. Part 1: Backsteps. AIAA J., 19(12), 1527–1530.

Smith, F.T. (1979a). Nonlinear stability of boundary layers for disturbances of different sizes. Proc. R. Soc. Lond. A, 368, 573–589.

Smith, F.T. (1979b). On the non-parallel flow stability of the Blasius boundary layer. Proc. R. Soc. Lond. A, 366, 91–109.

Smith, F.T. (1987). Non-linear effects and non-parallel flows: The collapse of separated motion. In D.L. Dwoyer and M.Y. Hussaini (Eds.), Stability of Time Dependent and Spatially Varying Flows (pp. 104–147). Berlin Heidelberg New York: Springer.

Smith, F.T. and Bodonyi, R.J. (1985). On short-scale inviscid instabilities in flow past surface-mounted obstacles and other nonparallel motions. J. R. Aeronaut. Soc., 89, 205–212.

Smith, A.M.O. and Gamberoni, N. (1956). Transition, pressure gradient and stability theory (Rep. ES 26388). El Segundo, Cal.: Douglas Aircraft Co.

Sommerfeld, A. (1908). Ein Beitrag zur hydrodynamischen Erklärung der turbulenten Flüssigkeitsbewegungen. In Proc. fourth internat. mathemat. congr. (Vol. 3, pp. 116–124). Rome.

Spalart, P.R. (1993). Numerical study of transition induced by suction devices. In R.M.C. So, C.G. Speziale and B.E. Launder (Eds.), Near-Wall Turbulent Flows (pp. 849–858). Tempe, Ariz., U.S.A: Elsevier.

Sparrow, E.M., Lin, S.H. and Lundgren, T.S. (1964). Flow development in the hydrodynamic entrance region of tubes and ducts. Phys. Fluids, 7(3), 338–347.

Spiridonov, A.N. and Chernorai, V.G. (2000). Generation and development of disturbances by a surface vibration in a boundary layer with a pressure gradient. In Proc. Seventh Internat. Conf. Stability of Homogeneous and Heterogeneous Fluids (pp. 165–167). Novosibirsk. (In Russian.)

Squire, H.B. (1933). On the stability for three-dimensional disturbances of viscous fluid between parallel walls. Proc. R. Soc. Lond. A, 142, 621–628.

Srokowski, A.J. and Orszag, S.A. (1977). Mass flow requirements for LFC wing design (AIAA Paper 77–1222).

Stewart, P.A. and Smith, F.T. (1987). Three-dimensional instabilities in steady and unsteady non-parallel boundary layers, including effects of Tollmien–Schlichting disturbances and cross flow. Proc. R. Soc. Lond. A, 409, 229–248.

Stewartson, K. (1969). On the flow near a trailing edge of a flat plate. Mathematika, 16, 106–121.

Stewartson, K. and Stewart, J.T. (1971). A nonlinear instability theory for a wave system in plane Poiseuille flow. J. Fluid Mech., 48, 529–545.

Strazisar, A.J., Reshotko, E. and Prahl, J.M. (1977). Experimental study of the stability of heated laminar boundary layers in water. J. Fluid Mech., 83, 225–247.

Streett, C.L. (1998). Direct harmonic linear Navier–Stokes methods for efficient simulation of wave packets (AIAA Paper 98–0784).

Struminskiy, V.V., Lebedev, Y.B. and Fomichev, V.M. (1986). Effect of streamwise pressure gradient on laminar boundary-layer extension in gas flow. Dokl. Akad. Nauk[11], 289(4), 813–816.

Stuart, J.T. (1960). On the nonlinear mechanisms of wave disturbances in stable and ubstable parallel flows. Part 1. The basic behaviour in plane Poiseuille flow. J. Fluid Mech., 9, 353–370.

Stuart, J.T. (1980). Stability and transition: Some comments on the problem. In R.Eppler and H.Fasel (Eds.), Laminar-Turbulent Transition (pp. 1–13). Berlin Heidelberg New York: Springer.

Suder, K.L., O'Brien, J.E. and Reshotko, E. (1988). Experimental study of bypass transition in a boundary layer (NASA TM 100913).

Swearingen, J.D. and Blackwelder, R.F. (1983). Parameters controlling the spacing of streamwise vortices on concave walls (AIAA Paper 83–0380).

Swearingen, J.D. and Blackwelder, R.F. (1987). The growth and breakdown of streamwise vortices in the presence of a wall. J. Fluid Mech., 182, 255–290.

[11] Translated in Phys. Dokl.

Tafti, D.K. and Vanka, S.P. (1991). A numerical study of flow separation and reattachment on a blunt plate. Phys. Fluids A, 3(7), 1749–1759.

Taghavi, H. and Wazzan, A.R. (1974). Spatial stability of some Falkner–Skan profiles with reversed flow. Phys. Fluids, 17(12), 2181–2183.

Takagi, S. and Itoh, N. (1994). Observation of traveling waves in the three-dimensional boundary layer along a yawed cylinder. Fluid Dyn. Res., 14, 167–189.

Takagi, S., Saric, W.S. and Radeztsky, R.H. (1991). Effect of sound and micro-sized roughness on crossflow dominated transition. Bull. Amer. Phys. Soc., 36(10), 2630.

Tam, C.K.W. (1978). Excitation of instability waves in a two-dimensional shear layer by sound. J. Fluid Mech., 89, 357–371.

Tam, C.K.W. (1981). The excitation of Tollmien–Schlichting waves in low subsonic boundary layers by free stream sound waves. J. Fluid Mech., 109, 483–501.

Tani, I. (1962). Production of longitudinal vortices in the boundary-layer along a curved wall. J. Geophys. Res., 67, 3075–3080.

Tani, I. (1964). Low-speed flows involving bubble separations. Progr. Aeronaut. Sci., 5, 70–103.

Tani, I. and Komoda, H. (1962). Boundary-layer transition in the presence of streamwise vortices. J. Aeronaut. Sci., 29, 440–444.

Taylor, G.I. (1936). Statistical theory of turbulence. V. Effects of turbulence on boundary layer. Theoretical discussion of relationship between scale of turbulence and critical resistance of sphere. Proc. R. Soc. Lond. A, 156, 307–317.

Taylor, M.J. and Peake, N. (1998). The long-time behaviour of incompressible swept-wing boundary layer subject to impulsive forcing. J. Fluid Mech., 355, 359–381.

Terent'ev, E.D. (1984). A linear problem on a vibrator oscillating harmonically at supercritical frequencies in a subsonic boundary layer. Prikl. Mat. Mekh.[12], 48(2), 264–272.

Theofilis, V. (2000). Global linear instabilities in laminar separated boundary layer flow. In H.F. Fasel and W.S. Saric (Eds.), Laminar-Turbulent Transition (pp. 663–668). Berlin Heidelberg New York: Springer.

Thomas, L.H. (1953). The stability of plane Poiseuille flow. Phys. Rev., 91, 780–783.

Thomas, A.S.W. (1983). The control of boundary-layer transition using a wave-superposition principle. J. Fluid Mech., 137, 233–250.

Thomas, A.S.W. and Lekoudis, S.G. (1978). Sound and a Tollmien–Schlichting wave in a Blasius boundary layer. Phys. Fluids, 21(11), 2112–2113.

[12] Translated in Appl. Math. Mech.

Tollmien, W. (1929). Über die Entstehung der Turbulenz. 1. Mitteilung. In Math. Phys. Klasse (pp. 21–44). Nachr. Ges. Wiss. Göttingen. (Translated as NACA TM 609, 1931.)

Trefethen, L.N. (1997). Pseudospectra of linear operators. SIAM Rev., 39, 383–406.

Trefethen, L.N., Trefethen, A.E., Reddy, S.C. and Driscoll, T.A. (1993). Hydrodynamic stability without eigenvalues. Science, 261, 578–584.

Troutt, T.R., Scheelke, B. and Norman, T.R. (1984). Organized structures in a reattaching separated flow field. J. Fluid Mech., 143, 413–427.

Tsao, T., Liu, C., Tai, Y.-C. and Ho, C.-M. (1994). Micromachined magnetic actuator for active fluid control. In P.R. Bandyopadhyay, K.S. Breuer and C.J. Blechinger (Eds.), Application of Microfabrication to Fluid Mechanics (Vol. FED–197, pp. 31–38). ASME.

Tumin, A. (1996). Receptivity of pipe Poiseuille flow. J. Fluid Mech., 315, 119–137.

Tumin, A. and Reshotko, E. (2001). Spatial theory of optimal disturbances in boundary layers. Phys. Fluids, 13(7), 2097–2104.

Tumin, A.M. and Shepelev, V.E. (1980). Numerical analysis of disturbance development in incompressible flat plate boundary layer. Chisl. Met. Mekhan. Splosh. Sredy, 1(3), 141–152. (In Russian.)

Ustinov, M.V. (1995). Secondary instability modes generated by Tollmien–Schlichting wave scattering from a bump. Theoret. Comput. Fluid Dyn., 7, 341–354.

Van Dam, C.P. and Elli, S. (1992). Simulation of nonlinear Tollmien–Schlichting wave growth through a laminar separation bubble. In M.Y. Hussaini, A.Kumar and C.L. Streett (Eds.), Instability, Transition, and Turbulence (pp. 311–321). Berlin Heidelberg New York: Springer.

Van Dyke, M. (1964). Perturbation methods in fluid mechanics. New York: Academic.

Van Ingen, J.L. (1975). On the calculation of laminar separation bubbles in two dimensional incompressible flow. In AGARD-CP–168.

Van Ingen, J.L. (1991). Research on laminar separation bubbles at Delft University of Technology. In V.V. Kozlov and A.V. Dovgal (Eds.), Separated Flows and Jets (pp. 537–556). Berlin Heidelberg New York: Springer.

Vasudeva, B.R. (1967). Boundary-layer instability experiment with localized disturbance. J. Fluid Mech., 29, 745–763.

Vasudevan, K.P., Dey, J. and Prabhu, A. (2001). Spots propagation characteristics in laterally strained boundary layers. Exp. Fluids, 30(5), 488–491.

Vatsa, V.N. and Carter, J.E. (1984). Analysis of airfoil leading-edge separation bubbles. AIAA J., 22(12), 1697–1704.

Vlasov, E.V., Ginevskiy, A.S. and Karavosov, R.K. (1977). Reaction of unstable laminar boundary layer to acoustic disturbances. In V.V.

Struminskiy (Ed.), Turbulent Flows (pp. 90–96). Moscow: Nauka. (In Russian.)

Voke, P.R. and Yang, Z.Y. (1995). Numerical study of a bypass transition. Phys. Fluids A, 7(9), 2256–2264.

Volodin, A.G. (1973). Stability of plane boundary layer with account of nonparallelity. Izv. Sib. Otd. Akad. Nauk SSSR, Ser. Tekhn. Nauk[13], 8(2), 14–17.

Volodin, A.G. and Gaponov, S.A. (1970). Stability of an incompressible boundary layer. Izv. Sib. Otd. Akad. Nauk SSSR, Ser. Tekhn. Nauk[13], 2(8), 55–58.

Volodin, A.G. and Zelman, M.B. (1978). Three-waves resonance interaction of disturbances in a boundary-layer. Fluid Dyn., 13(5), 698–703.

Waleffe, F. (1995a). Hydrodynamic stability and turbulence: Beyond transients to self-sustaining process. Stud. Appl. Math., 95, 319–343.

Waleffe, F. (1995b). Transition in shear flows. Nonlinear normality versus non-normal linearity. Phys. Fluids, 7(12), 3060–3066.

Walsh, M.J. (1983). Riblets as a viscous drag reduction technique. AIAA J., 21(4), 485–486.

Wang, B.-J., Bogucki, D.J. and Redekopp, L.G. (2000). Transition in separated flows via global instability. In H.F. Fasel and W.S. Saric (Eds.), Laminar-Turbulent Transition (pp. 669–674). Berlin Heidelberg New York: Springer.

Ward, J.W. (1963). The behaviour and effects of laminar separation bubbles on airfoils in incompressible flow. J. R. Aeronaut. Soc., 67, 783–790.

Watmuff, J.H. (1991). An experimental invetigation of boundary layer transition in an adverse pressure gradient. In D.C. Reda, H.L. Reed and R.Kobayashi (Eds.), Boundary Layer Stability and Transition to Turbulence (Vol. FED–114, pp. 129–136). ASME.

Watmuff, J.H. (1999). Evolution of a wave packet into vortex loops in a laminar separation bubble. J. Fluid Mech., 397, 119–169.

Watson, J. (1960). On the nonlinear mechanisms of wave disturbances in stable and ubstable parallel flows. Part 2. The development of a solution for plane Poiseuille flow and Couette flow. J. Fluid Mech., 9, 371–389.

Wazzan, A.R., Okamura, T.T. and Smith, A.M.O. (1970). The stability and transition of heated and cooled incompressible laminar boundary layers. In V.Grigull and E.Hahne (Eds.), Fourth Internat. Heat Transfer Conf. (Vol. 2). Amsterdam: Elsevier. (Paper No. FC-14.)

Weibust, E., Bertelrud, A. and Ridder, S.O. (1987). Experimental investigation of laminar separation bubbles and comparison with theory. J. Aircraft, 24(5), 291–297.

Westin, K.J.A. (1997). Laminar–turbulent boundary layer transition influenced by free stream turbulence (TRITA-MEK TR 1997:10). Stockholm: Royal Institute of Technology. (Doctoral Thesis.)

[13] Translated in Siberian Phys. Techn. J. (former Sov. J. Appl. Phys.)

Westin, K.J.A., Boiko, A.V., Klingmann, B.G.B., Kozlov, V.V. and Alfredsson, P.H. (1994). Experiments in a boundary layer subjected to free stream turbulence. Part 1. Boundary layer structure and receptivity. J. Fluid Mech., 281, 193–218.

Westin, K.J.A. and Henkes, R.A.W.M. (1997). Application of turbulence models to bypass transition. Trans. ASME Ser. I: J. Fluids Eng., 119(4), 859–866.

Westin, K.J.A., Bakchinov, A.A., Kozlov, V.V. and Alfredsson, P.H. (1998). Experiments on localized disturbances in a flat plate boundary layer. Part 1: The receptivity and evolution of a localized free stream disturbance. European J. Mech. B Fluids, 17(6), 823–846.

Wortmann, F.X. (1953). Eine Methode zur Beobachtung und Messung Wasserströmung mit Tellur. Z. Angew. Phys., 5(6), 200–206.

Wortmann, F.X. (1964). Experimental investigation of vortex occurence at transition in unstable boundary-layers (AFOSR 64–1280).

Wray, A. and Hussaini, M.Y. (1984). Numerical experiments in boundary-layer stability. Proc. R. Soc. Lond. A, 392, 373–389.

Wu, X. (2001). Receptivity of boundary layers with distributed roughness to vortical and acoustic disturbances: A second-order asymptotic theory and comparison with experiments. J. Fluid Mech., 431, 91–133.

Wu, X., Jacobs, R.G., Hunt, J. C.R. and Durbin, P.A. (1999). Simulation of boundary layer transition induced by periodically passing wakes. J. Fluid Mech., 398, 109–153.

Wygnanski, I.J. and Champagne, F.H. (1973). On transition in a pipe. Part 1. The origin of puffs and slugs and the flow in a turbulent slug. J. Fluid Mech., 59, 281–335.

Wygnanski, I.J., Sokolov, M. and Friedman, D. (1975). On transition in a pipe. Part 2. The equilibrium puff. J. Fluid Mech., 69, 283–304.

Wygnanski, I.J., Haritonidis, J.H. and Kaplan, R.E. (1979). On a Tollmien–Schlichting wave packet produced by a turbulent spot. J. Fluid Mech., 92, 505–528.

Wygnanski, I.J., Haritonidis, J.H. and Zilberman, M. (1982). On the spreading of a turbulent spot in the absence of a pressure gradient. J. Fluid Mech., 123, 69–90.

Yang, Z.Y. and Voke, P.R. (1991). Numerical simulation of transition under turbulence (Tech. Rep. ME–FD/91.01). Department of Mechanical Engineering, University of Surrey, U.K.

Yu, X. and Liu, J. T.C. (1991). The secondary instability in Goertler flow. Phys. Fluids A, 3(7), 1845–1847.

Yudovich, V.V. (1965). On stability of stationary flows of viscous incompressible fluid. Dokl. Akad. Nauk[14], 161(5), 1037–1040.

Zahn, J.P., Toomre, J., Spiegel, E.A. and Gough, D.O. (1974). Nonlinear cellular motions in Poiseuille channel flow. J. Fluid Mech., 64, 319–345.

[14] Translated in Phys. Dokl.

Zaman, K.B.M.Q. (1992). Effect of acoustic excitation on stalled flows over an airfoil. AIAA J., 30(6), 1492–1499.

Zaman, K.B.M.Q. and Hussain, A.K.M.F. (1981). Turbulence suppression in free shear flows by controlled excitation. J. Fluid Mech., 103, 133–159.

Zaman, K.B.M.Q. and McKinzie, D.J. (1991). Control of laminar separation over airfoils by acoustic excitation. AIAA J., 29(7), 1075–1083.

Zanin, B.Y. (1997). Hysteresis of a separated variable-velocity flow about a straight-wing model. J. Appl. Mech. Tehn. Phys., 38(5), 724–727.

Zavolskiy, N.A., Reutov, V.P. and Rybushkina, G.V. (1983). Excitation of Tollmien–Schlichting waves at scattering of acoustical and vortical disturbances in a boundary layer over a wavy wall. J. Appl. Mech. Tehn. Phys., 24(3), 79–86. (In Russian.)

Zelman, M.B. and Maslennikova, I.I. (1993). Tollmien–Schlichting-wave resonant mechanism for subharmonic-type transition. J. Fluid Mech., 252, 449–478.

Zelman, M.B. and Smorodski, B.V. (1991). On the influence of inflexion in mean velocity profile on the resonant interaction of disturbances in boundary layer. J. Appl. Mech. Tehn. Phys., 32(2), 200–204.

Zhigulev, V.N. and Sidorenko, N.V. (1985). On the issue of the nonlinear theory of the instability waves. In V.V. Kozlov (Ed.), Laminar-Turbulent Transition (pp. 81–86). Berlin Heidelberg New York: Springer.

Zhigulev, V.N. and Tumin, A.M. (1987). Onset of turbulence. Dynamical theory of exitation and development of instabilities in boundary layers. Novosibirsk: Nauka. (In Russian.)

Zhigulev, V.N., Sidorenko, N.V. and Tumin, A.M. (1980). Generation of instability waves in a boundary layer by external turbulence. J. Appl. Mech. Tehn. Phys., 21(6), 774–780.

Zhuk, V.I. and Ryzhov, O.S. (1980). Self-induced interaction and instabilty of an incompressible boundary layer. Dokl. Akad. Nauk[15], 253(6), 1326–1329.

[15] Translated in Phys. Dokl.

Index

GPSR Compliance
The European Union's (EU) General Product Safety Regulation (GPSR) is a set of rules that requires consumer products to be safe and our obligations to ensure this.

If you have any concerns about our products, you can contact us on

ProductSafety@springernature.com

In case Publisher is established outside the EU, the EU authorized representative is:

Springer Nature Customer Service Center GmbH
Europaplatz 3
69115 Heidelberg, Germany

www.ingramcontent.com/pod-product-compliance
Ingram Content Group UK Ltd.
Pitfield, Milton Keynes, MK11 3LW, UK
UKHW021831190726
13853UKWH00003B/1277

* 9 7 8 3 6 6 2 0 4 7 6 6 8 *